T0189135

Robust Methods for Dense Monocular Non-Rigid 3D Reconstruction and Alignment of Point Clouds

Vladislav Golyanik

Robust Methods for Dense Monocular Non-Rigid 3D Reconstruction and Alignment of Point Clouds

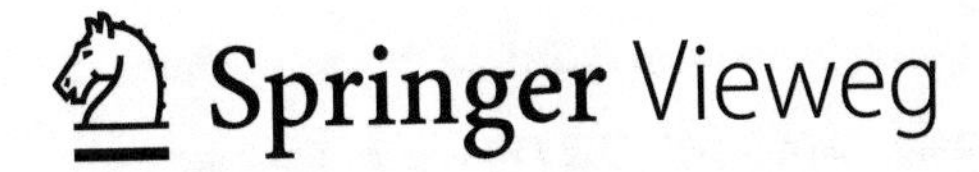

Vladislav Golyanik
Computer Graphics D4
Max Planck Institute for Informatics
Saarbruecken, Germany

Vom Fachbereich Informatik der Technischen Universität Kaiserslautern zur Verleihung des akademischen Grades Doktor der Ingenieurwissenschaften (Dr.-Ing.) genehmigte Dissertation

Datum der wissenschaftlichen Aussprache: 20. November 2019
Dekan: Prof. Dr. Stefan Deßloch
Vorsitzender der Promotionskommission: Prof. Dr. Hans Hagen
Erster Berichterstatter: Prof. Dr. Didier Stricker
Zweiter Berichterstatter: Prof. Dr. Reinhard Koch
Dritter Berichterstatter: Prof. Dr. Antonio Agudo

Technische Universität Kaiserslautern, 2019
D386

ISBN 978-3-658-30566-6 ISBN 978-3-658-30567-3 (eBook)
https://doi.org/10.1007/978-3-658-30567-3

This Springer Vieweg imprint is published by the registered company Springer Fachmedien Wiesbaden GmbH part of Springer Nature.
The registered company address is: Abraham-Lincoln-Str. 46, 65189 Wiesbaden, Germany

Acknowledgments

I am deeply and sincerely thankful to my doctoral supervisor Didier Stricker, the head of the Augmented Vision Department at the German Research Centre for Artificial Intelligence (DFKI) and a professor at the University of Kaiserslautern. He has always supported me in my research intentions, appreciated my independent work and strategically contributed to my development. He has also encouraged me to complete a research internship in a company.

The AV group has celebrated its tenth anniversary in August 2018. Its warm and collaborative environment has, without doubt, contributed to the accomplishment of this work. Therefore, I would like to acknowledge all colleagues and visiting researches of the AV group, who have been working in the AV group from 2014 until 2018, including my office mates, Adytia Tewari, Jilliam Maria Diaz Barros and Sarvenaz Salehi. A special thank goes to Leivy Michelly Kaul and Keonna Cunningham, for helping me to struggle through the office jungles and plenty of formalities.

I am deeply thankful to Bertram Taetz who was my postdoctoral colleague during my first months at DFKI. It was a unique experience to work with him which has effected my mindset in a way that I began to perceive computer vision in the continuous domain and from the perspective of applied mathematics. Since that time, I highly appreciate working with excellent mathematicians. I am deeply thankful to Gerd Reis who has often been helping me with improving paper drafts. Discussions with him have always been inspiring and rewarding.

During my sabbatical at NVIDIA Research in Santa Clara, I had the great fortune to work with Robert Maier who was an intern at NVIDIA Research at the same time as me, and Matthias Nießner who has been a visiting assistant professor at Stanford Univerity back then. I am thankful to my supervisors Kihwan Kim and Jan Kautz from NVIDIA Research for giving me the opportunity to work on a challenging topic and sharing the ideas. Moreover, working with them has helped me to strengthen my leadership qualities.

I am deeply thankful to all my further co-authors: Gabriele Bleser, Torben Fetzer, Soshi Shimada, Mitra Narsi, Oliver Wasenmüller, Jan C. Peters, Aman Shankar Mathur, Sk Aziz Ali, Mohammad Dawud Ansari, Tomonari Yosida, Kiran Varanasi, André Jonas and Christian Theobalt. I have enjoyed working with all of you, and I have learned from all of you. Furthermore, I would like to thank Stella Graßhof, Hanno Ackermann, Bodo Rosenhahn, Jörn Ostermann, Antonio Agudo, Francesc Moreno-Noguer, Daniel Cremers, Willi Frieden, Gabriele Steidl, Yongzhi Su, Alain Pagani, Norbert Schmitz, Marco Paladini, Deqing Sun and Thomas Breuel for collaborations, fruitful discussions and advice.

Last but not least, I am genuinely and sincerely grateful to my parents and my family. Without their patience and support, endurance and apprehension this work would have been barely possible. At all times, the music of Ludwig van Beethoven was helping me to make decisions and served as an inexhaustible source of patience, persistence and inspiration.

Vladislav Golyanik

Contents

List of Figures

List of Tables

Abstract

An accurate acquisition and processing of 3D point cloud data is an active research area in computer vision encompassing various unsolved problems. The thesis at hand addresses the jointly studied domains of dense non-rigid 3D reconstruction from monocular image sequences and point set registration under rigid as well as non-rigid transformations. Monocular non-rigid 3D reconstruction, which is in the focus of this work — known as non-rigid structure from motion (NRSfM) — relies on weak assumptions about the feasible deformation modes imposed on top of the motion and deformation cues. NRSfM and non-rigid point set registration are highly ill-posed problems in the sense of Hadamard.

The proposed dense NRSfM methods address the broad range of research questions including occlusion handling, scalability, interactive yet accurate processing as well as dense structure compression. For the occlusion handling and dealing with inaccurate point tracks, we propose a shape prior obtained on-the-fly and a new spatial regulariser — the coherency term. We also introduce a new model for NRSfM, which allows representing the recovered structure compactly.

The proposed point set registration methods aim at the enhanced registration accuracy for noisy data and samples with clustered outliers. For that reason, we embed prior correspondences into probabilistic point set registration and introduce a previously unexplored class of methods relying on principles of particle dynamics with simulated gravitational forces.

The thorough experimental evaluation confirms the efficiency and high accuracy of the proposed methods as well as the validity of the new ideas. By using the new principles, we advance the state of the art in dense monocular non-rigid 3D reconstruction and alignment of noisy point sets. Applications of the proposed NRSfM methods include (but are not limited to) 3D recovery and analysis of human and animal faces, endoscopic scenes and various other deformable surfaces. The proposed point set registration methods can be applied in robotics, automotive driving, face and shape recognition, and other areas. Apart from the abovementioned applications, we show how both method classes can be used for human appearance transfer, multiframe scene flow estimation from RGB-D as well as monocular image sequences. The developed methods offer numerous avenues for further investigation.

Zusammenfassung

Die genaue Eingabe und Verarbeitung von 3D Punktwolken ist ein aktives Forschungsfeld im maschinellen Sehen, das viele ungelöste Probleme umfasst. Die vorliegende Doktorarbeit befasst sich mit den in Zusammenwirkung erforschten Bereichen der dichten nicht-starren 3D Rekonstruktion aus monokularen Bildsequenzen, mit sowohl starren als auch nicht-starren Punktwolkenregistrierung. Die monokulare 3D Rekonstruktion, die im Fokus dieser Arbeit steht und die als nicht-rigide Struktur aus Bewegung (NRSaB) bekannt ist, wertet, einerseits, Bewegungen und Deformationen aus, und, andererseits, verknüpft diese mit den zusätzlichen Annahmen und Vorwissen über die Szene und die Art der zulässigen Zustände.

Die eingeführten Verfahren zur dichten NRSaB gehen auf mehrere offene Fragen ein, und zwar auf die Behandlung von Verdeckungen, die Skalierbarkeit und die Anpassungsfähigkeit auf unterschiedliche Szenarien und Größenordnungen der Szenen, interaktive und präzise Verarbeitung, sowie die Komprimierung dichter 3D Geometrie. Zwecks der Behandlung von Verdeckungen und fehlerhafter Punktkorrespondenzen werden Verfahren mit dem am Anfang einer Bildsequenz gewonnenen Formvorwissen sowie einem neuen räumlichen Kohärenz Regularisierer vorgestellt. Darüber hinaus, leiten wir ein neues NRSaB Verfahren her, das die gewonnene Geometrie in eine kompakte Repräsentation überführt.

Die entwickelten Verfahren zur Punktwolkenregistrierung verfolgen das Ziel, verrauschte und partielle Eingabedaten mit höherer Präzision zu verarbeiten als die Vorgängermethoden. Dementsprechend schlagen wir vor, die im Vorfeld hergestellten Korrespondenzen ins probabilistische Framework für die Punktwolkenregistrierung zu integrieren, und, zweitens, präsentieren wir eine neue und bisher unerforschte Verfahrensklasse, welche die Teilchenbewegungen unter virtuellen Schwerkräften simuliert.

Durch gründliche und zahlreiche Experimente ist es uns gelungen, die Geltung der neuen Ideen sowie die Präzision und Robustheit der entwickelten Verfahren zu bestätigen. Dank der neuen Prinzipien und Verfahren waren wir imstande, den Stand der Technik in beiden Bereichen der monokularen nicht-rigiden 3D Rekonstruktion sowie Punktwolkenregistrierung zu verbessern. Zu den Anwendungen neuer NRSaB Verfahren zählen die 3D Rekonstruktion von Menschen, Tieren und endoskopischer Aufnahmen sowie die Erfassung dünner Strukturen unterschiedlicher Herkunft. Die entwickelten Verfahren zur Punktwolkenregistrierung können unter anderem in Robotik, selbstfahrender Fahrzeugtechnik sowie Gesichts- und Formerkennung angewendet werden. Neben der erwähnten Gebieten wird in dieser Arbeit gezeigt, wie die beiden Verfahrensklassen zwecks der Übertragung des äußeren Erscheinungsbildes von Menschen sowie der Berechnung vom Szenenfluss aus Tiefenkamerabildern und monokularen Bildern angepasst werden können. Ferner, bieten die entwickelten Verfahren verschiedene Wege und Möglichkeiten zur Verbesserung und Weiterentwicklung, auf die am Schluss eingegangen wird.

1 Introduction

ONE of the objectives of computer vision is an accurate sensing of the real world and robust processing of the acquired data. Along with the material properties, knowledge of 3D geometry is a key component of complete scene description. 3D machine perception is the foundation for multiple applications which involve scene replication, scene understanding, localisation as well as a realistic superimposition of virtual contents, among others.

There are multiple sensors which come into question while designing a vision-based system including time-of-flight cameras, stereo cameras, lidars and sonars. A lightweight alternative to those is a single monocular camera. The advantages of a monocular camera are different designs and form-factors, affordability, relatively low electric energy consumption but also pervasiveness in modern electronic devices and wide acceptance in society. There are monocular cameras embedded in augmented reality glasses, helmets, mobile phones and tablets. Monocular cameras are central components in endoscopic surgery systems, surveillance systems, person identification systems, unmanned aerial and underwater vehicles, mobile robots, rovers for planetary explorations and autonomous cars. Thus, methods using monocular cameras for 3D sensing are of high relevance in a broad variety of systems and applications. Moreover, techniques for processing and analysis of the recovered raw 3D representations — often point sets and point clouds — are increasingly gaining relevance.

3D reconstruction is an extensively studied inverse problem in computer vision consisting in the recovery of the depth dimension of a scene lost during the imaging (together, the scene geometry), from single or multiple views. *Point set registration* is a computer vision problem of recovery the transformations aligning one or multiple point sets (raw 3D representations or 3D reconstructions) into a common coordinate frame or deforming the point sets so that their appearances match.

Depending on the available input and in many practical situations, 3D reconstruction can also be an ill-posed problem in the sense of Hadamard. Thus, 3D reconstruction from a single image is ill-posed, as multiple 3D scenes can result in the same 2D image. Additional prior knowledge is required to disambiguate the reconstruction such as a known object class, symmetry prior or a geometric prior. Starting from two views, additional constraints can be used ranging from epipolar geometry prior and trilinear constraints to the consistency constraints over multiple

V. Golyanik, *Robust Methods for Dense Monocular Non-Rigid 3D Reconstruction and Alignment of Point Clouds*,

views. Moreover, the rigidity assumption disambiguates the problem well, and impressive results were achieved in 3D reconstruction under the rigidity assumption, both from multiple views and the sequences of monocular views.

If several different views of the same scene at the same time frame are available, the subclass of the techniques is referred to as *multiple view geometry reconstruction*. In contrast, if multiple monocular views of the scene over several time frames are available, the subclass of the techniques is called *structure from motion* (SfM). In a general context, the input data corresponds to an image sequence, and some methods operate on a set of tracked points over the input views. The difference between multiple view geometry and structure from motion under rigidity is often subtle. In many cases, the techniques can be applied interchangeably, though the information about whether the cameras are static or moving can be advantageous (*e.g.*, motion blur prone to a moving camera can be accounted for). Compared to multi-view reconstruction, SfM often assumes smaller frame-to-frame displacements, as those observed in a video sequence. Video sequences also allow for stronger priors such as temporal smoothness.

Apart from predominantly static environments and sceneries, our surroundings are inhabited by living species including ourselves which move and deform. Besides, there are rigidly moving manmade instruments and products violating the assumption about the static and conserved ambience. Thus, capturing and processing of dynamic scenes is a core capability of robust vision-based systems.

1.1 Monocular Non-Rigid Dynamic 3D Reconstruction

The situation changes considerably, if the rigidity assumption does not hold anymore, *i.e.*, the scene undergoes non-rigid deformations. In the case of multiple views, the observations are captured at the same moment of time, and the geometry is still related by spatial rigidity between the views. If captured at different time frames and with different camera poses, the scene is observed in different states and the temporal rigidity does not hold anymore. The class of methods assuming non-rigid scenes over a temporal sequence of views is specified as *non-rigid structure from motion* (NRSfM). In NRSfM, the camera is moving, whereas the scene is moving and deforming. Similar to rigid SfM, the input of NRSfM is a set of tracked points over the available views.

Though NRSfM is a highly ill-posed problem which is sometimes said to be equivalent to the reconstruction from a single view, additional constraints can help to disambiguate it. *Real-world objects do not deform arbitrarily and rather follow a certain deformation pattern.* The deformation pattern is also often associated with periodicity, which implies that the scene states are repeated in a temporally-disjoint manner. Moreover, an average or middle state can be distinguished among all observed states. Additionally, it is more probable that the states in neighbouring frames are more similar than states in frames that are temporally further apart.

There is a substantial difference between single view rigid reconstruction and NRSfM which has crystallised out. In the single view rigid reconstruction, it is valid and common to assume a specific object class, and supervised learning methods are often applied. NRSfM, in contrast, assumes that no prior shape information about the observed scene is available, and solely relies on motion and deformation cues to obtain 3D surface reconstructions from monocular image sequences. This makes NRSfM capable of handling equally well — depending on the accuracy of correspondences — thin surfaces of different kinds (flags, sails, *etc.*), human and animal faces, clothes and body tissues in medical contexts.

Several new method classes have emerged which constrain the context of NRSfM, such as those assuming an accurate reconstruction of at least one of the frames in the sequence (template-based methods) and those assuming a pre-defined deformation model but different material properties.

1.2 Point Set Registration

When 3D surface recovery is complete, there are multiple ways how the dynamic reconstruction can be processed and analysed. One of the essential pre-processing steps is changing the reference frame or pose of the reconstruction for the further comparison, deformation transfer or recognition. This operation can be performed with *rigid point set registration* if the orientation of the reference frame or object is known. The comparison and deformation transfer can be accomplished with *non-rigid point set registration.*

The objective of point set registration is to align two or several point sets, *i.e.*, to recover a transformation which registers a *template* point set to an unaltered *reference.* A point set is an unordered set of coordinates (2D or 3D), with no further information available. As a representation of a shape, it can contain noise and clustered outliers, and some parts can be missing. Point set registration should not be confused with mesh registration methods (meshes are more complete shape representations consisting of points, triangles, normals *etc.*). 3D reconstructions obtained with NRSfM often represent point sets.

In the rigid case, the transformation is parametrised by the variables of rigid body motion with six degrees of freedom (three for rotation and three for translation). During a rigid transformation, no deformation is happening, and all distances between the points are preserved. Transformation of every point is given by the same rotation and translation. In the non-rigid case, due to deformations, distances between the points are not preserved, and the transformation is described by a general per-point displacement field. As monocular deformable reconstruction, *non-rigid point set registration relies on prior knowledge that real-world objects and scenes do not deform arbitrarily but rather follow some deformation rules and patterns.* One of the most commonly used and reasonable constraints in non-rigid point set registration is the topology the preserving constraint which states that point topology must be preserved despite the distances between the points are changing. It prevents intersections between the displacements, and, as a consequence, self-intersections of the surfaces represented by the points sets (though, point sets can also represent volumetric structures). Similarly to NRSfM, non-rigid point set registration is an ill-posed problem in the sense of Hadamard.

Despite the progress in point set registration which enabled various practical applications, one of the central research questions in point set registration remains improvement of the robustness to noise and disturbing effects in the data (missing parts and clustered outliers). Moreover, processing of large point sets is an ever-relevant problem (in other words, point sets containing one-two orders of magnitude more points than what is considered as a standard nowadays; the contemporary standard in NRSfM is around 30*k* points). It is addressed with faster hardware, parallelisation as well as data structures for acceleration. In contrast to methods for processing of synthetic 3D data (computer graphics), methods for processing of raw sensor inputs have to cope and consider noise and incompleteness of the data.

3D reconstruction and point set registration exhibit similarities. Thus, common types of assumptions and constraints can be applied to disambiguate them (*e.g.*, rigidity assumption and shape priors), and for handling non-rigid deformations, regularisation of displacement fields is required. In this thesis, the study of 3D reconstruction and point set registration is conducted jointly. As will be shown throughout the thesis, both related research fields facilitate and enrich each other with ideas. Point set registration provides tools for 3D reconstruction (algorithmic and evaluation tools), 3D reconstruction provides data for optimal evaluation of point set registration, and multiple concepts can be borrowed from one field to another one (regularisation of the displacement fields).

1.3 Scope of the Thesis

This thesis focuses on robust methods for dense monocular non-rigid 3D reconstruction from uncalibrated views and alignment of point cloud data.

The methods for dense monocular non-rigid 3D reconstruction should not assume that the calibration is known, though if it is known, the algorithms could optionally use the calibration parameters. Moreover, the new approaches should reconstruct the scene per-point densely, and, optionally, allow sparse reconstruction. The requirements to the new methods include robustness to self- and external occlusions, scalability, higher accuracy and lower runtime compared to the existing methods. Some of the requirements are not necessary facilitating towards the other ones, *i.e.*, it is more challenging to develop a scalable, accurate and fast method at the same time. Efficient NRSfM methods in conjunction with robust methods for dense correspondences would enable new applications based on commodity hardware.

Thanks to the point set registration, the reconstructed scenes can be compared to some reference data or the recovered appearance can be transferred to some other representations usable in different application scenarios. Thus, both method classes can be used in a single 3D reconstruction and processing pipeline.

There is also another reason to study the fields of monocular 3D reconstruction and point cloud alignment jointly. Even though the underlying methods pursue different goals and assume different input data, both fields are still related to each other, so that cross-fertilisation and exploitation of synergies is possible. Thus, non-rigid registration can help in the joint evaluation of NRSfM and correspondence establishment approaches, as will be shown in §5. Moreover, due to the handling of deformable structures in both algorithm classes, we proposed a new spatial regulariser (coherency constraint, §6) for NRSfM which was previously used exclusively in non-rigid point set registration.

The work at hand was also inspired by the maturing research area of augmented reality. Augmented reality is an interdisciplinary research field on the intersection of computer vision, computer graphics and hardware systems (which include material science, physics, mechatronics and electronics). The goal of augmented reality is to extend and enhance the perceived reality through useful virtual contents. Virtual contents should be realistic and indistinguishable from the real ones. Along with the realistic rendering, accurate placement of virtual contents is one of the quality factors. The acquisition of geometry of deformable objects with efficient methods for processing of the reconstructions is highly relevant for augmented reality as well. Both method classes addressed in this thesis — NRSfM and point set registration — can be used in augmented reality systems in a pipeline for 3D reconstruction and data processing with a single monocular moving camera.

1.4 Overview of the Contributions

The primary subject of the dissertation is dynamic 3D reconstruction of non-rigidly deforming scenes from monocular image sequences as well as processing of point sets. The main considered algorithm classes are non-rigid structure from motion (NRSfM) and point set registration (PSR). NRSfM is a highly ill-posed inverse problem. The input of NRSfM is a set of point tracks over several unsynchronised and uncalibrated views, and the objective is the recovery of the observed non-rigid 3D geometry. Thus, NRSfM uses motion and deformation cues as well as additional weak prior assumptions about the type of valid deformations for 3D recovery. In PSR, the inputs are two point sets with a different number of points, and the objective is the alignment of those into a common reference frame (in the rigid case) or the recovery of the displacements and correspondences non-rigidly aligning the inputs (in the non-rigid case). The two fields were studied jointly and complemented each other.

NRSfM and Monocular Surface Recovery

In the field of NRSfM, the thesis features the following contributions:

- First, a new dense variational NRSfM technique for handling large occlusions and inaccuracies in the data was proposed — *Shape Prior Variational Approach* (SPVA). SPVA estimates a shape prior from several first unoccluded frames of the sequence on-the-fly and guides the reconstruction by the occlusion tensor. The occlusion tensor is computed from the initial dense flow fields and indicates occlusion probabilities for every frame. The method allows for the reconstruction of scenes where large occlusions are expected (*e.g.*, in medical contexts). The method is parallelisable and is implemented on a GPU. The experimental results show the state-of-the-art accuracy on challenging sequences for which a shape prior can be obtained.

- Second, a new method with an intrinsic dynamic shape prior for 3D reconstruction and compression of sequences with temporally-disjoint rigidity is introduced. Temporally-disjoint rigidity occurs in most real video sequences, *i.e.*, the phenomenon of state reoccurrence. The repeating states can be separated by an arbitrary number of other states and can reappear in different poses. Our *Dynamic Shape Prior Reconstruction* (DSPR) approach takes advantage of temporally-disjoint rigidity and allows for dense reconstructions with low latencies. Experiments demonstrate that DSPR can operate on inaccurate correspondences.

- Third, a new spatial regulariser — the *coherency term* — for dense NRSfM is proposed which has allowed handling of large occlusions without a shape prior. The coherency term was adopted from the motion coherence theory. Before, the coherency term was used in non-rigid point set registration. We have shown how to minimise energy with the coherency term in the context of NRSfM.
- Fourth, we have addressed the problem of structure compressibility in the sense of data compression theory in the context of NRSfM and proposed a new *High-Dimensional Space Model* (HDSM) for NRSfM. In HDSM, non-rigid geometry in 3D is encoded as multiple projections of a single high dimensional structure onto different 3D subspaces. The proposed representation in combination with the factorisation-based (decoupled) formulation for camera pose and shape recovery allows compressing the structure in the high dimensional space. The resulting method encompassing handling of inaccurate point tracks with the coherency term and structure compression is known as *Lifted Coherent Depth Fields* (L-CDF).
- Fifth, we propose a new fast technique for dense NRSfM — *Accelerated Metric Projections* (AMP) — which allows to factorise dense batches of point tracks in seconds on a CPU. At the moment of publication, AMP was the fastest dense NRSfM method delivering high reconstruction accuracy. We have shown in AMP how to minimise a quadratic function on a set of orthonormal matrices using an efficient semidefinite programming solver. The method allows an arbitrary reshuffling of the per-frame measurements which can be advantageous in the cases when temporal information cannot be maintained.
- Sixth, we have addressed the question of scalability in the context of NRSfM. The core characteristic of the resulting robust NRSfM technique — *Scalable Monocular Surface Reconstruction* (SMSR) — is the steady high accuracy across a large variety of dense and sparse datasets with reasonable runtime and linear scalability w.r.t. the number of points. In SMSR, the camera pose is updated with singular value thresholding and proximal gradient techniques, whereas the surface is estimated by alternating direction method of multipliers.
- Seventh, we found a new way to regress non-rigid geometry with a trained encoder-decoder deep neural network. In the *Hybrid Deformation Model Network* (HDM-Net), the deformation model is learned from synthetic data in a supervised manner. Among contributions of HDM-Net is a new way to perform a convolution on a point set instead of a volumetric representation, an isometric loss and a contour loss. Moreover, the inference of a surface with over $5k$ points takes around 5 ms. Results on real images demonstrated the potential of the proposed architecture for augmented reality applications.

Besides, the thesis features several other contributions such as new joint evaluation methodology for correspondences and monocular non-rigid 3D reconstruction, a new dataset with ground truth geometry, dense correspondences and images, a new dataset for training a neural network for monocular non-rigid reconstruction, and a new method for monocular scene flow estimation based on NRSfM.

Point Set Registration

In the field of PSR, the thesis features the following contributions:

- First, we show how to embed prior correspondences into probabilistic PSR. The resulting algorithm is called *Extended Coherent Point Drift* (ECPD). The registration procedure with a set of sparse prior correspondences is more robust to noise and clustered outliers in the data compared to the state of the art. Non-rigid ECPD is implemented on a GPU and uses correspondence preserving subsampling of point clouds for a speed-up while processing large point sets. In the rigid case, ECPD performs joint pre-alignment and rigid registration with one or two prior correspondences which leads to an enhanced convergence basin compared to the separate processing. ECPD enables new applications with the processing of noisy human scans (*e.g.*, in the treatment of social pathologies).
- Second, we demonstrate how principles of collisionless particle dynamics can be used for alignment of point sets. Our discovery leads to a new class of point set registration methods relying on particle interactions under gravitational forces. The initially proposed *Gravitational Approach* (GA) for the rigid alignment by solving ordinary differential equations (ODE) of second order has demonstrated higher robustness to uniform noise, clustered outliers, as well as missing data, compared to widely used and state-of-the-art methods. The core statement of GA is that a locally optimal alignment corresponds to the locally minimal gravitational potential energy (GPE) of the joint system of particles.
- Third, in *Barnes-Hut Rigid GA* (BH-RGA), the particles interact in the world with altered laws of physics. This allows to perform a more stable optimisation with standard non-linear least squares compared to the solution with ODEs and achieve a broader basin of convergence compared to the state of the art. The core contribution of BH-GRA is fast multiply-linked registration of large point sets through adopting a tree data structure, *i.e.*, Barnes-Hut quadtree in 2D or octree in 3D, used in astrodynamical simulations and N-body problems. Through a mass normalisation technique, BH-RGA registers LIDAR point clouds without subsampling. Further applications of BH-RGA include alignment of cultural heritage data.

- Fourth, we develop *Non-Rigid GA* (NRGA) for non-rigid alignment of point sets using principles of particle dynamics. We pose non-rigid registration as multiple local rigid interactions of overlapping point groups. The individual displacements are then recovered with the new in the context of PRS *coherent collective motion* (CCM) regulariser. NRGA shows unprecedented registration accuracy in the presence of uniform noise and structured outliers in the data. NRGA balances well global topology and capture of local details in the template. It enables new applications with appearance transfer and 3D face matching. Note that GA, BH-RGA and NRGA allow for embedding of weak prior correspondences through point mass hierarchies.
- Fifth, a new method for scene flow estimation from RGB-D images is introduced. The method considers a short temporal window of frames and is called *Multiframe Scene Flow* (MSF) with piecewise rigid motion. In contrast to previous variational RGB-D scene flow approaches, MSF is formulated as an energy optimisation problem, and the energy is optimised with non-linear least squares. The energy includes, among other terms, the projective iterative closest point term operating directly on 3D points and the lifted segment pose regulariser term distinguishing between local coherent and independent motions. The experiments have shown the validity and high accuracy of the method.

1.5 Thesis Structure

This work is partitioned in eleven chapters including Introduction. §2 describes preliminaries such as frequently used constructs, optimisation methods as well as methods used as parts in the proposed approaches. §3.1 and §3.2 review NRSfM and point set registration method classes, respectively. §4 describes scalable NRSfM methods and §5 and 6 discusses dense NRSfM with explicit occlusion handling. Monocular non-rigid reconstruction with a learned deformation model is discussed in §7. §8 elaborates on probabilistic point set registration methods with prior correspondences whereas §9 expands on point set registration relying on principles of gravitational interaction. §10 shows applications of NRSfM and point set registration to scene flow estimation from different inputs (monocular and RGB-D inputs), and §11 concludes the thesis with a summary and a discussion of future directions.

Table 1.1: List of supporting publications with relevance attribution to the thesis chapters. † marks approximately equal contributions. See Table 1.2 for the conference abbreviations.

Publication	Chapter
V. Golyanik and D. Stricker. Dense Batch Non-Rigid Structure from Motion in a Second. In *WACV*, 2017. M. D. Ansari, **V. Golyanik** and D. Stricker. Scalable Dense Monocular Surface Reconstruction. In *3DV*, 2017.	§4
V. Golyanik, T. Fetzer and D. Stricker. Accurate 3D Reconstruction of Dynamic Scenes from Monocular Image Sequences with Severe Occlusions. In *WACV*, 2017. **V. Golyanik**, A. Jonas, D. Stricker and C. Theobalt. Intrinsic Dynamic Shape Prior for Dense Monocular Non-Rigid 3D Reconstruction and Compression of Sequences with Temporally-Disjoint Rigidity. In *ArXiv.org*, 2019.	§5
V. Golyanik, T. Fetzer and D. Stricker. Introduction to Coherent Depth Fields for Dense Monocular Surface Recovery. In *BMVC*, 2017. **V. Golyanik** and D. Stricker. High Dimensional Space Model for Dense Monocular Surface Recovery. In *3DV*, 2017.	§6
V. Golyanik, S. Shimada, K. Varanasi and D. Stricker. HDM-Net: Monocular Non-rigid 3D Reconstruction with Learned Deformation Model. In *EuroVR*, 2018.	§7
V. Golyanik, B. Taetz and D. Stricker. Joint Pre-Alignment and Robust Rigid Point Set Registration. In *ICIP*, 2016. **V. Golyanik**†, B. Taetz†, G. Reis and D. Stricker. Extended Coherent Point Drift Algorithm with Correspondence Priors and Optimal Subsampling. In *WACV*, 2016. **V. Golyanik**, G. Reis, B. Taetz and D. Stricker. A Framework for an Accurate Point Cloud Based Registration of Full 3D Human Body Scans. In *MVA*, 2017.	§8
V. Golyanik, S. A. Ali and D. Stricker. Gravitational Approach for Point Set Registration. In *CVPR*, 2016. **V. Golyanik**, C. Theobalt and D. Stricker. Accelerated Gravitational Point Set Alignment with Altered Physical Laws. In *ICCV*, 2019. S. A. Ali, **V. Golyanik** and D. Stricker. NRGA: Gravitational Approach for Non-Rigid Point Set Registration. In *3DV*, 2018.	§9
V. Golyanik, K. Kim, R. Maier, M. Nießner, D. Stricker, J, Kautz. Multiframe Scene Flow with Piecewise Rigid Motion. In *3DV*, 2017. **V. Golyanik**, A. S. Mathur and D. Stricker. NRSfM-Flow: Recovering Non-Rigid Scene Flow from Monocular Image Sequences. In *BMVC*, 2016.	§10

1.5.1 Supporting Publications

This section resolves the relation between the chapters of this thesis and published as well as submitted works. Table 1.1 provides a list of supporting publications, with a key for conference abbreviations given in Table 1.2. A comprehensive list of publications by the author, including the supporting and additional works as of 2020 (but now standing in the focus of this thesis) can be found in the Appendix.

Table 1.2: List of conference proceedings with abbreviations.

Conference	**Abbreviation**
IEEE Winter Conference on Applications of Computer Vision	WACV
IEEE Conference on Computer Vision and Pattern Recognition	CVPR
IEEE International Conference on Image Processing	ICIP
International Conference on 3D Vision	3DV
British Machine Vision Conference	BMVC
International Conference on Machine Vision Applications	MVA
International Conference on Virtual Reality and Augmented Reality	EuroVR

2 Preliminaries

THIS chapter reviews the concepts, mathematical techniques and computer vision methods which constitute the background knowledge relevant for the thesis at hand. The chapter is organised into three sections, each of which follows an individual common thread. In Sec. 2.1, starting from an overview of the imaging models, we review Hadamard problem classification, basics of the theory of inverse problems and non-linear least squares. In Sec. 2.2, we recap techniques relevant for NRSfM chapters including singular value decomposition, rigid body parametrisation, optical flow, rigid factorisation-based structure from motion approach of Tomasi-Kanade [281] and NRSfM approach of Bregler *et al.* [53]. Sec. 2.3 is devoted to the foundation of the chapters on point set registration. It elaborates on N-body simulations, expectation-maximisation and Gaussian mixture models as well as outlines two methods for point set registration — Iterative Closest Point (ICP) [40] and Coherent Point Drift (CPD) [203].

2.1 Computer Vision Primer

In this section, we review models of image formations, Hadamard problem classification and non-linear least squares as a powerful optimisation technique for inverse problems in computer vision.

2.1.1 Perspective and Orthographic Projections

Computer vision encompasses techniques for understanding and inferring from visual signals or images. One of the essential tools of low-level vision is an image formation formalism, *i.e.*, a mathematical model relating a physical 3D scene with the observed 2D images. The underlying imaging process is inherently non-linear. Objects of the same size placed at different ranges from the observer appear to be of different sizes. More distant objects appear smaller than the closer ones, and the apparent size is inversely proportional to the distance. Besides, the depth influences where the objects are observed in 2D. *Perspective* is the spatial scene configuration attached to the observer. Among several effects (occlusions, colour desaturation), it includes distance-dependent object size relativity of objects. The

V. Golyanik, *Robust Methods for Dense Monocular Non-Rigid 3D Reconstruction and Alignment of Point Clouds*,

perspective projection is an accurate mapping model from 3D to 2D which describes well the majority of imaging processes encountered in everyday life. The human visual system also operates according to the perspective projection, *i.e.*, it models how humans see. A simple yet accurate *pinhole camera model* relates 3D object coordinates with the coordinates of the observed points in 2D in an idealised imaging process (*e.g.*, without lens distortion). Many computer vision algorithms assume that the input images are undistorted (the undistortion usually happens in a pre-processing step) so that the pinhole camera is an accurate model in practice.

In the perspective camera model, every visible 3D point $\mathbf{P} = (P_x, P_y, P_z)$ is projected to the image plane with the projection operator $\pi : \mathbb{R}^3 \rightarrow \mathbb{R}^2$:

$$\mathbf{p}(x,y) = \pi(\mathbf{P}) = \left(f_x \frac{P_x}{P_z} + c_x, f_y \frac{P_y}{P_z} + c_y \right)^{\mathsf{T}}, \tag{2.1}$$

with the 2D projection $\mathbf{p}(x,y)$, the *focal lengths* of the camera $\{f_x, f_y\}$ and the *principal point* $(c_x, c_y)^{\mathsf{T}}$. The principal point is the point where the optical axis intersects the image plane. In high-quality sensors, the principal point is located very close to the centre of the sensor. Focal length is the measure of the optical power, *i.e.*, how strong a lens refracts the light. It is defined as the distance from the vertical axis of the lens to the convergence point of the initially parallel incoming rays. The degree of sharpness at the focal length depends on the quality of the lens.

The inverse projection operator $\pi^{-1} : \mathbb{R}^2 \times \mathbb{R} \rightarrow \mathbb{R}^3$ maps a 2D image point to a 3D scene point along the preimage given a depth value z as follows:

$$\pi^{-1}(\mathbf{p}(x,y), z) = \left(z \frac{x - c_x}{f_x}, z \frac{y - c_y}{f_y}, z \right)^{\mathsf{T}}. \tag{2.2}$$

If depth variation of a scene is smaller than the average depth (or the distance to the camera) along with a small field of view, the weak-perspective projection is an accurate approximation of the imaging process. In the weak-perspective projection, the depth P_z is replaced by the average depth P_{mean}:

$$\pi_{\text{weak perspective}}(\mathbf{P}) = \left(f_x \frac{P_x}{P_{\text{mean}}} + c_x, f_y \frac{P_y}{P_{\text{mean}}} + c_y \right)^{\mathsf{T}}. \tag{2.3}$$

By setting $P_{\text{mean}} = 1$ in Eq. (2.3), we obtain orthographic projection — a further simplified linear approximation of the imaging process:

$$\pi_{\text{orthographic}}(\mathbf{P}) = \left(s_x P_x + c_x, s_y P_y + c_y \right)^{\mathsf{T}}, \tag{2.4}$$

with the scale vector $(s_x, s_y)^\top$. In orthographic projection, the depth of a point P_z does not influence its projected position in the image plane, *i.e.*, irrespective of the depth, all parts of the scene are projected using the same rule. This implies that the lengths are preserved irrespective of the depth, and no perspective effects are present. By analogy with the perspective projection, the 2D-to-3D back projections with weak-perspective and orthographic models can be readily derived.

Orthographic projection is the main tool in technical drawing. The objective of a technical drawing is to depict the technical characteristics of a product, suppose an aeroplane, and enable its reproducibility. Thus, the lengths have to be conveyed in an unaltered way and without distortions specific to the perspective projection. In computer vision, the orthographic camera model is used when the assumptions for the weak-perspective camera model are fulfilled and when intrinsic camera parameters are not known and cannot be reliably recovered due to the ill-posedness of the problem. Monocular non-rigid reconstruction addressed in this thesis adopts the orthographic camera model.

2.1.2 Problem Classification in the Sense of Hadamard

According to J. Hadamard [135], a problem is *well-posed* if and only if the following three conditions hold simultaneously: a solution *exists*, the solution is *unique* and *changes gradually with the gradual change of the inputs*. If one or several of the above-listed conditions are violated, the problem becomes *ill-posed*. Originally, the three conditions mentioned above were formulated for boundary value problems in mathematical physics. Regularisation applied to an ill-posed problem targets at mimicking some physical properties of the underlying problem.

In computer vision, most of the problems are ill-posed. First, there can be multiple solutions to the problem, and additional prior knowledge is required to choose one solution out of many, *i.e.*, the most suitable one in the solution space according to the selected optimality criterion. Second, the solution does not always depend continuously on the input data. Further developing the notion of the solution continuity, we can deduce that not only sudden changes can occur. Additionally, the solution can evince different properties (faster or slower changes, more or less revealed noise, singularities, corruption) with the different accuracy and noise level of the input data. The ill-posedness of computer vision problems stems to a large extent from their nature, *i.e.*, because they are inverse problems.

2.1.3 Inverse Problems in Computer Vision

For computer vision, with no exception for the thesis at hand, the theory of inverse problems is of high relevance. The term *inverse problem* encompasses several aspects. *The objective of an inverse problem is to infer the causing physical state or phenomena, given implicit measurements of the latter.* In other words, the implicit observations represent the consequence, and the physical state or phenomena represent the origin. The implicit observations contain partial information about the cause. In practical situations, measured data contains noise (due to the imperfection of the sensors and external disturbing effects) and does not always strictly obey a mathematical model (first, due to the noise and, second, due to possible simplifications and assumptions in the modelling of the direct process).

Theory of inverse problems is an area of applied mathematics. It always assumes that the data is noisy, the observations are partial, the model of the direct process can be inaccurate and that the source is unknown. In many cases, the unknown variable or function is high dimensional [97]. A *direct problem* — in contrast to an inverse problem — consists in the derivation or simulation of the cause given the source or model of the causing physical state or phenomenon. Physics operate "in a direct way", *i.e.*, it solves direct problems according to the laws of physics. Inversion in physics if possible, for instance, if a body is heated and then cooled down and returned to its initial temperature. In this case, the inverse process has direct formulation. Speaking in mathematical analogies, an inverse problem in physics is well-posed and has a direct formulation. It is known that it is not possible to invert the flow of time in non-symmetric processes (*e.g.*, diffusion in physics). Thus, most time-dependent processes and problems in computer vision are non-invertible.

An example of a direct problem is physical prediction and forecasting. Starting from an initial state, it is possible to simulate physical laws and forecast future states of the system. Inverse problem consists in inverting the direct process beginning from the consequences.

An example of a direct problem in computer vision is a projection of a known 3D scene by a virtual camera to a virtual image plane. This is a standard step in visual odometry and RGB-D scene flow estimation (see chapter 10). In computer vision, images are implicit observations. Examples of inverse problems are image denoising, optical flow and scene flow estimation, 3D reconstruction, recovery of a displacement field between several shapes or several states of the same deformed shape (non-rigid point set registration), ultrasound imaging and inverse kinematics. Thus, computer vision, following its purpose, naturally deals with inverse and ill-posed problems.

Ill-posed problems can be regularised to obtain a reasonable solution. Regularisation means altering the objective in such a way that it becomes less ill-posed or well-posed. At the same time, the solution to the new problem should correspond or suit as a solution to the original problem.

One of the frequently encountered examples of regularisation techniques is the least squares method. Instead of solving an ill-posed $\mathbf{Ax} = \mathbf{b}$ problem, least squares replaces it by $\mathbf{A}^\mathsf{T}\mathbf{Ax} = \mathbf{A}^\mathsf{T}\mathbf{b}$ problem which has a closed-form solution. When opting for the least squares, it has to be taken into account that the corresponding least squares problem can be rank defficient and not always well-posed. Rank deficient least squares problems are severely ill-posed, as a small perturbation of the data can result in a significant change of the solution violating the Hadamard's condition of the gradual solution change (see Sec. 2.1.2). Whether a least-squares system is well- or ill-posed can be seen from the system's condition number. Non-linear least squares (NLLS) is a regularisation methodology for problems involving non-linear operators. The next section describes NLLS in detail.

2.1.4 Non-Linear Least Squares

C. F. Gauss has shown that a least squares estimate is equivalent to a maximum-likelihood estimation under zero-mean Gaussian noise. Thus, least squares and quadratic functions suit well for problems where the Gaussian noise models well a process. In many problems in computer vision, normally distributed noise is a reasonable assumption which leads to accurate results in practice. In Chapter 10, we show that NLLS suits well for optimisation of compound energy functional in the context of RGB-D scene flow estimation.

Target functionals in the least squares problems can be highly anisotropic and — due to non-linearities — non-convex. The main idea of non-linear least squares is linearisation around the current solution estimate and solving a series of linear least squares problems. Suppose in an NLLS problem, x is a set of M unknowns to be estimated and $f_i(\mathrm{x}),\, i \in \{1,\ldots,N\}$ is a set of known residuals parametrised by x. We stack the residuals into a single multivariate vector-valued function $\mathbf{F}(\mathrm{x}) : \mathbb{R}^M \to \mathbb{R}^N$:

$$\mathbf{F}(\mathrm{x}) = [f_1(\mathrm{x}), f_2(\mathrm{x}), \ldots, f_N(\mathrm{x})]^\mathsf{T}. \tag{2.5}$$

An NLLS problem can now be compactly written as

$$\min_{\mathrm{x}} \mathbf{E}(\mathrm{x}) = \min_{\mathrm{x}} ||\mathbf{F}(\mathrm{x})||_2^2. \tag{2.6}$$

In Eq. (2.6), $||\mathbf{F}(\mathrm{x})||_2^2$ is a compact notation for a sum of squared residuals written down as a squared ℓ_2-norm. The aim of NLLS is to find an optimal parameter set x' so that

$$\mathrm{x}' = \arg\min_{\mathrm{x}} ||\mathbf{F}(\mathrm{x})||_2^2. \tag{2.7}$$

The optimality condition for the optimisation (2.7) is generally given by

$$2\sum f_i(\mathrm{x})\frac{\partial f_i}{\partial \mathrm{x}} = 0, j \in \{1,\ldots,M\}. \tag{2.8}$$

Usually, Eq. (2.8) is not directly solvable in practice, and we resort to an iterative technique.

As the problem in Eq. (2.7) is non-linear, we iteratively linearise the objective around the current solution x_t and find an update through minimisation of the linear objective function. The first-order Taylor expansion of (2.7) leads to

$$\mathbf{F}(\mathrm{x}+\Delta\mathrm{x}) \approx \mathbf{F}(\mathrm{x}) + \mathbf{J}(\mathrm{x})\Delta\mathrm{x}, \tag{2.9}$$

where $\mathbf{J}(\mathrm{x})_{\mathrm{M}\times(\mathrm{K}^{\mathrm{N}}C_2+n_{\mathrm{pp}})}$ is the Jacobian of $\mathbf{F}(\mathrm{x})$ at point x. Consequently, we define the objective for $\Delta\mathrm{x}$:

$$\min_{\Delta\mathrm{x}} \|\mathbf{J}(\mathrm{x})\Delta\mathrm{x} + \mathbf{F}(\mathrm{x})\|^2. \tag{2.10}$$

Problem (2.10) is convex, and the minimum is achieved when

$$\mathbf{J}(\mathrm{x})\Delta\mathrm{x} = -\mathbf{F}(\mathrm{x}). \tag{2.11}$$

Since (2.10) is overconstrained, Eq. (2.11) has a solution in the least-squares sense. By projecting $\mathbf{F}(\mathrm{x})$ into the column space of $\mathbf{J}(\mathrm{x})$, we obtain the corresponding normal equations

$$\mathbf{J}(x)^{\mathsf{T}}\mathbf{J}(\mathrm{x})\,\Delta\mathrm{x} = -\mathbf{J}(\mathrm{x})^{\mathsf{T}}\,\mathbf{F}(\mathrm{x}) \tag{2.12}$$

which has a unique solution.

2.1.4.1 Levenberg-Marquardt Algorithm

For an enhanced convergence, we introduce a Tikhonov-type regulariser resulting in the Levenberg-Marquardt (LM) method [175, 189]:

$$\left(\mathbf{J}(x)^{\mathsf{T}}\mathbf{J}(\mathrm{x}) + \lambda\,\mathrm{diag}(\mathbf{J}(x)^{\mathsf{T}}\mathbf{J}(\mathrm{x}))\right)\Delta\mathrm{x} = -\mathbf{J}(\mathrm{x})^{\mathsf{T}}\,\mathbf{F}(\mathrm{x}), \tag{2.13}$$

where λ is a damping parameter. Beginning with a substantial λ value, LM starts as a Gradient Descent method and changes over to the Gauss-Newton in the vicinity of the local minimum. To solve the system of linear equations (2.13), Conjugate Gradient on the Normal Equations (CGNE) can be applied. CGNE is a method from the Krylov subspace class. The main idea of CGNE consists in casting a solution of a linear system as a minimisation of a quadratic form and performing an update step in the direction accounting for the space stretching. CGNE takes advantage of the sparse matrix structure and is suitable for large systems. Though $\mathbf{J}(\mathrm{x})^{\mathsf{T}}\mathbf{J}(\mathrm{x})$ can be substantially large depending on the parameter number, $\mathbf{J}(\mathrm{x})$ is highly sparse, as every individual residual depends only on few parameters from x. Thus, it neither stores nor computes $\mathbf{J}(\mathrm{x})$ and $\mathbf{J}(\mathrm{x})^{\mathsf{T}}\mathbf{J}(\mathrm{x})$ explicitly. Non-linear conjugent gradient (CG) can also be used to directly solve non-linear optimisation problems. The convergence rate of CG is, however, only linear in this case (and CG is similar to gradient descent with momentum) [136].

2.1.4.2 Huber Norm

It is possible to use a Huber loss in NLLS defined as

$$\left\|a^2\right\|_{\varepsilon} = \begin{cases} \frac{1}{2}a^2, & \text{for } |a| \leq \varepsilon \\ \varepsilon(|a| - \frac{1}{2}\varepsilon), & \text{otherwise,} \end{cases} \tag{2.14}$$

with a non-negative scalar threshold ε. Huber loss allows to reduce the influence of large outliers, *i.e.*, if the residual exceeds ε, the values are not squared. As a result, the optimisation is more robust against outliers and noise in the input data.

2.2 From Sparse Rigid Structure from Motion to Sparse Non-Rigid Structure from Motion

In this chapter, we review the essentials for monocular non-rigid 3D reconstruction including eigenvalue and singular value matrix decompositions, two classical factorisation-based rigid and non-rigid structure from motion methods, techniques for parametrisation of rotations, matrix projection into the rotation group $SO(3)$ as well as optical flow techniques for computing dense correspondences from image sequences.

2.2.1 Eigenvalue Decomposition of a Matrix

This section is partially based on [260] and partially on [94]. Consider a square diagonalisable matrix $\mathbf{A}$ of dimension D. For any such $\mathbf{A}$ there exists a decomposition of the form

$$\mathbf{A} = \mathbf{Q}\boldsymbol{\Lambda}\mathbf{Q}^{-1}, \tag{2.15}$$

where $\mathbf{Q}$ is a square matrix and $\boldsymbol{\Lambda}$ is a diagonal matrix. Moreover, the columns $\mathbf{q}_i$ of $\mathbf{Q}$ and the non-zero elements λ_i of $\boldsymbol{\Lambda}$ are related to each other as follows:

$$\mathbf{A}\mathbf{q}_i = \lambda_i \mathbf{q}_i, \;\; \forall i \in \{1, \ldots, D\}. \tag{2.16}$$

In the eigenvalue equation (2.16), the unknown $\mathbf{q}_i$ are called *eigenvectors* of $\mathbf{A}$ and the unknown λ_i are called *eigenvalues* of $\mathbf{A}$. From Eq. (2.16), we see that $\mathbf{A}\mathbf{q}_i$

are the scaled versions of $\mathbf{q}_i$ with the scaling factors λ_i, *i.e.*, $\mathbf{A}$ applied to $\mathbf{q}_i$ does not change the directions of $\mathbf{q}_i$. An arbitrary vector $\mathbf{Ax}$, in general, comes out in a different direction compared to $\mathbf{x}$. Eigenvalues of $\mathbf{A}$ can be repeated (though still correspond to different eigenvectors). In the general case, eigenvalues can be complex. If $\mathbf{A}$ is symmetric, the eigenvalues are real and $\mathbf{Q}$ can be chosen to be orthogonal:

$$\mathbf{A} = \mathbf{Q}\boldsymbol{\Lambda}\mathbf{Q}^{\mathsf{T}}. \tag{2.17}$$

Eigenvalues and eigenvectors are primary characteristics of a matrix. Thus, the sum of eigenvalues is equal to the trace of the matrix:

$$\mathrm{trace}(\mathbf{A}) = \sum_i \lambda_i, \tag{2.18}$$

and the product of eigenvalues equals its determinant:

$$\det(\mathbf{A}) = \prod_i \lambda_i. \tag{2.19}$$

Note that in Eq. (2.18)–(2.19), all eigenvalues are included along with the repeated values (the total number of λ_i has to be equal to D). Next, eigenvalues of $\mathbf{A}^{-1}$ are $\frac{1}{\lambda_i}$, and the eigenvectors are the same.

An eigendecomposition of $\mathbf{A}$ can be found by considering

$$(\mathbf{A} - \lambda\mathbf{I})\mathbf{x} = \mathbf{0}. \tag{2.20}$$

Thus, $\mathbf{A}$ shifted by $\lambda\mathbf{I}$ has to be singular — as it maps a non-zero vector $\mathbf{x}$ to a zero vector — implying the *characteristic equation*

$$\det(\mathbf{A} - \lambda\mathbf{I}) = 0 \tag{2.21}$$

A solution to Eq. (2.21) is a set of λ_i. The corresponding set of eigenvectors $\mathbf{q}_i$ can be found from Eq. (2.20). For every substituted λ_i, the corresponding eigenvector $\mathbf{q}_i$ is a null-space of Eq. 2.20.

Some approaches in computer vision require only selected eigenvalues and eigenvectors to be computed. Numerical methods for finding selected eigenvalues include the *QR-method* with its several variants. The corresponding eigenvectors are found using different types of *inverse power iteration* method [94]. For large sparse matrices, *Arnoldi iteration* method is a frequent choice. Arnoldi iteration is an indirect method which computes eigenvalues from maps of test vectors. In chapter 8, we show how Implicitly Restarted Arnoldi Method [174] is used in point set registration. In the next section, a more general matrix decomposition is discussed — the singular value decomposition (svd). In contrast to eigenvalue decomposition, svd exists for every arbitrary rectangular matrix.

2.2.2 Singular Value Decomposition

This section is partially based on [260] and partially on [94]. Singular value decomposition (svd) is a matrix decomposition of the form

$$\mathbf{A} = \mathbf{U}\boldsymbol{\Sigma}\mathbf{V}^{\mathsf{T}}, \tag{2.22}$$

with $\mathbf{A} \in \mathbb{R}^{m \times n}$, $\mathbf{U} \in \mathbb{R}^{m \times m}$, $\mathbf{V} \in \mathbb{R}^{n \times n}$ and $\boldsymbol{\Sigma} \in \mathbb{R}^{m \times n}$.

$\mathbf{A}$ is a real matrix, $\mathbf{U}$, $\mathbf{V}$ are real orthogonal matrices, and $\boldsymbol{\Sigma}$ is a diagonal matrix with the elements denoted by σ_i, $i \in \{1,,\ldots,\min(m,n)\}$. In contrast to eigenvalue decomposition, svd exists for every matrix including non-square and complex matrices. In the following, we assume $\mathbf{A}$ to be real. σ_i are non-negative real numbers called singular values arranged in the descending order so that σ_1 is the largest singular value.

A closer look to svd reveals that $\mathbf{U}$ and $\mathbf{V}$ represent orthonormal bases for the column and row spaces of $\mathbf{A}$, respectively. Moreover, svd is also a matrix diagonalisation and decomposition with two different bases. The singular values serve as scaling factors between the respective basis vectors of column and row spaces. For every pair of basis vectors, we can write

$$\mathbf{A}\mathbf{v}_i = \sigma_i \mathbf{u}_i, \tag{2.23}$$

with $\mathbf{V} = [\mathbf{v}_1, \ldots, \mathbf{v}_k]$, $\mathbf{U} = [\mathbf{u}_1, \ldots, \mathbf{u}_k]$, $k = \min(m,n)$.

svd reveals multiple properties of a matrix. A rank of a matrix is a number its non-zero singular values. A numerical rank of a matrix is a number of its non-zero singular values, up to a threshold ε. Thus, svd allows determining whether a matrix is full rank. Next, svd allows computing a low-rank approximation of $\mathbf{A}$ by setting the smallest singular values to zero and reassembling the matrix. The spectral norm of $\mathbf{A}$ equals to σ_1, *i.e.*, the largest singular value of $\mathbf{A}$. The nuclear norm of $\mathbf{A}$ equals to the sum of singular values of $\mathbf{A}$. Frobenius norm of a matrix is a length of a vector consisting of all singular values of a matrix:

$$\|\mathbf{A}\|_{\mathfrak{F}} = \sqrt{\sum_{i,j} a_{i,j}^2} = \sqrt{\sum_i \sigma_i^2}. \tag{2.24}$$

A known svd of $\mathbf{A}$ allows calculating an inverse or a pseudo-inverse of a matrix by inverting the singular values and reassembling the decomposed matrix. A pseudo-inverse is required for obtaining a solution to a linear least squares problem in a closed form. The so-called normal equations can be solved if a pseudo-inverse $\mathbf{A}^{\dagger}$ of $\mathbf{A}$ is known. If $\mathbf{A}$ has a full rank, then a solution to $\mathbf{A}\mathbf{x} = \mathbf{b}$ that minimises the sum of squared differences between the left and right sides is given by

$$\mathbf{x} = (\mathbf{A}^{\mathsf{T}}\mathbf{A})^{-1}\mathbf{A}^{\mathsf{T}}\mathbf{b}, \tag{2.25}$$

and $\mathbf{A}^\dagger = (\mathbf{A}^\top\mathbf{A})^{-1}\mathbf{A}^\top$. Note that a more robust solution to linear least squares, especially for systems with low condition numbers, is obtained by an orthogonal decomposition method (with QR-decomposition). The orthogonal decomposition method avoids direct computation of the product $\mathbf{A}^\top\mathbf{A}$. At the same time, it is slower than a solution through normal equations.

The relation between an eigenvalue decomposition and svd arises from the consideration of $\mathbf{A}^\top\mathbf{A}$:

$$\mathbf{A}^\top\mathbf{A} = \mathbf{V}\Sigma\mathbf{U}^\top\mathbf{U}\Sigma\mathbf{V}^\top = \mathbf{V}\Sigma^2\mathbf{V}^\top, \tag{2.26}$$

since $\mathbf{A}^\top\mathbf{A}$ is symmetric, square and positive semidefinite matrix. Thus, $\mathbf{V}$ contains eigenvectors of $\mathbf{A}^\top\mathbf{A}$ and Σ^2 contains eigenvalues of $\mathbf{A}^\top\mathbf{A}$ on the diagonal. Eigenvalues are positive since $\mathbf{A}^\top\mathbf{A}$ is positive semidefinite, and the eigenvectors are orthogonal to each other.

The relationship between eigenvalue decomposition and svd in Eq. (2.26) offers the means for the computation of svd. First, singular values of $\mathbf{A}$ and $\mathbf{V}$ can be found as square roots of eigenvalues of $\mathbf{A}^\top\mathbf{A}$ and orthonormal eigenvectors of $\mathbf{A}^\top\mathbf{A}$, respectively. $\mathbf{U}$ can be found similarly by considering an eigenvalue decomposition of $\mathbf{A}\mathbf{A}^\top$. This results in a slow and, in many cases, not accurate svd, especially for large matrices. In practice, svd is computed numerically by one-sided Jacobi method or a zero-shift downward sweep algorithm by Demmel and Kahan [77].

svd plays a vital role in factorisation-based structure from motion techniques [53, 281]. It allows to obtain an initial estimate of the camera poses and the deformable geometry and serves as a starting point in several algorithms proposed in this thesis (see chapters 4–5).

2.2.3 Rigid Structure from Motion by Factorisation

A classic SfM pipeline consists of several steps including finding reliable correspondences between the input view, camera pose initialisations using the epipolar constraint (with the estimation of either essential or fundamental matrix) and a subsequent global refinement of the geometry and camera poses over multiple views with bundle adjustment. Either a calibrated or an uncalibrated perspective camera model is usually assumed. An orthographic camera is a linear approximation of an imaging process which is accurate when the distance to the scene exceeds ca. the fivefold of the scene extent (see Sec 2.1.1). In the case of an orthographic camera, SfM can be performed avoiding the estimation of the essential or fundamental matrix. Instead, a rank constraint on the measurement matrix can be used to recover camera poses and the scene structure. Tomasi and Kanade proposed a factorisation-based method for rigid SfM [281]. In this section, we will refer to their approach as Tomasi-Kanade factorisation (TKF).

Suppose P points are tracked over F frames so that their pixel coordinates are known for every $f \in \{1,\ldots,F\}$. The measurement matrix $\mathbf{W} \in \mathbb{R}^{2F\times P}$ encompasses all coordinates over F frames:

$$\mathbf{W} = \begin{bmatrix} \mathbf{X} \\ \hline \mathbf{Y} \end{bmatrix}, \tag{2.27}$$

with $\mathbf{X} = [x_{fp}]$, $\mathbf{Y} = [y_{fp}]$, $f \in \{1,\ldots,F\}$, $p \in \{1,\ldots,P\}$ the absolute x and y coordinates of the p-th point in the f-th frame, respectively.

In the first step, TKF rectifies $\mathbf{W}$ so that the centroid of the scene coincides with the origin of the coordinate system. This step resolves translation and ensures that the recovered camera poses are relative to the origin of the coordinate system:

$$\tilde{\mathbf{w}}_{f,p} = \mathbf{w}_{f,p} - \mathbf{c}_f, \tag{2.28}$$

where $\tilde{\mathbf{w}}_{f,p}$ is the entry of the rectified measurement matrix $\tilde{\mathbf{W}}$ for frame f and 2D point p, and $\mathbf{c}_f = \frac{1}{P}\sum_{p=1}^{P} \mathbf{w}_{f,p}$ is the mean of all coordinates $\mathbf{W}_f = [\mathbf{w}_{f,p}]$, $p \in \{1,\ldots,P\}$ at frame f.

Next, TKF factorises $\tilde{\mathbf{W}}$ into the product of camera poses and a rigid shape:

$$\tilde{\mathbf{W}} = \mathbf{RS}, \tag{2.29}$$

with camera poses $\mathbf{R} \in \mathbb{R}^{2F\times 3}$ and the 3D structure $\mathbf{S} = \begin{bmatrix} x_p & y_p & z_p \end{bmatrix}^\mathsf{T} \in \mathbb{R}^{3\times P}$, $p \in \{1,\ldots,P\}$. Since x and y coordinates in $\tilde{\mathbf{W}}$ are grouped together, the rows of camera poses $\mathbf{R}_f = \begin{bmatrix} \mathbf{i}_f \\ \mathbf{j}_f \end{bmatrix} \in \mathbf{R}^{2\times 3}$ are also grouped in $\mathbf{R}$:

$$\mathbf{R} = \begin{bmatrix} \mathbf{i}_1^\mathsf{T} & \ldots & \mathbf{i}_F^\mathsf{T} & | & \mathbf{j}_i^\mathsf{T} & \ldots & \mathbf{j}_F^\mathsf{T} \end{bmatrix}^\mathsf{T}. \tag{2.30}$$

From Eq. (2.29) we see that $\tilde{\mathbf{W}}$ is of rank 3 if both $\mathbf{R}_{2F\times 3}$ and $\mathbf{S}_{3\times P}$ are of the full rank (of rank 3 in both cases). In real applications, noise in the measurements can spuriously increase the rank of $\tilde{\mathbf{W}}$. Tomasi and Kanade have shown that the best possible shape and a set of camera poses are obtained after performing a rank-3 approximation of $\tilde{\mathbf{W}}$. Accordingly, in the next step TKF calculates the initial estimate of $\mathbf{R}$ and $\mathbf{S}$ by svd (see Sec. 2.2.2):

$$\tilde{\mathbf{W}} = \mathbf{U}\sqrt{\mathbf{\Sigma}}\sqrt{\mathbf{\Sigma}}\mathbf{V}^\mathsf{T} \tag{2.31}$$

and performs a rank-3 approximation of the registered measurement matrix $\tilde{\mathbf{W}}$ in the decomposed form:

$$\tilde{\mathbf{W}}' = \underbrace{\mathbf{U}'\sqrt{\Sigma'}}_{\tilde{\mathbf{R}}}\underbrace{\sqrt{\Sigma'}\mathbf{V}'^{\mathsf{T}}}_{\tilde{\mathbf{S}}} \tag{2.32}$$

(note that $\tilde{\mathbf{W}}'$ is not explicitly reassembled since we are searching for the factorisation of $\tilde{\mathbf{W}}$). Moreover, we keep only the first three columns of $\mathbf{U}$ and the first three rows of $\mathbf{V}^{\mathsf{T}}$. The decomposition $\tilde{\mathbf{W}}' = \tilde{\mathbf{R}}\tilde{\mathbf{W}}$ is not unique, since

$$\tilde{\mathbf{R}}\tilde{\mathbf{W}} = \tilde{\mathbf{R}}\mathbf{Q}\mathbf{Q}^{-1}\tilde{\mathbf{W}}, \tag{2.33}$$

where $\mathbf{Q} \in \mathbb{R}^{3\times 3}$ is an arbitrary invertible matrix. Nevertheless, despite $\tilde{\mathbf{R}}$ and $\tilde{\mathbf{S}}$ are different from the sought $\mathbf{R}$ and $\mathbf{S}$, respectively, there must exist linear corrective transformations between $\tilde{\mathbf{R}}$ and real $\mathbf{R}$ and $\tilde{\mathbf{S}}$ and real $\mathbf{S}$. For simplicity, we denote the corrective transformation by $\mathbf{Q}$.

Factorisation as shown in Eq. (2.33) does not take into account that $\mathbf{R}$ contains rotation matrices with orthonormal vectors. Imposing orthonormality constraints allows disambiguating the factorisation in Eq. (2.33) and find a valid $\mathbf{R}$ and an undistorted $\mathbf{S}$. The corrective transformation $\mathbf{Q}$ can be found by solving the over-constrained system of linear equations:

$$\begin{cases} \hat{\mathbf{i}}_f^{\mathsf{T}}\,\mathbf{Q}\mathbf{Q}^{\mathsf{T}}\,\hat{\mathbf{i}}_f^{\mathsf{T}} = 1 \\ \hat{\mathbf{j}}_f^{\mathsf{T}}\,\mathbf{Q}\mathbf{Q}^{\mathsf{T}}\,\hat{\mathbf{j}}_f^{\mathsf{T}} = 1 \\ \hat{\mathbf{i}}_f^{\mathsf{T}}\,\mathbf{Q}\mathbf{Q}^{\mathsf{T}}\,\hat{\mathbf{j}}_f^{\mathsf{T}} = 0 \end{cases}, \forall f \in \{1,\ldots,F\}. \tag{2.34}$$

In Eq. (2.34), $\hat{\mathbf{i}}_f$ and $\hat{\mathbf{i}}_f$ denote the row pairs in $\tilde{\mathbf{R}}$ corresponding to the same rotation matrix. The details of the solution of the system of equations (2.34) with least squares are outlined in Morita and Kanade [199].

Using the property of a valid rotation matrix $\mathbf{R}\mathbf{R}^{\mathsf{T}} = \mathbf{I}$, a valid rotation matrix in 3D $\mathbf{R}_f^{3D}$ can be obtained from $\mathbf{R}_f$ as

$$\mathbf{R}_f^{3D} = \begin{pmatrix} \mathbf{i}_f \\ \mathbf{j}_f \\ \mathbf{i}_f \times \mathbf{j}_f \end{pmatrix}, \tag{2.35}$$

where $\times$ denote a vector cross product.

There are several remaining ambiguities in TKF method. First, the depth is recovered up to the sign, since svd is prone to the sign ambiguity of singular vectors. Second, the camera poses are recovered relative to each other. The orthonormality constraints in the equation system (2.34) ensure that $\mathbf{R}_f$ are orthonormal matrices, but still, an arbitrary orthogonal matrix $\bar{\mathbf{Q}}$ can be inserted into Eq. (2.29):

$$\tilde{\mathbf{W}}_f = \mathbf{R}_f \bar{\mathbf{Q}} \bar{\mathbf{Q}}^\mathsf{T} \mathbf{S}, \ \forall f \in \{1,\ldots,F\} \tag{2.36}$$

while the orthonormality constraints are satisfied and both $\mathbf{R}_f\bar{\mathbf{Q}}$ and $\bar{\mathbf{Q}}^\mathsf{T}\mathbf{S}_f$ are the valid camera poses and structure, respectively. As a linear transformation, $\bar{\mathbf{Q}}$ preserves the inner product and can express rotation, reflection or rotation and reflection simultaneously (an improper rotation).

2.2.4 Non-Rigid Structure from Motion by Factorisation with Low-Rank Subspace Model

Factorisation based SfM under orthography has been extended to the non-rigid case by Bregler *et al.* [53]. They proposed to model the time-varying 3D geometry $\mathbf{S}_f$, $f \in \{1,\ldots,F\}$ as a linear combination of a few K unknown basis shapes $\mathbf{B}_k$, $k \in \{1,\ldots,K\}$, *i.e.*,

$$\mathbf{S}_f = \sum_{k=1}^{K} l_k \mathbf{B}_k. \tag{2.37}$$

We largely follow the notation introduced in Sec. 2.2.3. In this section, P stands for the total number of points tracked throughout F frames, and all points are combined into the measurement matrix $\mathbf{W} \in \mathbb{R}^{2F \times P}$:

$$\mathbf{W} = \begin{bmatrix} \mathbf{W}_1 \\ \mathbf{W}_2 \\ \vdots \\ \mathbf{W}_F \end{bmatrix} = \begin{bmatrix} \begin{bmatrix} x_{11} & x_{12} & \cdots & x_{1P} \\ y_{11} & y_{12} & \cdots & y_{1P} \end{bmatrix}_{f=1} \\ \begin{bmatrix} x_{21} & x_{22} & \cdots & x_{2P} \\ y_{21} & y_{22} & \cdots & y_{2P} \end{bmatrix}_{f=2} \\ \vdots \\ \begin{bmatrix} x_{F1} & x_{F2} & \cdots & x_{FP} \\ y_{F1} & y_{F2} & \cdots & y_{FP} \end{bmatrix}_{f=F} \end{bmatrix}. \tag{2.38}$$

Similar to rigid SfM, the measurement matrix is assumed to be centred at the origin, and the translation is resolved. We search for the factorisation of the form

$$\mathbf{W}_f = \mathbf{R}_f \sum_{k=1}^{K} l_k \mathbf{B}_k = \begin{bmatrix} l_1\mathbf{R}_f & l_2\mathbf{R}_f & \ldots & l_K\mathbf{R}_f \end{bmatrix} \begin{bmatrix} \mathbf{B}_1 \\ \mathbf{B}_2 \\ \vdots \\ \mathbf{B}_K \end{bmatrix}, \ \forall f \in \{1,\ldots,F\}. \tag{2.39}$$

For all $\mathbf{W}_f$ we can write

$$\mathbf{W} = \mathbf{M}\hat{\mathbf{B}} = \begin{bmatrix} \mathbf{q}_1 \\ \mathbf{q}_2 \\ \vdots \\ \mathbf{q}_{2F} \end{bmatrix} \begin{bmatrix} \hat{\mathbf{B}}_1 \\ \hat{\mathbf{B}}_2 \\ \vdots \\ \hat{\mathbf{B}}_K \end{bmatrix} = \begin{bmatrix} l_{11}\mathbf{R}_1 & l_{12}\mathbf{R}_1 & \dots & l_{1K}\mathbf{R}_1 \\ l_{21}\mathbf{R}_2 & l_{22}\mathbf{R}_2 & \dots & l_{2K}\mathbf{R}_2 \\ \vdots & \vdots & & \vdots \\ l_{F1}\mathbf{R}_F & l_{F2}\mathbf{R}_F & \dots & l_{FK}\mathbf{R}_F \end{bmatrix} \begin{bmatrix} \hat{\mathbf{B}}_1 \\ \hat{\mathbf{B}}_2 \\ \vdots \\ \hat{\mathbf{B}}_K \end{bmatrix}, \quad (2.40)$$

with $\mathbf{M} \in \mathbb{R}^{2F\times 3K}$ and $\hat{\mathbf{B}} \in \mathbb{R}^{3K\times P}$, and $\mathbf{q}_i$ denoting rows of $\mathbf{M}$. We see that $\mathbf{W}$ is at most of rank $3K$. The initial $\mathbf{M}$ and $\mathbf{B}$ can be estimated by svd:

$$\mathbf{W} = \underbrace{\mathbf{U}\sqrt{\Sigma}}_{\mathbf{M}}\underbrace{\sqrt{\Sigma}\mathbf{V}^{\top}}_{\mathbf{B}}. \quad (2.41)$$

Note that first $3K$ columns and rows of $\mathbf{U}$ and $\mathbf{V}^{\top}$ and the $3K$ largest singular values are sufficient. Next, the weights l_{fk} and the camera poses $\mathbf{R}_f$ have to be extracted from $\mathbf{M}$.

Consider pairs of rows in $\mathbf{M}$ (the frame index is dropped for convenience):

$$\begin{bmatrix} \mathbf{m}_{2f-1} \\ \mathbf{m}_{2f} \end{bmatrix}_f = \begin{bmatrix} l_1 r_{11} & l_1 r_{12} & l_1 r_{13} \dots l_K r_{11} & l_K r_{12} & l_K r_{13} \\ l_1 r_{21} & l_1 r_{22} & l_1 r_{23} \dots l_K r_{21} & l_K r_{22} & l_K r_{23} \end{bmatrix}_f, \quad (2.42)$$

where $\begin{bmatrix} r_{11} & r_{12} & r_{13} \\ r_{21} & r_{22} & r_{23} \end{bmatrix}_f = \mathbf{R}_f$. After reordering the right side of Eq. (2.42), we obtain

$$\begin{bmatrix} \bar{\mathbf{m}}_{2f-1} \\ \bar{\mathbf{m}}_{2f} \end{bmatrix}_f = \begin{bmatrix} l_1 r_{11} & l_1 r_{12} & l_1 r_{13} & l_1 r_{21} & l_1 r_{22} & l_1 r_{23} \\ l_2 r_{11} & l_2 r_{12} & l_2 r_{13} & l_2 r_{21} & l_2 r_{22} & l_2 r_{23} \\ \vdots & \vdots & \vdots & \vdots & \vdots & \vdots \\ l_K r_{11} & l_K r_{12} & l_K r_{13} & l_K r_{21} & l_K r_{22} & l_K r_{23} \end{bmatrix}_f = \begin{bmatrix} l_1 \\ l_2 \\ \vdots \\ l_K \end{bmatrix} \mathrm{vec}(\mathbf{R}^{\top})^{\top}, \quad (2.43)$$

with vec$(\cdot)$ denoting a vectorisation operator which stacks columns of a matrix into a column vector. Thus, $\begin{bmatrix} \bar{\mathbf{m}}_{2f-1} \\ \bar{\mathbf{m}}_{2f} \end{bmatrix}_f$ is of rank 1, $\forall f \in \{1,\dots,F\}$. Hence, by applying svd to $\begin{bmatrix} \bar{\mathbf{m}}_{2f-1} \\ \bar{\mathbf{m}}_{2f} \end{bmatrix}_f$ we can obtain the initial estimates for the decoupled l_{fk} and the camera poses $\hat{\mathbf{R}}_f$. Similarly to TKF [281], orthonormality constraints have to be imposed to recover the corrective transformation $\mathbf{Q}$ rectifying $\hat{\mathbf{R}}_f$ and $\hat{\mathbf{B}}_k$:

$$\begin{aligned} \mathbf{R}_f &= \mathbf{Q}\hat{\mathbf{R}}_f, \ \forall f \\ \mathbf{B}_k &= \mathbf{Q}^{-1}\hat{\mathbf{B}}_k, \ \forall k. \end{aligned} \quad (2.44)$$

The system of linear equations for finding $\mathbf{Q}$ reads

$$\begin{cases} \begin{bmatrix} r_{11} & r_{12} & r_{13} \end{bmatrix}_f \mathbf{Q}\mathbf{Q}^\top \begin{bmatrix} r_{11} & r_{12} & r_{13} \end{bmatrix}_f^\top = 1 \\ \begin{bmatrix} r_{21} & r_{22} & r_{23} \end{bmatrix}_f \mathbf{Q}\mathbf{Q}^\top \begin{bmatrix} r_{21} & r_{22} & r_{23} \end{bmatrix}_f^\top = 1 \\ \begin{bmatrix} r_{11} & r_{12} & r_{13} \end{bmatrix}_f \mathbf{Q}\mathbf{Q}^\top \begin{bmatrix} r_{21} & r_{22} & r_{23} \end{bmatrix}_f^\top = 0 \end{cases}, \forall f \in \{1,\dots,F\}. \tag{2.45}$$

The system of linear equations (2.45) is equivalent to the equation system (2.34) and can be solved by least squares [199].

The outlined NRSfM method has a free parameter K which has to be set depending on the scene complexity and the expected accuracy of point tracks in $\mathbf{W}$. Thus, an optimal K is case-specific and is not known in advance. The proposed low-rank subspace model can naturally cope with small and moderate deformations. Many techniques were inspired by the Bregler *et al.*'s method. One of the early extensions can handle missing data [283]. Multiple follow-up techniques addressed the questions of basis ambiguity, scene reconstructability (*e.g.*, integrating of additional constraints, handling of large deformations, alternative deformation models) and dense reconstructions.

2.2.5 Parametrisation of Rotations

Rotation and translation entirely describe rigid body transformation with six degrees of freedom in total. A convenient form of rotation representation is a rotation matrix

$$\mathbf{R} = \begin{bmatrix} r_{11} & r_{12} & r_{13} \\ r_{21} & r_{22} & r_{23} \\ r_{31} & r_{32} & r_{33} \end{bmatrix}. \tag{2.46}$$

Eq. (2.46) is an example of a 3D rotation. It is an element of a special orthogonal or rotational group $SO(3)$. For the arbitrary dimension n, the rotational group is defined as

$$SO(n) = \{\mathbf{R} \in \mathbb{R}^{n\times n} | \mathbf{R}\mathbf{R}^\top = \mathbf{R}^\top\mathbf{R} = \mathbf{I},\ \det(\mathbf{R}) = 1\}. \tag{2.47}$$

An arbitrary 3D point $\mathbf{P} = (P_x, P_y, P_z)$ can be rotated around the origin of the coordinate system by a single matrix-vector multiplication:

$$\mathbf{P}' = \mathbf{R}\mathbf{P}. \tag{2.48}$$

In computer vision, one often optimises for an optimal rotation. One of the ways to find an optimal rotation is to update nine parameters of the candidate rotation matrix in an unconstrained manner and subsequently project the update into the $SO(3)$ space with the method outlined in Sec. 2.2.6. Another way is to optimise in the space without the redundant degrees of freedom. Often used rotation parametrisations are Euler angles, axis-angle and quaternions. In the following, we review two latter representations.

2.2.5.1 Axis-Angle Representation

The motivation behind axis-angle representation is that every rotation can be equivalently represented as a rotation around a single axis by a single angle. In this section, we will use the $\mathfrak{so}(3)$ space is the Lie algebra corresponding to the $SO(3)$ group. Elements of $\mathfrak{so}(3)$ are axis-angle representations of rotations in the 3D space. In the angle-axis representation, a rotation is encoded by a vector $\boldsymbol{\alpha} = (\alpha_x, \alpha_y, \alpha_z)$. The direction (a unit vector) of $\boldsymbol{\alpha}$ indicates the axis of rotation $\boldsymbol{\alpha}_n$ (it is obtained by normalisation of $\boldsymbol{\alpha}$) and the length of $\boldsymbol{\alpha}$ indicates the angle of rotation θ around $\boldsymbol{\alpha}_n$ according to the right-hand rule. To rotate a segment, we convert $\boldsymbol{\alpha}$ to the corresponding rotation matrix $\mathbf{R}$ using a corollary of the Rodrigues' rotation formula leading to the following exponential map:

$$\mathbf{R} = \exp(\theta\,\mathbf{K}) = \mathbf{I} + \sin\theta\,\mathbf{K} + (1-\cos\theta)\mathbf{K}^2, \tag{2.49}$$

where $\mathbf{K} = \begin{pmatrix} 0 & -\alpha_z & \alpha_y \\ \alpha_z & 0 & \alpha_x \\ -\alpha_y & -\alpha_x & 0 \end{pmatrix} \in \mathfrak{so}(3)$ is given by a skew-symmetric cross-product matrix of $\boldsymbol{\alpha}_n$.

Given a rotation matrix $\mathbf{R}$, the corresponding axis-angle representation can be found as

$$|\boldsymbol{\alpha}| = \theta = \cos^{-1}\left(\frac{\mathrm{tr}(\mathbf{R})-1}{2}\right), \tag{2.50}$$

$$\frac{\boldsymbol{\alpha}}{|\boldsymbol{\alpha}|} = \boldsymbol{\alpha}_n = \frac{1}{2\cos\theta}\begin{bmatrix} r_{32}-r_{23} \\ r_{13}-r_{31} \\ r_{21}-r_{12} \end{bmatrix}. \tag{2.51}$$

The conversion according to (2.50) and (2.51) is not unique since the cosine is a symmetric function and $\cos\theta = \cos(-\theta)$. Thus, for two possible angles θ and $-\theta$ there are two possible corresponding axes $\boldsymbol{\alpha}_n$ or $-\boldsymbol{\alpha}_n$. Nevertheless, both axis-angles represent the same rotation in 3D space.

If the updates always occur on the rotational manifold (in the parametric space), the described ambiguity does not affect the optimisation. Care has to be taken if there are intermediate updates or initialisations of axis-angles from rotation matrices. If a unique and unambiguous representation of rotations is necessary, one option is to utilise quaternions.

2.2.5.2 Quaternions

Quaternions are four-dimensional numbers denoted by $\mathbb{H}$ which extend complex numbers with two additional imaginary basis elements. In total, there are three imaginary basis elements commonly denoted by i, j and k and one real basis element 1. It is postulated that

$$i^2 = j^2 = k^2 = ijk = -1, \tag{2.52}$$

with further non-commutative multiplication rules derived from (2.52). A quaternion can be represented as a vector with four elements:

$$\mathbf{q} = \begin{bmatrix} q_0 & q_1 & q_2 & q_3, \end{bmatrix}^\mathsf{T} = \begin{bmatrix} q_0 \\ \mathbf{v} \end{bmatrix}, \tag{2.53}$$

with q_0 denoting the real part and $\mathbf{v} = \begin{bmatrix} q_1 & q_2 & q_3 \end{bmatrix}^\mathsf{T}$ denoting the imaginary i, j and k parts, respectively.

Given two quaternions $\mathbf{q}_1 = \begin{bmatrix} q_0^1 \\ \mathbf{v}^1 \end{bmatrix}$ and $\mathbf{q}_2 = \begin{bmatrix} q_0^1 \\ \mathbf{v}^2 \end{bmatrix}$, their sum is calculated as

$$\mathbf{q}_1 + \mathbf{q}_2 = \begin{bmatrix} q_0^1 + q_0^2 \\ \mathbf{v}^1 + \mathbf{v}^2 \end{bmatrix}, \tag{2.54}$$

and the product is given by

$$\mathbf{q}_1\mathbf{q}_2 = \begin{bmatrix} q_0^1 q_0^2 - \mathbf{v}^1 \cdot \mathbf{v}^2 \\ \mathbf{v}^1 \times \mathbf{v}^2 + q_0^1 \mathbf{v}^2 + q_0^2 \mathbf{v}^1 \end{bmatrix}, \tag{2.55}$$

which can also be written as a matrix-vector product:

$$\mathbf{q}_1\mathbf{q}_2 = \mathbf{Q}\mathbf{q}_2 = \begin{bmatrix} q_0 & -\mathbf{v}^1 \\ \mathbf{v}^1 & \mathbf{B} \end{bmatrix} \mathbf{q}_2 = \begin{bmatrix} q_0 & -\mathbf{v}^1 \\ \mathbf{v}^1 & \begin{bmatrix} q_0^1 & -q_3^1 & q_2^1 \\ q_3^1 & q_0^1 & -q_1^1 \\ -q_2^1 & q_1^1 & q_0^1 \end{bmatrix} \end{bmatrix} \mathbf{q}_2. \tag{2.56}$$

The product $\mathbf{q}_2\mathbf{q}_1$ can be expressed in a similar way, with the transposed $\mathbf{B}$ from Eq. (2.56), namely as

$$\mathbf{q}_2\mathbf{q}_1 = \mathbf{Q}^\sharp \mathbf{q}_2 = \begin{bmatrix} q_0 & -\mathbf{v}^1 \\ \mathbf{v}^1 & \mathbf{B}^\mathsf{T} \end{bmatrix} \mathbf{q}_2. \tag{2.57}$$

A conjugate of $\mathbf{q}$ is obtained by inverting the sign of the imaginary part as $\bar{\mathbf{q}} = \begin{bmatrix} q_0 \\ -\mathbf{v} \end{bmatrix}$. For unit quaternions, a conjugate of the quaternion coincides with its multiplicative inverse.

Rotations with Quaternions

Consider the map $L_{\mathbf{q}}(\mathbf{p}) = \mathbb{H} \times \mathbb{R}^3 \rightarrow \mathbb{R}^3$:

$$L_{\mathbf{q}}(\mathbf{p}) = \mathbf{q}\mathbf{p}\bar{\mathbf{q}} = (\mathbf{Q}\mathbf{p})\bar{\mathbf{q}} = \bar{\mathbf{Q}}^{\sharp}(\mathbf{Q}\mathbf{p}) = \left(\mathbf{Q}^{\sharp^{\mathsf{T}}}\mathbf{Q}\right)\mathbf{p}, \tag{2.58}$$

with a unit quaternion $\mathbf{q}$, a quaternion $\mathbf{p}$ with a zero real part and

$$\mathbf{Q}^{\sharp^{\mathsf{T}}}\mathbf{Q} = \begin{bmatrix} |\mathbf{q}|^2 & \mathbf{0}^{\mathsf{T}} \\ \mathbf{0} & \mathbf{R} \end{bmatrix} = \begin{bmatrix} |\mathbf{q}|^2 & 0 & 0 & 0 \\ 0 & q_0^2+q_1^2-q_2^2-q_3^2 & 2(q_1q_2-q_0q_3) & 2(q_1q_3+q_0q_2) \\ 0 & 2(q_1q_2+q_0q_3) & q_0^2-q_1^2+q_2^2-q_3^2 & 2(q_2q_3-q_0q_1) \\ 0 & 2(q_1q_3-q_0q_2) & 2(q_2q_3+q_0q_1) & q_0^2-q_1^2-q_2^2+q_3^2 \end{bmatrix}. \tag{2.59}$$

$L_{\mathbf{q}}(\mathbf{p})$ maps purely imaginary quaternions (3D vectors) to purely imaginary quaternions while preserving their norms. $\mathbf{Q}^{\sharp}$, $\mathbf{Q}$ and $\mathbf{Q}^{\sharp^{\mathsf{T}}}\mathbf{Q}$ and $\mathbf{R}$ are orthogonal matrices. Moreover, $\mathbf{R}$ is a rotation matrix, and $\mathbf{q}$ can be used to represent a rotation of a 3D point $\mathbf{p}$ in 3D space. The real part q_0 encodes the angle of rotation θ (as $\cos\frac{\theta}{2}$), and the direction of $\mathbf{q}$ encodes the rotation axis. Eq. (2.59) reveals how a unit quaternion can be converted to the equivalent rotation matrix. To convert a rotation matrix to an equivalent quaternion, we can write down nine equations relating every entry of $\mathbf{R}$ and four elements of $\mathbf{q}$. An alternative way is to express ten different $\mathbf{q}_x\mathbf{q}_y$ combinations of $\mathbf{R}$ in Eq. (2.59) through the elements of $\mathbf{R}$ and solve a system of ten linear equations. As $\mathbf{R}$ contains squared q_1, q_2, q_3 and q_4, both negative and positive numbers are possible in $\mathbf{q}$ while representing the same rotation. The fact that $\mathbf{q}$ and $-\mathbf{q}$ represent the same rotation is not a problem in practice. When performing transformation of a quaternion to a rotation matrix, negative solutions are discarded as rotation angle in quaternions does not exceed π. For angles $> \pi$, a corresponding configuration with another axis and an angle $\leq \pi$ can always be found. Thus, most transformation routines choose the sign of the quaternion to be positive.

Averaging of Rotations with Quaternions

Often in computer vision, it is required to compute an (perhaps weighted) average rotation from several given rotations. Quaternions enable an elegant norm-preserving solution to this problem [188]. A straightforward weighted averaging of quaternion elements is not optimal as there is no guarantee that the average is a normalised quaternion. Let

$$\mathbf{M} = \sum_{i}^{K} w_i \mathbf{q}_i \mathbf{q}_i^{\mathsf{T}}. \tag{2.60}$$

An average quaternion can be found by solving the optimisation problem

$$\mathbf{q}' = \underset{\mathbf{q}\in\mathbb{S}^3}{\arg\max}\, \mathbf{q}^\mathsf{T}\mathbf{M}\mathbf{q}, \tag{2.61}$$

where $\mathbb{S}^3$ is a three-dimensional unit sphere representing the space of normalised quaternions. The solution to (2.61) is given by a unit eigenvector corresponding to the largest eigenvalues of $\mathbf{M}$. To conform to the convention, the sign of $\mathbf{q}'$ is set to be positive. If the largest eigenvalue if $\mathbf{M}$ is distinct, then the average quaternion is also distinct, and we are done [188]. Note that since only the eigenvector corresponding to the largest eigenvalue is required, efficient iterative methods can be used which avoid the complete decomposition (see Sec. 2.2.1).

2.2.6 Finding a Closest Rotation Matrix to a Given A

The problem of projection of a given arbitrary matrix $\mathbf{A} \in \mathbb{R}^{D\times D}$ to the SO(3) group, *i.e.,* to the rotation manifold in 3D, often arises in computer vision. For instance, it is possible to perform an unconstrained update of a rotation matrix and then compute the closest rotation matrix. This problem can be solved with svd (see Sec. 2.2.2). Suppose $\mathbf{A} = \mathbf{U}\Sigma\mathbf{V}^\mathsf{T}$. The closest rotation matrix $\mathbf{R}$ to $\mathbf{A}$ is those which minimises the squared Frobenius norm of their differences $\|\mathbf{A}-\mathbf{R}\|^2_{\mathscr{F}}$. $\mathbf{R}$ can be obtained as

$$\mathbf{R} = \mathbf{U}\,\mathrm{diag}\left(1\ 1\ \ldots\ 1\ \det(\mathbf{U}\mathbf{V}^\mathsf{T})\right)\mathbf{V}^\mathsf{T}. \tag{2.62}$$

In Eq. (2.62), $\mathbf{R}$ can be ambiguous if $\mathbf{A}$ is a rank deficient matrix or when $\det(\mathbf{A})$ is negative and σ_D (the smallest singular value) is not distinct [202].

2.2.7 Optical Flow Estimation

The problem of apparent point displacement or velocity estimation based on 2D images is called *optical flow estimation* problem. Optical flow is the observed motion of a scene arising due to the motion of the scene relative to the camera. Optical flow estimation is an ill-posed inverse problem since camera motion the 3D scene configuration and the intrinsic camera parameters are not known.

One of the fundamental assumptions in optical flow estimation is the brightness constancy assumption. It states that

$$\mathbf{I}(\mathbf{x}(f), f) = \text{const},\ \textit{i.e.,} \tag{2.63}$$

intensities of the same physical points observed in 2D are constant throughout an image sequence. Brightness constancy assumption is a reasonable assumption though it only approximately holds. How well it holds depends on many factors (*e.g.*, lighting, shading, specular properties of the surface *etc.*). Thus, optical flow methods require additional regularisation.

Consider the case with two images. Let $u = \frac{dx}{dt}$ and $v = \frac{dy}{dt}$ denote the flow components (optical flow velocities) in the continuous domain, and let $\mathbf{I}_x$, $\mathbf{I}_y$ and $\mathbf{I}_t$ be partial derivatives of point brightnesses w.r.t the coordinates x, y and time t, respectively. Horn and Schunck proposed smoothness of optical flow as an additional constraint [145]. Their method optimises the following energy functional:

$$\mathscr{E}^2 = \int\int \left(\alpha^2 \Bigg(\underbrace{\left(\frac{\partial u}{\partial x}\right)^2 + \left(\frac{\partial u}{\partial y}\right)^2}_{\nabla u} + \underbrace{\left(\frac{\partial v}{\partial x}\right)^2 + \left(\frac{\partial v}{\partial y}\right)^2}_{\nabla v} \Bigg) + \underbrace{\mathbf{I}_x u + \mathbf{I}_y v + \mathbf{I}_t}_{\text{brightness constancy}} \right) dx\,dy, \tag{2.64}$$

with α^2 being a weight balancing the relative influence of the smoothness and data (intensity) terms. In Eq. (2.64), ∇u and ∇v are the magnitudes of the flow gradients.

To minimise the energy functional in Eq. (2.64), $\mathscr{E}^2$ has to be differentiated w.r.t. unknown u and v which eventually results in a system of two linear equations [145]. The method of Horn and Schunck is known to oversmooth boundaries and is not well suitable for scenes with fine-grained and various motions, as even moderate details can be overregularised and lost [145]. The method can be improved by considering multiple frames for computing time derivatives with higher accuracy.

2.2.8 Multiframe Optical Flow with Subspace Constraints and Occlusion Handling

For well-textured scenes depicting non-rigid motion, several multiframe optical flow (MFOF) methods were recently proposed.

Let $\mathbf{I}(\mathbf{x}, f) : \Omega \times \{1, \ldots, F\} \rightarrow \mathbb{R}$ be an image sequence with N images, $f \in \{1, \ldots, F\}$, and $\mathbf{x} \in \Omega \in \mathbb{R}^2$. Suppose we choose a reference view at the moment $f = 1$. For every point of the reference frame, we define 2D point trajectories of every $\mathbf{x}$ throughout the remaining $F - 1$ frames:

$$\mathbf{u}(\mathbf{x}, f) : \Omega \times \{2, \ldots, F\} \rightarrow \mathbb{R}^2. \tag{2.65}$$

Thus, $\mathbf{u}(\mathbf{x}, f)$ are the coordinates of every physical point $\mathbf{x}$ (observed in the reference frame $f = 1$) in frame $f \neq 1$. Instead of the absolute coordinates, the trajectories $\mathbf{u}(\mathbf{x}, f)$ can be interchangeably represented by point displacements in the image plane relative to the reference frame.

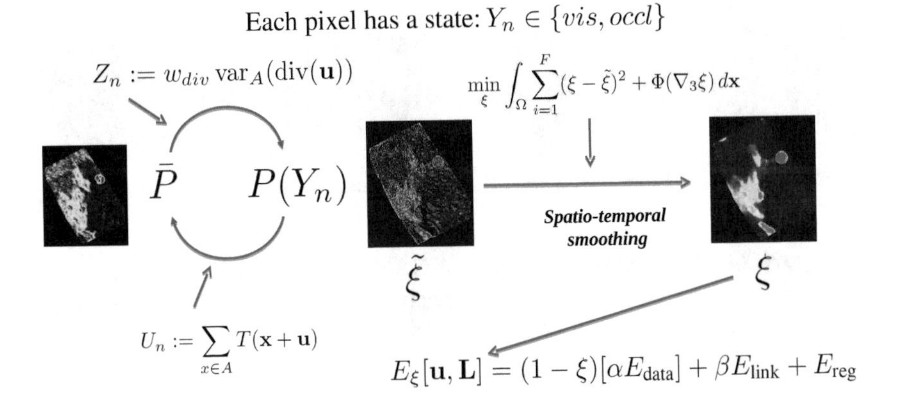

Figure 2.1: Overview of the energy function of the multiframe optical flow approach with explicit occlusion handling by Taetz *et al.* [267].

One of the key ideas of MFOF methods is to regularise point trajectories $\mathbf{u}(\mathbf{x}, f)$ over multiple frames. Multiframe subspace flow (MFSF) [105] constrains point $\mathbf{u}(\mathbf{x}, f)$ to lie in a linear space spanned by a few basis trajectories $\mathbf{q}_1, \mathbf{q}_2, \ldots, \mathbf{q}_R$.

During optimisation, every $\mathbf{x}$ obtains a set of weights $\mathbf{L}(\mathbf{x})$ with R elements determining the contribution of every $\mathbf{q}_r$, $r \in \{1, \ldots, R\}$. The energy functional of MFSF can be written as

$$\mathbf{E}(\mathbf{u}(\mathbf{x}, f), \mathbf{L}(\mathbf{x})) = \alpha\, \mathbf{E}_{\text{data}} + \beta\, \mathbf{E}_{\text{link}} + \gamma \mathbf{E}_{\text{reg}}, \text{ with} \tag{2.66}$$

$$\mathbf{E}_{\text{data}} = \int_{\Omega} \sum_{f=2}^{F} \Phi\Big(\mathbf{I}(\mathbf{x} + \mathbf{u}(\mathbf{x}, f), f) - \mathbf{I}(\mathbf{x}, f_{\text{ref}})\Big)\, \mathrm{d}\mathbf{x}, \tag{2.67}$$

$$\mathbf{E}_{\text{link}} = \int_{\Omega} \sum_{f=2}^{F} \left\| \mathbf{u}(\mathbf{x}, f) - \sum_{r=1}^{R} \mathbf{L}_r(\mathbf{x})\, \mathbf{q}_r(f) \right\|^2 \mathrm{d}\mathbf{x}, \text{ and} \tag{2.68}$$

$$\mathbf{E}_{\text{reg}} = \int_{\Omega} \sum_{r=1}^{R} g(\mathbf{x}) \Phi\Big(\nabla \mathbf{L}_r(\mathbf{x})\Big)\, \mathrm{d}\mathbf{x}. \tag{2.69}$$

In Eq. (2.67) and (2.69), $\Phi(\cdot)$ stands for a robust norm (ℓ_1 or Huber norm). Besides, $g(\mathbf{x})$ in Eq. (2.69) is a space-varying weight which encourages flow discontinuities to coincide with the contours of the reference frame. Thus, $g(\mathbf{x})$ lowers the influence of the regularisation near the contours. Note that in Eq. (2.67)–(2.69), the

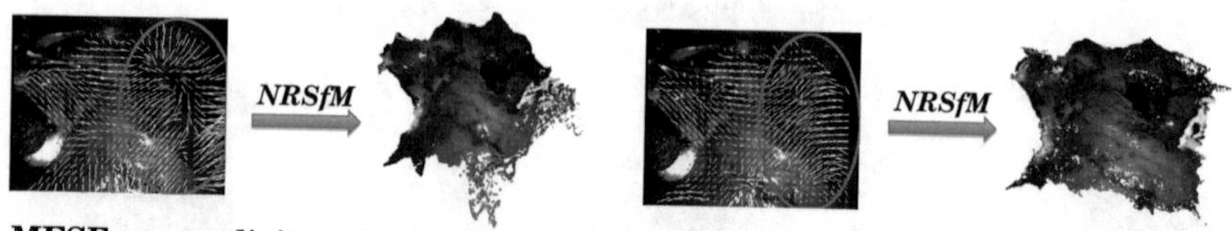

Figure 2.2: Qualitative comparison of NRSfM reconstructions of a human heart (the sequences originates from [259]) obtained on the uncorrected (left) and corrected tracks (right).

continuous notation is used along with the discrete notation which is a common practice in computer vision if one wishes to express a process on a high level (*i.e.*, decoupled from the discretisation effects at the respective stage). The energy functional in Eq. (2.66) can be optimised alternatingly w.r.t. $\mathbf{u}(\mathbf{x}, f)$ and $\mathbf{L}(\mathbf{x})$ with the primal-dual algorithm proposed by Chambolle and Pock [64].

MFSF is used in chapters 4–6 for computing dense correspondences from real images for NRSfM. For the cases with short-time occlusions, we also proposed a modification of MFSF which automatically computes an occlusion map $\xi(\mathbf{x}, f)$ for every frame and dynamically adjusts the influence of the data term in Eq. (2.66) [267]. An occlusion map is a matrix which indicates for every pixel of the reference frame its probability of being occluded (the object is either externally occluded or self-occluded) in every consecutive frame of the image sequence. The modified energy functional of Taetz *et al.* [267] reads

$$\mathbf{E}_{\xi(\mathbf{x},f)}(\mathbf{u}(\mathbf{x}, f), \mathbf{L}(\mathbf{x})) = (1 - \xi(\mathbf{x}, f))[\alpha\, \mathbf{E}_{\text{data}}] + \beta\, \mathbf{E}_{\text{link}} + \gamma \mathbf{E}_{\text{reg}}, \tag{2.70}$$

with $\mathbf{E}_{\text{data}}$, $\mathbf{E}_{\text{link}}$ and $\mathbf{E}_{\text{reg}}$ as in Eq. (2.67)–(2.69), respectively. In Eq. (2.70), $\xi(\mathbf{x}, f)$ is the reliability measure for the data term, see Fig. 2.1 for an overview of the method (more details about the notations and steps of the method can be found in [267]).

The modified multlframe optical flow (MFOF) [267] with occlusion handling operates in multiple steps. First, it computes the initial flow estimates by setting $\xi(\mathbf{x}, f) = 0$, $\forall \mathbf{x}$ and $\forall f$. The first step is thus equivalent to MFSF [105]. Second, MFOF estimates $\xi(\mathbf{x}, f)$ from the initial flow fields by evaluating weighted divergence variances computed in local neighbourhoods of each $\mathbf{x}$. Negative divergence operator was shown as a reliable detector of self-occlusions in optical flow estimation [30]. Here, we extended the occlusion detector by also considering positive divergence accounting for external occlusions, and proposed a Bayesian measurement model with a state transition (with occluded and non-occluded states) and a Bayesian smoother for $\xi(\mathbf{x}, f)$. In the third step, MFOF minimises the energy functional in Eq. (2.70) with a known binarised $\xi(\mathbf{x}, f)$. Similar to MFSF [105], minimisation in performed with the primal-dual algorithm [64]. Fig. 2.2 demonstrates the qualitative difference in the surface reconstruction obtained by the SPVA method (with the static shape prior term switched off, see Sec. 5.1) on uncorrected and corrected dense point tracks, respectively.

2.3 Local Refinement and Probabilistic Approaches for Point Set Registration

In this section, we review a method for estimation of an optimal transformation given point correspondences and two classic methods for point set registration. Moreover, we discuss N-body simulations, Gaussian mixture models and probabilistic expectation-maximisation approach for unsupervised learning.

2.3.1 Estimation of an Optimal Transformation

Suppose correspondences between two point sets $\mathbf{X} \in \mathbb{R}^{N \times D}$ and $\mathbf{Y} \in \mathbb{R}^{N \times D}$ are known, with the number of points N (the same in both point sets) and the dimensionality of the point sets D. The objective of transformation estimation is finding an optimal rigid transformation between $\mathbf{X}$ and $\mathbf{Y}$. $\mathbf{X}$ is assumed to be fixed and $\mathbf{Y}$ is being transformed. Optimal transformation estimation with rotation matrices is a special case of the constrained Procrustes problem.

One of the techniques for transformation estimation is based on svd (see 2.2.2) and was described by W. Kabsch [158]. First, both point sets are translated so that their centroids are located in the origin of the coordinate system. This step resolves translation and ensures that the recovered rotation is defined around the origin and preserve the origin. Next, a cross-covariance matrix $\mathbf{C}$ for $\mathbf{X}$ and $\mathbf{Y}$ is computed as

$$\mathbf{C} = \mathbf{Y}^{\mathsf{T}}\mathbf{X}. \tag{2.71}$$

$\mathbf{C}$ is a measure of similarity between $\mathbf{X}$ and $\mathbf{Y}$. Next, C is decomposed by svd:

$$\mathbf{C} = \mathbf{U}\boldsymbol{\Sigma}\mathbf{V}^{\mathsf{T}}. \tag{2.72}$$

Suppose $\mathbf{R}$ is the sought rotation matrix, with the following properties:

$$\begin{aligned} \det(\mathbf{R}) &= 1, \\ \mathbf{R}^{\mathsf{T}}\mathbf{R} = \mathbf{R}^{-1}\mathbf{R} &= \mathbf{I}. \end{aligned} \tag{2.73}$$

Next, using the properties listed in (2.73), we can reassemble $\mathbf{C}'$ so that it is a valid rotation matrix as:

$$\mathbf{C}' = \mathbf{U}\,\mathrm{diag}\left(1\ 1\ \ldots\ 1\ \det(\mathbf{U}\mathbf{V}^{\mathsf{T}})\right)\mathbf{V}^{\mathsf{T}}. \tag{2.74}$$

The reassembled matrix $\mathbf{C}'$ is a valid rotation matrix and it equals to an optimal rotation aligning $\mathbf{Y}$ to $\mathbf{X}$. The correction $\det(\mathbf{U}\mathbf{V}^{\mathsf{T}})$ in Eq. (2.74) is required to ensure the property $\det(\mathbf{R}) = 1$.

2.3.2 Iterative Closest Point

Iterative Closest Point (ICP) is an iterative local refinement technique for point set registration introduced by Besl and McKay [40]. It adopts transformation estimation for point set registration by operating on correspondences obtained in local neighbourhoods. In the first step of every iteration, ICP computes nearest neighbours for every template point. In the second step of every iteration, ICP estimates the local transformation of the template on the estimated local correspondences. ICP requires an accurate initialisation. Otherwise, it often converges to local minima and is known to be sensitive to noise.

Suppose $\mathbf{X} \in \mathbb{R}^{N \times D}$ and $\mathbf{Y} \in \mathbb{R}^{M \times D}$ are the reference and the template with N and M points, respectively. Both point clouds are assumed to contain points of the same dimensionality D. We seek a rigid transformation $\mathbf{T}$ aligning $\mathbf{Y}$ to $\mathbf{X}$.

For every point $\mathbf{y}_m \in \mathbf{Y}$, we denote its corresponding point in $\mathbf{X}$ in every iteration with $\mathbf{y}_m^{\mathrm{corr}}$. In ICP, $\mathbf{y}_m^{\mathrm{corr}}$ is obtained as the nearest neighbour for every $\mathbf{y}_m$. Various ways for obtaining nearest neighbours in ICP were proposed in the literature so far, ranging from general-purpose techniques like a k-d tree to case-specific ones taking advantage of point set parametrisations (*e.g.*, if point clouds are obtained from an RGB-D camera, they are parametrised through a regular 2D grid). With available correspondences, ICP solves the following optimisation problem in every iteration:

$$\mathbf{T}_{i+1} = \arg\min_{\mathbf{T}} \sum_{m \in M} \|g(\mathbf{T}, \mathbf{y}_m) - \mathbf{y}_m^{\mathrm{corr}}\|^2, \tag{2.75}$$

where $\mathbf{T}$ is a rigid transformation and $g(\mathbf{T}, \mathbf{y}_m)$ denotes the rigid transformation operator rigidly transforming point $\mathbf{y}_m$ with $\mathbf{T}$ in iteration i. Eq. (2.75) can be solved in a closed form by transformation estimation techniques or using non-linear optimisation. When $\mathbf{T}_{i+1}$ is found, it is applied to $\mathbf{Y}$, and ICP proceeds with updating the correspondences $\mathbf{y}_m^{\mathrm{corr}}$. If $\mathbf{y}_m^{\mathrm{corr}}$ is not altered after the update of

$\mathbf{T}$, the algorithm converges, since the transformation estimation step will likely output a transformation close to identity matrix for rotation and a null vector for the translation. The final rigid transformation $\mathbf{T}'$ after L iterations equals to the concatenation of all transformations $\mathbf{T}_i$, $i \in \{1,\ldots,L\}$:

$$\mathbf{T}' = \mathbf{T}_L \circ \ldots \circ \mathbf{T}_2 \circ \mathbf{T}_1, \tag{2.76}$$

with $\circ$ denoting concatenation operator for rigid transformations.

Chen and Medioni [66] proposed a variant of ICP which takes into account point normals. For the point set registration case, the transformation estimation problem with point normals can be expressed as

$$\mathbf{T}_{i+1} = \arg\min_{\mathbf{T}} \sum_{m \in M} \left\| \left(g(\mathbf{T}, \mathbf{y}_m) - \mathbf{y}_m^{\text{corr}}\right) \cdot \mathbf{n}(\mathbf{y}_m^{\text{corr}}) \right\|^2, \tag{2.77}$$

where $\mathbf{n}(\cdot)$ denotes a normal of $\mathbf{X}$. The formulation in Eq. (2.77) stabilises registration of planes and surfaces. In general, the spatial diversity of $\mathbf{X}$ facilitates more accurate registrations.

2.3.3 N-Body Simulations

The objective of an N-body problem in celestial mechanics is the prediction of future positions and trajectories of celestial bodies (stars, planets, comets, and so on) interacting under gravitational forces [80]. Most frequently, the masses in N-body problems are assumed to be concentrated in an infinitely small volume of space, and the system consists of particles. In this section, we use arrow notation for vectors which is more common in physics. A superimposed gravitational field induced by individual particles exerts gravitational force $\vec{F}_i$ to every particle $i \in \{1,\ldots,|A| = N\}$ from the set A in the system [284]:

$$\vec{F}_i = -\nabla \phi_i(\vec{r}_i) - \nabla \phi_{ext}(\vec{r}_i) = -G m_i \sum_{i \neq j} \frac{m_j(\vec{r}_i - \vec{r}_j)}{\left\|\vec{r}_i - \vec{r}_j\right\|^3} - \nabla \phi_{ext}(\vec{r}_i), \tag{2.78}$$

where G is the gravitational constant determining the gravitational strength between two bodies of unit masses separated by a unit distance, m_i, m_j and $\vec{r}_i$, $\vec{r}_j$ are particle masses and position vectors, respectively, $\phi_i(\cdot)$ and $\phi_{ext}(\cdot)$ are the internal and external gravitational potentials, respectively, ∇ denotes the gradient operator, and $\|.\|$ denotes the ℓ^2-norm. Newton's Second Law of motion relates the forces exerted on particles with their accelerations by the second-order ordinary differential equations (ODE):

$$\ddot{\vec{r}}_i = \frac{\vec{F}_i}{m_i}, \forall i \in A, \tag{2.79}$$

where $\ddot{\vec{r}}_i$ is the particle's acceleration. There exists a unique solution to the equation system (2.79) as long as initial conditions are specified, *i.e.,* an initial position and velocity of every particle are known. A solution is obtained by double numerical integration since no analytical solution for $N > 3$ exists. Depending on the assumptions and objectives, N-body simulations can be classified into *collisional* and *collisionless*. While the former class allows the particles (the bodies) to merge, no merging, splitting and re-distribution of masses are allowed in a collisionless simulation. In computer vision, collisionless simulations are used almost exclusively, as the number of particles (irrespective of what they are modelling) needs to be preserved in most of the cases. To avoid gravitational collapses and numerical instabilities (as the masses are concentrated in infinitesimal volumes), a near-field regularisation is required. For more details on classical N-body problems, the reader may refer to [5,6].

With the advent of the Einstein's general relativity (GR) [87], the perspective to the interaction between masses has changed. According to modern GR, celestial masses bend the space-time continuum and influence the space-time curvature. A metric on the 4D space-time defines its geometrical properties (such as Ricci tensor and scalar curvature) appearing in GR equations. The celestial masses placed in the curved space-time move along curved lines (geodesics). The latter phenomenon causes the effect known in Newtonian mechanics as gravitational force, whereas in GR, gravitational force does not exist. Thus, in the presence of mass, the space-time is a flat four-dimensional manifold $\mathbb{M}^4$ (Minkowski space). In GR, space-time curvature depends on the mass and energy distribution in space, the speed of light c and G [310]. Even though gravitational force in GR is replaced by the curvature of the 4D space-time, it is useful to use the notion of *relativistic gravitational force* (RGF), *i.e.*, a gravitational force parameterised by the Gaussian curvature of the force inducing particles. Eventually, the N-body problem was extended to the case of relativistic motion. For the simulation in a space with a constant Gaussian curvature, Eq. (2.78) is adjusted for RGF. Substituted to Eq. (2.79), it yields [81]

$$\ddot{\vec{r}}_i = -G \sum_{i \neq j} \frac{m_j \left[\vec{r}_j - \left(1 - \frac{\kappa \left\| \vec{r}_i - \vec{r}_j \right\|^2}{2} \right) \vec{r}_i \right]}{\left\| \vec{r}_i - \vec{r}_j \right\|^3 \left(1 - \frac{\kappa \left\| \vec{r}_i - \vec{r}_j \right\|^2}{4} \right)^{\frac{3}{2}}}, \tag{2.80}$$

with $\kappa \in \mathbb{R}$ denoting constant Gaussian curvature of the space. Similar to the classic N-body problem, new $\vec{r}_i$ are updated as double integrals of $\ddot{\vec{r}}_i$ w.r.t. time. If $\kappa = 0$, Eq. (2.80) reduces to the analogous equation for the non-relativistic case (*cf.* Eqs. (2.78)–(2.79)).

2.3.3.1 Acceleration Techniques for N-body Simulations

Methods considering every particle in an N-body problem individually require $\mathcal{O}(N^2)$ operations. Several methods were proposed to reduce the computational complexity and enable simulation of large-scale N-body systems. Ahmad-Cohen (AC) neighbour scheme employs two time scales for each particle [18]. Thereby, force evaluations for neighbouring particles occur more frequently than for distant particles. For smaller time steps, contributions from the distant points are approximated. Various strategies for the neighbourhood selection were proposed [6]. Though the AC scheme achieves a speedup, the complexity class remains $\mathcal{O}(N^2)$. Barnes and Hut proposed a tree data structure for hierarchical computation of forces and gravitational potential energies (GPE) [31]. The main idea of the Barnes-Hut octree is to cluster the particles dependent on the distance to other particles and accumulate the contributions to the force or GPE at a given point $\vec{r}_i$. The algorithm achieves $\mathcal{O}(N \log N)$ complexity. The fast multipole method extends this idea by also allowing collective contributions to groups of points [130]. It also employs hierarchical space decomposition taking advantage of multipole expansions. Thus, adjacent particles in the near-field tend to accelerate similarly under forces exerted by particles in the far field. This class of algorithms exploits the idea of rank-deficiency of the $N \times N$ interaction matrix considering the nature of the far-field interactions. As the distance to a particle group increases, the group expands. This strategy results in an $\mathcal{O}(N \log N)$ algorithm, whereby the multiplicative constant depends on the approximation accuracy. An $\mathcal{O}(N)$ algorithm is also possible but leads to the further increase in the multiplicative constant.

2.3.4 Gaussian Mixture Models and Expectation-Maximisation

Suppose A and B are events in the sample space Ω. The probability of event A given that event B has occurred (the probability of A conditioned upon the event B) is called conditional probability. Conditional probability $P(A|B)$ can be calculated as

$$P(A|B) = \frac{P(A \cap B)}{P(B)}, \tag{2.81}$$

where $P(A \cap B)$ is the set notation for a joint probability. Essentially, Eq. (2.81) restricts Ω to the subspace of event B. Similarly, for $P(B|A)$ we can write

$$P(B|A) = \frac{P(A \cap B)}{P(A)}, \tag{2.82}$$

where Ω is narrowed to the subspace of event A. From Eq. (2.81) and (2.82), we can derive the Bayes rule — the expression for $P(A|B)$ if $P(B|A)$ known:

$$P(A|B) = \frac{P(B|A)P(A)}{P(B)}. \tag{2.83}$$

The Bayes rule is the foundation for many probabilistic models and methods in computer vision. It allows performing the inference of the unknown inverse conditional probability $P(A|B)$ from the known conditional probability $P(B|A)$.

2.3.4.1 Gaussian Mixture Models

Probability density function (PDF) of a Gaussian Mixture Model (GMM) is a linear convex combination of individual Gaussian PDFs. GMMs are used in computer vision for clustering problems and probabilistic point set registration, among other domains. A GMM can be formalised through the underlying generative process. Suppose $z \in \{1,\dots,K\}$ is a vector-valued random variable indicating the mixture component. Suppose $P(z=k) = \pi_k$ are the probabilities of the individual mixture components $k \in \{1,\dots,K\}$. π_k are called mixing coefficients, for the reasons to become clear. Given $z = k$, the random variable $\mathbf{x}$ is distributed normally:

$$\mathbf{x} \sim \mathcal{N}(\mu_k, \Sigma_k), \tag{2.84}$$

where $\mu_k \in \mathbb{R}^D$ are the means and $\Sigma \in \mathbb{R}^{D\times D} \succeq 0$ are the positive semi-definite covariance matrices of the Gaussian distribution $\mathcal{N}(\mu_k, \Sigma_k)$. Thus, z is a latent, not directly observed variable of the mixture of Gaussians. Consider the marginal distribution on $\mathbf{x}$:

$$P(\mathbf{x}) = \sum_z P(z)P(\mathbf{x}|z) = \sum_{k=1}^{K} P(z=k)P(\mathbf{x}|z=k) = \sum_{k=1}^{K} \pi_k \mathcal{N}(\mathbf{x}|\mu_k, \Sigma_k). \tag{2.85}$$

The right component of Eq. (2.85) is the PDF of GMM and a convex combination of Gaussian PDFs. Using the Bayes rule, we can express the inverse conditional probability $P(z=k|\mathbf{x})$ as

$$P(z=k|\mathbf{x}) = \frac{P(\mathbf{x}|k)P(k)}{P(\mathbf{x})} = \frac{\pi_k \mathcal{N}(\mathbf{x}|\mu_k, \Sigma)}{\sum_{l=1}^{K} \pi_l \mathcal{N}(\mathbf{x}|\mu_l, \Sigma_l)}. \tag{2.86}$$

In the next section, we show how the parameters of a GMM can be automatically inferred in an unsupervised manner, *i.e.*, when the data labels are unknown.

2.3.4.2 Expectation-Maximisation Algorithm

Expectation-Maximisation (EM) is an unsupervised learning technique with soft assignments for clustering. EM operates in two steps — expectation (E-)step and maximisation (M-)step. In the E-step, the latent hidden variables are updated while the model parameters are fixed, whereas in the M-step, the model parameters are updated while the hidden variables are kept fixed. We follow the notation introduced in Sec. (2.3.4.1). In the E-step of EM for GMM, the probabilities of latent variables $P(z = k|\mathbf{x})$ (cf. Eq. (2.86)) are updated while $\theta = \pi_k$, μ_k and Σ_k are fixed:

$$P(z_i = k|\mathbf{x}_i) \propto \pi_k \mathscr{N}(\mathbf{x}_i|\mu_k, \Sigma_k), \tag{2.87}$$

where $\mathbf{x}_i \in \{1, \dots, N\}$ is the sample with index i out of N samples. In the M-step, π_k, μ_k and Σ_k are updated while $P(z = k|\mathbf{x}_i)$ are fixed:

$$\pi_k = \frac{\sum_{i=1}^N P(z_i = k|\mathbf{x}_i)}{N}, \tag{2.88}$$

$$\mu_k = \frac{\sum_{i=1}^N P(z_i = k|\mathbf{x}_i)\mathbf{x}_i}{\sum_{i=1}^N P(z_i = k|\mathbf{x}_i)}, \tag{2.89}$$

$$\Sigma_k = \frac{\sum_{i=1}^N P(z_i = k|\mathbf{x}_i)(\mathbf{x}_i - \mu_k)(\mathbf{x}_i - \mu_k)^\mathsf{T}}{\sum_{i=1}^N P(z_i = k|\mathbf{x}_i)}. \tag{2.90}$$

The algorithm repeats the E- and M-steps and converges when no changes are occurring. At the beginning, the parameters π_k, μ_k and Σ_k can be chosen randomly unless some prior information is available.

The advantages of EM algorithm include numerical stability (the likelihood increases with every iteration), reliable global convergence under general conditions, a low per-iteration cost (well counterbalancing perhaps a large number of iterations) and applicability in scenarios with missing data [160]. On the contrary, EM does not automatically provide an estimate of the covariance matrix for the parameters (in this case, an appropriate methodology is required to offset this disadvantage [160, 192]), it can converge slow and in some problems, both the E- and M-steps can be analytically intractable [160].

2.3.5 Coherent Point Drift

Coherent Point Drift (CPD) method [203] considers alignment of two D-dimensional point sets $\mathbf{X}_{N \times D} = (\mathbf{x}_1, \dots, \mathbf{x}_N)^T$ and $\mathbf{Y}_{M \times D} = (\mathbf{y}_1, \dots, \mathbf{y}_M)^T$ as a probability density estimation problem where one point set represents the GMM centroids ($\mathbf{Y}_{M \times D}$)

and the other one represents the data points ($\mathbf{X}_{N\times D}$). According to different deformation models between the two point sets, an appropriate rigid or non-rigid transformation can be selected. At the optimum two point sets become aligned and the correspondences are obtained using the maximum of the GMM posterior probability for a given data point. Core to CPD method is to force the GMM centroids to move coherently as a group to preserve the topological structure of the point sets. The GMM probability density function of the CPD method can be written as

$$p(\mathbf{x}) = \sum_{m=1}^{M+1} P(m)p(\mathbf{x}|m). \tag{2.91}$$

To explicitly model outliers, the density function is split in the following form

$$p(\mathbf{x}) = w\frac{1}{N} + (1-w)\sum_{m=1}^{M} P(m)p(\mathbf{x}|m), \tag{2.92}$$

with $p(\mathbf{x}|m) = \frac{1}{(2\pi\sigma^2)^{D/2}}\exp(-\frac{\|\mathbf{x}-\mathbf{y}_m\|^2}{2\sigma^2})$ and $0 \leq w \leq 1$ — a weight parameter that reflects the assumption about the amount of outliers in the reference point set. The GMM centroids are adjusted by a set of transformation parameters θ that can be estimated by minimising the negative log-likelihood function

$$E(\theta,\sigma^2) = -\sum_{n=1}^{N} \log \sum_{m=1}^{M+1} P(m)p(\mathbf{x}_n|m). \tag{2.93}$$

An EM algorithm is used to find θ and σ^2. The E-step constructs a guess of the parameter values based on the previous ("old") values and then uses the Bayes' theorem to compute a posteriori probability distribution $P^{old}(m|\mathbf{x}_n)$. The M-step updates the parameters by minimising an upper bound of the negative log-likelihood (2.93). After leaving out the terms constant w.r.t. σ^2 and θ the objective function reads:

$$Q(\theta,\sigma^2) = \frac{1}{2\sigma^2}\sum_{n=1}^{N}\sum_{m=1}^{M} P^{old}(m|\mathbf{x}_n)\|\mathbf{x}_n - T(\mathbf{y}_m,\theta)\|^2 + \frac{N_P D}{2}\log\sigma^2 \tag{2.94}$$

where $T(\mathbf{y}_m,\theta)$ is a transformation applied to $\mathbf{Y}$, $N_p = \sum_{n=1}^{N}\sum_{m=1}^{M} P^{old}(m|\mathbf{x}_n)$ and

$$P^{old}(m|\mathbf{x}_n) = \frac{\exp(-\frac{1}{2}\|\frac{\mathbf{x}_n - T(\mathbf{y}_m,\theta^{old})}{\sigma^{old}}\|^2)}{\sum_{k=1}^{M}\exp(-\frac{1}{2}\|\frac{\mathbf{x}_n - T(\mathbf{y}_k,\theta^{old})}{\sigma^{old}}\|^2) + c} \tag{2.95}$$

with $c = (2\pi\sigma^2)^{D/2}\frac{w}{1-w}\frac{M}{N}$. Based on the formulas above the transformation T can be specified for rigid, affine and non-rigid point set registration.

3 Review of Previous Work

THE core problems in this thesis, *i.e.,* monocular non-rigid 3D reconstruction and point set registration, are reviewed in this chapter. For every problem, we discuss method classes for solving them and relevant previous works. In some following chapters, we also additionally elaborate on most related methods for the respective proposed approaches.

3.1 Non-Rigid Structure from Motion

The problem of monocular non-rigid 3D reconstruction consists in the recovery of time-varying geometry of a scene — poses and shapes or surfaces for every frame — viewed by a monocular camera. This challenging problem is also known as dynamic or 4D monocular reconstruction, and it is an **ill-posed inverse problem** in the sense of Hadamard, since the uniqueness of the solution is violated, see Fig. 3.1 for an example (the same projections in the image plane can originate from different shapes). To obtain a unique and reasonable solution, the problem must be regularised, *i.e.,* additional assumptions about the observed scene and types of valid deformations must be introduced to constrain the solution space.

Efficient techniques for monocular non-rigid surface recovery will enable 4D reconstruction systems to be built based on ubiquitously available and mature imaging technology avoiding specialised and expensive hardware (*e.g.,* stereo cameras, multi-view setups and laser scanners) as well as overheads related to their set up and operation (*e.g.,* space requirements, imperfection of sensors, calibration and synchronisation between the views). Another significant factor is the reduced data volume and the accosiated storage and throughput requirements. In the light of turning conventional monocular recording devices into a general sensing technology for non-rigid scenes, the following application domains appear to be promising:

- *Analysis of deforming surfaces and thin objects in engineering.* Engineers often wish to analyse elastic, bending and aerodynamic surface properties of wings, sails, skis, bridges and dampers. This information would be helpful to improve their design. Also in such scenarios, a full 3D reconstruction can be superfluous and prohibitively costly to obtain.

V. Golyanik, *Robust Methods for Dense Monocular Non-Rigid 3D Reconstruction and Alignment of Point Clouds*,

Figure 3.1: Depth ambiguity is the reason for the ill-posedness of NRSfM. All three reconstructions of the image shown on the left result in a low reprojection error.

- *Augmented and mixed reality systems*. In augmented and virtual reality devices, knowledge about scene geometry will help to place virtual objects and enable retexturing in the environments with non-rigidly deforming parts.
- *Post factum reconstruction in cultural and historic heritage*. An extensive legacy of monocular historic recordings was consigned to us from the past. Inference of 3D geometry from those recordings is of wide scientific interest, and therefore, structure from motion (both rigid and non-rigid) could be used.
- *Scene understanding*. Reconstructed geometry provides stronger cues for scene understanding (*e.g.*, human behaviour and emotions), compared to 2D analysis.
- *Minimally invasive surgery (MIS)*. The current trend in surgery is minimising operational interventions in human bodies with smaller volumes as well as less and smaller insicions. The advantages include shorter postoperative periods, the smaller probabilities of infections, faster recovery times and shorter anaesthetisation times, among others. The insicions must be sufficiently large to place the actuators and laparoscopic devices with sensors in the operational area. One of the unobstrusive imaging sensors is a monocular camera. A real-time video helps surgeons to localise the instruments, and an additional 3D representation could generate virtual and augmented views, and eventually facilicate operational presicion and reduce the overal time of the surgical invasion.

3.1.1 Approaches to Monocular Non-Rigid Surface Recovery

Two main method classes for non-rigid surface recovery from monocular views are known as template-based monocular non-rigid 3D reconstruction (TBR) and non-rigid structure from motion (NRSfM). If a 3D reconstruction for at least a single frame is available and used as a prior for non-rigid tracking, a method falls into TBR [246], [187]. Availability of a template is a very strong assumption which harms the generalisability and restricts the usage of this method class in practice. The latter class of methods — NRSfM — is template-free and allows to reconstruct arbitrary surfaces while solely relying on the observed motion and deformation cues [53, 216]. Additionally, NRSfM requires accurate point correspondences throughout a video stream or a set of views as an input.

Most of TBR and NRSfM methods assume that the observed scene consists of a single moving and non-rigidly deforming object. Moreover, in NRSfM camera motion is assumed to be unknown and the views are often assumed to be uncalibrated. During the tracking phase, there is one reference view of the scene determining points of interest. Only points visible in the reference view are tracked and eventually reconstructed. In most cases, it is more preferable to chose a frontal or a reoccurring view of a scene as a reference view. *Batch* NRSfM operates on correspondences obtained over all expected views at a time.

Thus, batch NRSfM takes advantage of the global optimisation over all available views. *Sequential* NRSfM processes correspondences of the incoming views upon availability. Since not all observations covering the entire variety of deformations are available from the beginning, the reconstructions can be less accurate compared to batch NRSfM. However, apart from the incremental accuracy improvement with more and more views, an incremental refinement of already performed reconstructions is also possible.

The core of every NRSfM approach is a deformation model. Deformation model determines the space of valid deformations, *i.e.*, it encodes which states are reachable and more probable than the other ones. An exact configuration of the deformation model is usually learned from the data, as NRSfM operates in an unsupervised manner. The input to NRSfM is a set of sparse or dense optical flows between the reference view and every other view of the image sequence. A measurement matrix combining input correspondences is either obtained through a sparse keypoint tracking or a dense tracking of all visible points with optical flow. In this setting, we call the set of optical flows with the same reference view *multiframe optical flow* (MFOF). MFOF can be computed either naively by a unification of multiple two-frame optical flows [229] or jointly and more accurately with additional constraints over multiple frames reflecting the properties of the observed scene (in many cases, the scene contains a single deformable object or a surface).

Initialisations in NRSfM can be performed in different ways — as the most expected or an intermediate state upon the deformation model, as a planar structure, by a backprojection of the measurements into the 3D space or under the rigidity assumption. In the case of small deformations, initialisation under rigidity assumption also enforces the state closest to the most expected state upon the deformation model. Moreover, knowledge about the origin of the views in NRSfM is advantageous. Thus, if the observations originate from a video stream, it is reasonable to assume temporal smoothness in the scene. On the other hand, if the ordering is not known, arbitrary reshuffling of the views is possible, without affecting the reconstruction accuracy.

To summarise, the pipeline for dense monocular non-rigid 3D reconstruction with NRSfM consists of dense point tracking, solution initialisation, shape recovery and post-processing (such as registration, recognition, stitching *etc.*). In the following, we consider NRSfM for the detailed discussion, as this method class is in focus of this thesis. Specifically, we are interested in dense NRSfM, *i.e.*, recovery of densely tracked 2D points throughout temporally-consistent image sequences of scenes consisting of a single object or a single surface.

3.1.2 Previous Work in Non-Rigid Structure from Motion

NRSfM methods made significant advances during the recent years in terms of the ability to reconstruct realistic non-rigid motion. The earliest works on 3D reconstruction from monocular video streams consider the rigid case, sparse representation and the orthogonal camera model. In rigid factorisation approach of Tomasi and Kanade [281], shape and rotation matrices are recovered by factorisation of the measurement matrix. The first solution for the non-rigid case was introduced in [53], while the deforming shape was assumed to lie in a linear subspace of a finite number of basis shapes. The basis shapes and the coefficients for every frame were assumed to be unknown. To obtain a unique solution, orthonormality constraints on rotation matrices were applied, similar to [281]. Without further constraints, the proposed method is sensitive to noise, and the first successor methods elaborated on orthonormality and rank constraints [51, 52, 283]. Multiple successor methods elaborated on the idea of constraining surfaces to lie in the linear subspaces spanned by a few basis shapes [20, 52, 216]. Torresani *et al.* [282] constrain 3D shapes to lie near a linear subspace in a probabilistic manner, with a Gaussian prior on every basis shape. The main difficulty in modelling with an explicit basis is the need to decide about the optimal basis cardinality since it varies across datasets exhibiting different degrees of deformations and non-linearities. Moreover, there is no guarantee that an optimal number of basis shapes exists capturing all shape variation while discarding the noise [282]. Despite the basis and coefficients are ambiguous, later it was proven that the shapes can be, nevertheless, reconstructed uniquely [20].

Since NRSfM is a highly ill-posed inverse problem in the sense of Hadamard, a prior knowledge is required to disambiguate the solution space and obtain a unique solution such as basis constraints [313], shape prior [56], temporal consistency and smoothness assumption [3, 58, 104, 126, 172, 226, 227, 282, 331], local rigidity assumption [228, 273], soft inextensibility constraint [12, 67, 293], shape prior [56, 270–272] or the assumption on a compliance with a physical deformation model [12,13]. Several approaches formulate NRSfM in the trajectory space [21,22] and use a pre-defined set of generic basis trajectories instead of basis shapes. As a result, the overall number of unknowns in NRSfM with trajectory space is reduced. A predefined basis is, at the same time, a disadvantage, and the performance can likewise vary considerably depending on data. A frequent choice for basis trajectories is a set of basis functions of discrete cosine transform. The first approach that used constraints in both metric and trajectory space was [125]. Another way to reduce the overall number of unknowns is surface modelling with physics-based priors [13, 14, 17]. Dai *et al.* [76] proposed an energy-based method with a minimal number of priors which was extended for dense reconstructions in the variational approach of Garg *et al.* [104]. The performance of Dai *et al.*'s method considerably varies with the dataset [76]. In NRSfM, we are always confronted with the dilemma which set of constraints is the most appropriate in a given scenario. There is no universal set of priors and regularisers covering all cases. Along with methods supporting an orthographic camera model [14,22,104,216,226–228,270,282], there are methods supporting a full perspective (in most of the cases calibrated) camera model [12, 33, 67, 138, 181, 293, 314, 331], dense reconstructions [11, 104, 238], sequential processing [9, 12, 13, 214, 272] and compound scenes [239].

In contrast to most NRSfM operating on the whole batch of 2D correspondences, sequential NRSfM processes the incoming frames upon arrival [9, 10, 14, 214]. Sequential methods are especially promising for real-time processing. However, without additional prior assumptions, sequential NRSfM either only supports sparse settings or do not reach the accuracy of NRSfM with global optimisation over a batch of frames. Piece-wise rigid approaches interpret the structure as locally rigid in the spatial domain [273]. In contrast to locally rigid models, several methods divide the structure into patches each of which can deform non-rigidly [89, 173]. High granularity level of operation allows these methods to reconstruct large deformations as opposed to methods relying on linear low-rank subspace models [89]. One of the recent methods by Lee *et al.* [173] achieves state-of-the-art accuracy on several sparse datasets including impressive results for several human

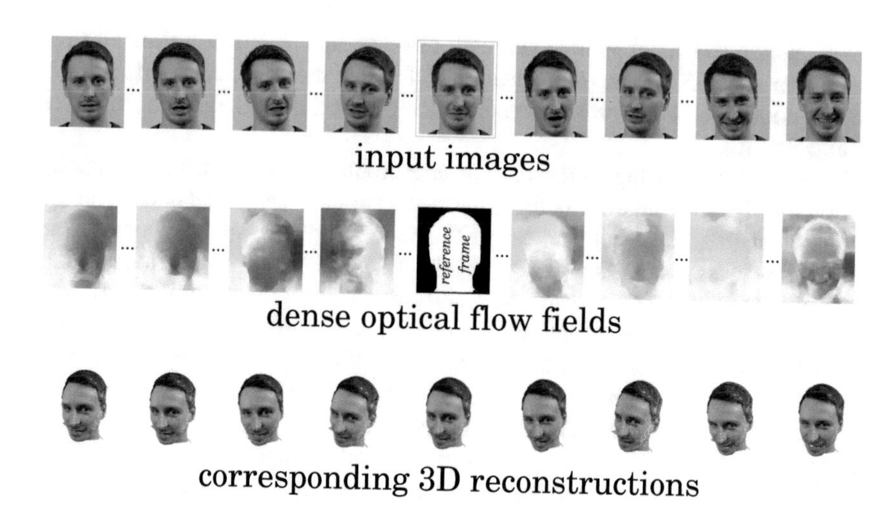

Figure 3.2: To reconstruct non-rigidly deforming surfaces, NRSfM uses motion and deformation cues contained in the set of dense point tracks over a monocular image sequence.

motion sequences. However, it scales poorly with the number of points and involves several computationally expensive steps.

Tresadern and Reid [285], Yan and Pollefeys [315] and Palladini *et al.* [216] address the articulated case with two rigid body movements and detect a hinge location. Later, an approach relying on spatial smoothness and dealing with an arbitrary number of rigid parts with simultaneous segmentation was proposed by Fayad *et al.* [90]. Next, Valmadre *et al.* [290] propose a dynamic programming approach for reconstruction of 3D articulated trees from input joint positions operating in linear time. Multibody SfM methods reconstruct multiple independent rigid body transformations [73] and non-rigid deformations in the same scene [171].

3.1.2.1 Dense Non-Rigid Structure from Motion

In the last several years, the advent of accurate dense multi-frame optical trackers [105, 267] paved the way for dense NRSfM. In dense NRSfM, points of interest usually constitute a connected region and are densely reconstructed from multiple views, see Fig. 3.2 for an overview. While many principles tested for sparse NRSfM can be directly generalised for the dense reconstruction, the transition from sparse to dense brings additional possibilities to constrain the solution space. One of such constraints is spatial consistency with topology preservation — instead of reconstructing disjoint points of interest, dense NRSfM recovers connected surfaces.

Several previous works addressed dense monocular non-rigid 3D reconstruction. Russel *et al.* [238] showed how existing piecewise NRSfM approaches could be improved for the scalability with the increased number of points and thus adapted for dense reconstructions. Garg *et al.* [104] proposed a variational approach with an energy functional minimised based on recent advances in optimisation techniques and the theory of inverse problems [64, 71]. The algorithm estimates the camera poses and shapes alternatingly, imposing spatial and temporal shape smoothness. The method was shown to be more robust against basis ambiguity peculiar to the low-rank shape model since it automatically estimates the optimal dimension of the subspace. Another technique is [239], where the scenes are segmented to foreground and background and consequently reconstructed. In [11], dense 3D shapes are recovered sequentially through completion of a downsampled sparse rest shape. All in all, only several methods can handle dense settings so far [11,104,238], and only a few dense datasets for quantitative evaluation are currently available.

In the following, chapters 4–5 address several open questions in the area of dense NRSfM including scalability with the number of points, occlusion handling with different prior assumptions, compressed representations, handling of topological changes, consolidation of multiple ideas and datasets for evaluation of dense NRSfM. In chapter 6, we propose a new class of methods for dense monocular reconstruction which uses a hybrid deformation model (HDM) which is closely related to TBR. On the one hand, HDM does not require point matches and operates directly on images. On the other hand, it requires pre-defined shape states for training a neural network for a deformation model. In this regard, a neural network uses learned deformation cues for the surface inference directly from images.

3.2 Point Set Registration

As 3D acquisition devices become ubiquitous, the demand for reliable point cloud processing and alignment algorithms grows. In this chapter, we define the problem of point set registration and review previous works. Point set registration is a key component in many domains such as 3D reconstruction, medical image registration, computer graphics and shape recognition. It also finds broad applications in engineering, computer-aided design, industrial quality control, autonomous driving and robotics.

3.2.1 Scope of Point Set Registration

The objective of *point set registration* (PSR) is the recovery of an optimal transformation and correspondences between two (generally several) point sets so that the point sets coincide according to some optimality criterion. In the most applications, one of the point sets is fixed and called a *reference*, and the point set which is transformed or aligned is called a *template*. A common special case of PSR is when point sets are related by parameters of rigid body motion, *i.e.*, rotation and translation applied to all points simultaneously. In other words, the objective of rigid PSR is the alignment of two point sets into a common reference frame. Due to the high practical relevance, rigid PSR became an active and standalone research field. Another significant special case arises when the displacement field is given by an affine transformation. There are multiple applications of rigid PSR in science and engineering including

- *Lidar data registration in autonomous driving*. Rigid PSR is used in the registration of lidar and sonar data to stitch the measurements into a complete map of the surroundings. Since at every time step, a vehicle is moving and the environment is changing, the individual measurements are partially overlapping and contain clustered outliers w.r.t each other.
- *Augmented reality*. In augmented reality devices, the environment is often scanned with RGB-D sensors. To build a complete 3D representation, point set registration is required, similar to the case with lidar data registration. Often, in augmented reality glasses or helmets, inertial sensors track the pose of the device in 3D space and provide pose initialisation for point set registration.
- *3D reconstruction*. Rigid alignment is widely used in 3D reconstruction to align partial reconstructions and to refine the camera poses for fusion. Thus, in 3D reconstruction with structured light and many other 3D multi-view reconstruction methods, alignment of partial scans is an essential step. 3D reconstruction is highly relevant for cultural heritage, where digitalisation of historical objects allows to capture the state, replicate and analyse the antiquities without exposing them to additional disturbing factors. A digital copy provides access to the relic for a broad audience.
- *Robotics*. In robotics, rigid point set registration is used for visual simultaneous localisation and mapping (SLAM).
- *Medicine*. In computer-assisted surgery, rigid PSR is used for many different instrument localisation and surface registration tasks. The examples are the alignment of a 3D model and the measurements, localisation of a 3D model in

a 3D representation of the human body, bringing the sensor measurement and the positions of the instruments in the common reference frame. Moreover, rigid PSR allows for avoiding artificial markers in some cases.

- *Product quality control.* In industrial production, rigid point set registration is used to measure the geometric similarity between a scan of a product and a 3D model. This allows detecting flaws in the products and machinery.

In the non-rigid case, the point set to point set transformation is described by a more general point displacement field, since non-rigid deformations imply individual displacements for every point. Therefore, the problem becomes highly ill-posed and regularisation of the displacement field is required to obtain a reasonable solution. Existing applications of non-rigid PSR are extensive and new applications are yet still to be discovered:

- *Pattern, shape and emotion recognition.* Non-rigid PSR can be used to compare how close a given 3D shape matches a pre-defined shape. Thus, spatial pattern recognition in biology, 3D facial expression and emotion recognition are facilitated by non-rigid PSR. Comparison of curves in characters is an application example of 2D non-rigid PSR in optical character recognition.
- *Appearance transfer and animation.* Non-rigid PSR can be used for appearance transfer between virtual characters, both covering facial expression transfer and morphological appearance.
- *Medical image registration (including volumetric images).* In medicine, non-rigid PSR is widely used for registration of volumetric images represented by point clouds. Moreover, 2D non-rigid PSR is used to register the parse representation of images which can be obtained by key point extraction.
- *Scientific visualisation.* Non-rigid PSR plays a significant role in scientific visualisation. Since PSR optionally recovers point displacements and affiliations, it can be used for visualisation of different types of deformations and shape and surface evolution in various scenarios. Having several states of the same scene at different moments of time, non-rigid PSR recovers transformations between those and signalises the differences.

A common byproduct of PSR is a set of correspondences relating points of the registered point sets. Correspondences can be often ambiguous, *i.e.,* not expressible by a bijective function (one-to-one correspondences). In many cases, correspondences are given by a non-injective surjective function (one-to-many correspondences). Not every template's point may possess a valid correspondence in the reference, and vice versa.

Captured point sets differ from the synthetic data (*e.g.*, 3D data created by a designer) in several ways. 3D acquisition devices often output noisy point clouds, perhaps with clustered outliers, and some parts can be missing — all the above-mentioned factors are encompassed under the notion of *disturbing effects*. Disturbing effects make the methods for manipulating artificial data in computer graphics and acquired real-world data different in several ways. Thus, general point set registration techniques differ from several other algorithm classes such as mesh registration methods [26] and matching techniques [240] by the unavailability of any further information except point coordinates, and, perhaps, sparse prior matches. The latter are assumed to be provided and not established by the algorithm.

A field related to PSR is image registration. The objective of image registration is to register two or several images. The images can be interpreted as regular 2D point sets, and the intensities, colours and gradients serve as the main alignment cues. PSR, in contrast, uses spatial arrangement and configuration of the point sets as the main registration cue (though, the usage of colour as an additional cue in PSR has been demonstrated [250]). It is possible to cast image registration to 2D point set registration by extracting the features from the images and performing PSR on them. This is one of the common use cases for 2D point set registration. An overview of registration methods can be found in [269]. Next, the previous work in rigid and non-rigid PSR will be discussed in detail.

3.2.2 Previous Work in Rigid Point Set Registration

Rigid alignment (RA) of point sets provided in different reference frames is an essential algorithmic component in multiple fields including but not limited to visual odometry [159], 3D reconstruction and augmented reality [208], [207], robot navigation and localisation [243], computer graphics [269], computer-aided design (CAD) modelling [316], cultural heritage [54] and medical technology [66], [129]. Even though rigid point set registration is a well-studied area, multiple challenges are remaining — on the top of the list are the broader basin of convergence, the robustness to disturbing effects in the data (noise, missing data, clustered outliers) and handling of massive point sets.

The seminal works of Besl and McKay [40] and Chen and Medoni [66] introduced iterative closest point (ICP) and opened the way for new applications. In practice, ICP has proven to be robust, stable and easy to implement. Since correspondences in the classic ICP are determined by deterministic nearest neighbour search, its accuracy was found to be sensitive to outliers and suboptimal initialisations. Through its simplicity and despite the disadvantages, it is one of the most widely used PSR algorithms.

Multiple improvements were proposed to overcome the shortcomings of ICP ranging from accelerating policies for nearest-neighbour search and relaxation of correspondences to more efficient optimisation schemes. Spectral registration methods operate on proximity matrices with distance measures between the points [251]. They are generally computationally expensive which narrows their scope. Gold *et al.* [113] proposed a combination of deterministic annealing and softassign with a relaxation of binary correspondences to probabilistic correspondences. Expectation-Maximisation (EM) ICP uses multiple matches weighted by normalised Gaussian weights and is more accurate under suboptimal initialisations [129]. In [93], the ICP energy is minimised using the non-linear least squares (NLLS) framework allowing integration of robust norms (such as Huber norm). In [39], optimisation is performed by iteratively reweighted least squares (robust M-estimation techniques). In both latter cases, energy optimisation with least squares resulted in a broader convergence basin and robustness against outliers. Nüchter *et al.* propose a cashed (partially reusable) *k*-d tree achieving a noticeable acceleration [209]. Several methods introduce an extra term to cope with outliers explicitly [203, 308] while making an additional assumption on the type of the noise distribution.

A milestone in RA is related to modelling of point sets by probability density functions, especially Gaussian mixture models (GMM). In probabilistic methods with GMM, EM algorithm is used for the likelihood function optimisation and transformation recovery. The idea of probabilistic assignments was further evolved in [69] and [203] where the template and the reference are interpreted as a Gaussian mixture and observed data, respectively. Thus, aligning the point sets is equivalent to explaining the data by the GMM — or finding maximum a posterior estimates of GMM — using maximum likelihood principle. The advantages of probabilistic methods such as coherent point drift (CPD) [203] is locally multiply-linked assignments and, often, explicit modelling of noise which results in the increased robustness to noise. CPD was extended in [302] with an additional term for outlier modelling. In [84], a wider convergence basin is achieved by Gaussian mixture decoupling. The multiply-linked Kernel Correlation (KC) approach minimises the Renyi's quadratic entropy of the joint system composed of the reference and the transformed template parameterised by $\mathbf{R}$ and $\mathbf{t}$ [286]. KC maximisation is equivalent to minimisation of a sum of distances in the sense of M-estimator. In practice, the KC cost function is approximated, and still, only local neighbourhoods are directly involved in one-to-many interactions. Being less sensitive to noise compared to ICP, KC is not widely used due to the high computational complexity. KC was further extended with GMM in [156]. The unified framework for probabilistic point set registration *GMM Registration* (GMR) [157] interprets both input point sets as GMMs and casts RA as a problem of mixture alignment, *i.e.*, minimising the discrepancy between two probability distributions measured by the ℓ_2 distance.

It was also shown in [157] that ICP and KC can be interpreted as special cases of maximising the similarity between probability distributions.

A subclass of the probabilistic algorithms takes advantage of particle filters [196, 248]. Only the latest particle filtering approach to point set registration [248] can be seen as a general-purpose point set registration algorithm, as it makes no assumptions on the point set density and is more accurate by a moderate computational cost. This method was developed to cope with partial rigid registrations. However, it includes ICP as an intermediate step.

Next, Campbell and Petersson [63] leveraged conversion of the input point sets to densified though lightweight GMM representations and introduced the support vector registration (SVR) algorithm. In SVR, the intermediate mapping to the continuous domain is performed by a support vector machine and accompanied by the selection of reliable points and outlier pruning. RA is then performed as ℓ_2 distance minimisation between the densified GMMs. SVR was demonstrated to be noise-resistant as well as exhibiting a broader basin of convergence than GMR, CPD and ICP.

Since lately, physics-based formulations are enriching the spectrum of available RA techniques. Recently, Deng *et al.* proposed a method for solving RA in the Schrödinger distance transform representation [78], where a minimal geodetic distance between two points on a unit sphere in a Hilbert space serves as an optimality criterion.

3.2.3 Previous Work in Non-Rigid Point Set Registration

In this section, we provide a concise overview of modern non-rigid point set registration methods. Though the body of literature on non-rigid point set registration is vast, current approaches can be classified into few categories. At large, the taxonomy of non-rigid PSR is similar to the taxonomy of rigid PSR. Whenever an idea had been proposed for the rigid case, it has been soon tested and possibly generalised for the non-rigid case (up to few exceptions).

Coarsely, the primary method classes either involve a nearest-neighbour search step [40, 236] (variants of ICP), employ various probabilistic principles for unsupervised learning of optimal transformation fields [156, 185], or are physics-based schemes which adopt or simulate natural and physical phenomena [78, 184]. Each category can be further refined through multiple subcategories differentiating between used optimisation techniques (deterministic annealing scheme in [70], expectation-maximisation [185]), acceleration techniques (spherical triangle constraint for nearest neighbour search in ICP [132]), types of probabilistic models and distributions (Gaussian mixture model of GMMReg [156] versus Student's-t mixture model [330]), types of borrowed physical laws (Schrödinger distance transform [78]) or types of displacement field regularisers (a thin-plate spline in [70] or variational regulariser of motion coherence in [203]). Moreover, auxiliary classifications distinguish between two point set (hereto belongs the majority of the approaches) and multi-set registration [247], whether methods allow to embed priors matches (akin to ICP methods) or require some additional information aside from 3D coordinates [24, 107].

Early works on non-rigid point set registration used probabilistic Gaussian Mixture Models (GMM) positioned along contours. The contours were modelled by splines allowing non-rigid transformations, but the method was restricted to contour-like registrations [141], [230]. Several extensions of the Iterative Closest Point (ICP) for the non-rigid case were proposed [93], [24]. They evince the same drawbacks as their rigid counterpart, namely high sensitivity to outliers. One of the widely used non-rigid point set registration methods is based on modelling the transformation with thin plate splines (TPS) [46] followed by robust point matching (RPM) and is known as the TPS-RPM [70]. It uses deterministic annealing and alternates between updates of the soft assignment and estimation of transformation parameters. The authors showed how the expectation-maximisation (EM) algorithm can be embedded into a deterministic annealing scheme [69]. An optimised implementation of the TPR-RPM on GPU for point sets with thousands of elements was addressed in [200]. A correlation-based approach was proposed in [286] and later improved in [156]. It tries to align two distributions whereby each of the point sets represents GMM centroids. CPD method was introduced in [204] and further improved in [203]. It employs an EM algorithm to optimise the GMM and the regularisation originating in the motion coherence theory [320], [321]. Compared to TPS-RPM, CPD offers superior accuracy and stability with respect to non-rigid deformations in the presence of outliers. An additional parameter for outlier modelling of CPD was introduced for a hybrid optimisation with the Nelder-Mead simplex method in [302]. CPD was also modified by imposing the Local Linear Embedding topological constraint to cope with highly articulated non-rigid deformations [108]. However, this extension performs poorly on data with inhomogeneous density and is more sensitive to noise than CPD. Recently, a non-rigid registration method based on Student's Mixture Model (SMM) showed to be even more robust and accurate on noisy data than the CPD approach [330].

Several methods were proposed for improving registration quality in challenging cases through embedding prior knowledge and constraining the solution space. Some are based on shape priors and address registration of a human template to human scans [143]. A recent approach for human shape registration [45] not only registers the geometric shape but also the appearance of shapes and the optical flow between shapes obtained from textured 3D scans. Some spectral differential geometry methods, constituting rather a separate class of algorithms, are able to regard prior correspondences [167]. Compared to point set registration methods, they are more noise sensitive and typically operate on meshes (*i.e.*, need surfaces and normals), whereas real-world scans are usually noisy point clouds. In the context of the related problem of medical image registration, several methods utilise correspondences between SIFT keypoints or hybrid detectors and couple them into the registration procedure [198], [325]. No further assumptions about the content of the image are made, which allows decoupling correspondence search and image registration.

For the acceleration purposes, subsampling of point sets can be applied. In the rigid case, the recovered transformation refers simultaneously to all points. Hence, the transformation, recovered for a subsampled point set can be directly generalised to the initial one [236]. In contrast, in the case of non-rigid registration, such transformation does not generalise to the initial dense point set in the general case.

4 Scalable Dense Non-Rigid Structure from Motion

THE focus of this chapter lies on scalable NRSfM methods. In the recent years, the scalability in NRSfM has gained increased attention. Thus, the goal is not only obtaining accurate reconstructions but also the results have to remain consistently accurate with the different number of input point tracks and for as many different scenarios as possible. In this chapter, these properties comprise the notion of *scalability*. Moreover, the desired property of a scalable NRSfM method is the support of dense settings at interactive rates and potential suitability for real-time applications.

With these limitations in mind, our motivation is to find a scalable NRSfM method, *i.e.*, a method supporting point track matrices of different sizes, to make NRSfM robust to self-occlusions and easy to implement. Among existing approaches, it is hard to find one which fulfils all the properties stated above. The proposed in this chapter methods are based on the direct factorisation of the input dense point tracks.

In the first section, we show how to minimise a quadratic function on a set of orthonormal matrices using an efficient semidefinite programming solver with application to dense NRSfM. Thanks to the proposed technique, a new form of the convex relaxation for the Metric Projections (MP) [216] algorithm is obtained. The modification results in an efficient single-core CPU implementation enabling dense factorisations of long image sequences with tens of thousands of points into camera pose and non-rigid shape in seconds, *i.e.*, at least two orders of magnitude faster than the runtimes reported in the literature so far. The proposed implementation can be useful for interactive or real-time robotic and other applications, where monocular non-rigid reconstruction is required. In a narrow sense, Sec. 4.1 complements research on MP, though the proposed convex relaxation methodology can also be useful in other computer vision tasks. The experimental part with runtime evaluation and qualitative analysis concludes the first section.

Next, we introduce Scalable Monocular Surface Reconstruction (SMSR) approach in Sec. 4.2. Existing techniques using motion and deformation cues rely on multiple prior assumptions, are often computationally expensive and do not perform equally well across the variety of datasets. In contrast, SMSR combines

V. Golyanik, *Robust Methods for Dense Monocular Non-Rigid 3D Reconstruction and Alignment of Point Clouds*,

the strengths of several algorithms, *i.e.,* it is scalable with the number of points, can handle sparse and dense settings as well as different types of motions and deformations. We estimate camera pose by singular value thresholding and proximal gradient. Our formulation adopts alternating direction method of multipliers which converges in linear time for large point track matrices. In the proposed SMSR, trajectory space constraints are integrated by smoothing of the measurement matrix. In the extensive experiments, SMSR is demonstrated to consistently achieve state-of-the-art accuracy on a wide variety of datasets.

4.1 Scalable NRSfM with Semidefinite Programming

4.1.1 Introduction

Template-free deformable surface reconstruction from monocular image sequences referred to as Non-Rigid Structure from Motion (NRSfM) experienced significant advances during the last ten years. Previously being able to retrieve sparse structures under small non-rigid deformations, NRSfM methods nowadays can recover dense surfaces exhibiting large deformations. While entering the realm of dense reconstructions, the computational time of NRSfM methods has increased significantly due to the inherent ill-posedness of the problem, higher complexity of the models, and high computational complexity of the optimisation methods. This tendency is aside from the increased pre-processing time for the dense correspondence establishment, which the vast majority of NRSfM methods rely on. At the same time, many robotics and medical application require not only dense reconstructions but also impose harsh timing constraints — either new reconstructions need to be obtained upon arrival of a new frame (real-time requirement), or reconstructions are required within an arguably reasonable time interval after the image sequence is acquired (requirement of interactivity).

Among NRSfM algorithms, Metric Projections (MP) [216] possesses properties which make it considerable for interactive, real-world applications such as high reconstruction accuracy, feasible computational complexity, availability of efficient and fast optimisation methods, notable scalability with the number of points as well as robustness to noisy and missing data. An important for this section aspect is that MP requires solving a quadratic optimisation problem on the set of orthonormal matrices of the form

$$\min_{\mathbf{q}=vec(Q)} \mathbf{q}^{\top}\mathbf{E}\mathbf{q}, \tag{4.1}$$

where $E \in \mathbb{R}^{6\times6}$ and $Q \in \mathbb{R}^{3\times2}$ is an orthonormal matrix. According to [216], the problem in Eq. (4.1) can be convex-relaxed and approximated as

$$\min_{\mathbf{q}=vec(Q)} \operatorname{tr}(\mathbf{E}\mathbf{q}\mathbf{q}^\top) = \min_{\mathbf{X}\in co(S)} \operatorname{tr}(E\mathbf{X}), \tag{4.2}$$

where $co(S)$ is approximated by a set of real symmetric matrices $X \in \mathbb{R}^{6\times6} = \begin{pmatrix} A & B \\ B^\top & C \end{pmatrix}$, whereby $A \in \mathbb{R}^{3\times3}$, $X \succcurlyeq 0$, $\operatorname{tr}(A) = \operatorname{tr}(C) = 1$, $\operatorname{tr}(B) = 0$, with an additional constraint on the combined matrix Y:

$$Y = \begin{pmatrix} I_3 - A - C & \mathbf{w} \\ \mathbf{w}^\top & 1 \end{pmatrix} \succcurlyeq 0, \tag{4.3}$$

with $\mathbf{w} = \begin{pmatrix} b_{23} - b_{32} \\ b_{31} - b_{13} \\ b_{12} - b_{21} \end{pmatrix}$ (see Sec. 4.1.3 for further details).

The problem in Eq. (4.2) is a Semidefinite Programming (SDP) problem. In an SDP, a linear objective function with linear constraints over a set of Positive Semidefinite (PSD) matrices needs to be optimised [140]. SDP is a field of convex optimisation which finds applications in various areas of science and engineering ranging from the approximation of NP-hard combinatorial problems (max-cut, graph bisection, min-max eigenvalue problems) [140] and quantum complexity [32], to the theory of automatic control [219]. An SDP problem can be solved using a Matlab SDP solver, *e.g.*, SeDuMi [262] which allows specifying constraints on a high level of abstraction directly in terms of traces, block matrices and values of particular elements. However, for an efficient C++ SDP solver such as CSDP [48], an SDP problem must be specified in a general form as

$$\begin{aligned} &\max \operatorname{tr}(C'U), \\ &\text{subject to } \left\{\mathscr{A}(U) = a;\ U \succcurlyeq 0\right\}, \end{aligned} \tag{4.4}$$

where

$$\mathscr{A}(U) = \begin{bmatrix} \operatorname{tr}(A_1U) & \operatorname{tr}(A_2U) & \ldots & \operatorname{tr}(A_mU) \end{bmatrix}^\top. \tag{4.5}$$

In the optimisation model given by Expr. (4.4), the matrices $C', A_1, A_2, \ldots, A_m$ as well as the vector a are known, and the operator $\mathscr{A}$ maps the unknown PSD matrix U to a vector.

Thus, the existing form of the convex relaxation as presented in Eq. (4.2) does not allow to employ CSDP and implement MP in C++ efficiently. A drawback is that robotic and medical applications do not often allow to combine Matlab and C++ source code on an embedded platform due to performance and memory issues. Thus, research on MP algorithm can not be considered as accomplished yet. Moreover, both academia and industry are interested in an efficient monocular NRSfM which supports real-time frame rates.

4.1.1.1 Contributions

Accordingly, our contributions in this section are as follows:

- We show how to formulate the convex optimisation problem given in Eq. (4.2) in terms of the standard SDP problem as accepted by an efficient SDP solver (Expr. (4.4)). The shown methodology can also be useful in a broad range of algorithms taking advantage of SDP. In this section, it is applied to an optimised CPU-only implementation of MP which we refer to as Accelerated Metric Projections (AMP).
- We demonstrate experimentally that AMP is suitable for dense factorisation of long image sequences with tens of thousands of points in seconds, and compare runtimes of AMP with an optimised implementation of the Variational NRSfM Approach (VA) [104].

In the performed experiments, AMP finishes up to 20 times faster than an optimised implementation of VA [116]. Considering the achieved performance, AMP can be useful as a building block in a broad range of real-time and interactive applications including medical and robotic ones which require monocular non-rigid reconstruction. AMP can also be used for initialisation of more accurate but computationally expensive, perhaps sequential NRSfM algorithms.

4.1.2 Related Work

SDP was extensively applied in computer vision [8, 161, 162, 177] including NRSfM [67, 76, 216]. We noticed, on the one hand, that a detailed consideration of SDP problems and their efficient solutions for computer vision tasks were rather sparsely discussed in the literature. On the other hand, research in the area of NRSfM was mainly focused on algorithmic improvements which were not always accompanied by an appropriate consideration of efficient optimisation methods. As a result, it is difficult to find high-performance implementations able to handle dense data in a few seconds among those shown in the literature, although several algorithms are potentially capable of it [11, 104, 238].

An efficient implementation is often not possible without changes in underlying optimisation methods. We discover that this is the case with the MP algorithm [216]. The optimisation proposed in this section allows to fully unfold the strength of the approach regarding the runtime. Whereas [216] reports around 30 seconds for sequences with 37 points tracked throughout 74 frames, AMP allows reconstructing $5 \cdot 10^4$ points tracked throughout 202 frames in 30 seconds. Despite the original MP implementation was performed in Matlab, the difference with our implementation can not be barely explained by a re-implementation in another programming language (C++ in our case) or an improved hardware during the last 48 months between [216] and the submission of our method to a conference in 2016 (recall that an efficient implementation in C++ was not earlier possible). Various runtimes for dense NRSfM were reported for other algorithms in the literature so far. Thus, [238] reports 600 seconds for 90 frames with $7.8 \cdot 10^4$ points per frame. [293] reports 720 seconds for 50 frames with 540 points in every frame. The sequential algorithm of Agudo *et al.* [11] achieves 62 seconds per frame for the dense flag sequence [105] with $\sim 10^4$ points. An efficient implementation of this method could potentially run in real-time. Nevertheless, the runtimes reported in the literature are still at least two orders of magnitude higher than the runtime reported in this section, considering comparable number of points and frames (*e.g.*, for the synthetic flag sequence [105]).

4.1.3 Accelerated Metric Projections (AMP) Approach

First, we summarise the core MP method. The input of MP is a measurement matrix $\mathbf{W}$ which combines image coordinates of the tracked points for all frames:

$$\mathbf{W}_{2F\times N} = [\mathbf{W}_1\mathbf{W}_2\ldots\mathbf{W}_F]^\top, \tag{4.6}$$

where $\mathbf{W}_i = [\mathbf{w}_{ij}] = [(u_{ij}v_{ij})^\top]$, $i \in \{1,\ldots,F\}$ is a frame index and $j \in \{1,\ldots,N\}$ is a point index. Values in $\mathbf{W}$ are registered to the centroid of the scene. MP adopts the low-rank shape model introduced in [53] so that every non-rigid shape $\mathbf{S}_i$ observed in frame i can be expressed as

$$\mathbf{S}_i = \sum_{d=1}^{k} l_{id}\mathbf{B}_d, \tag{4.7}$$

i.e., a linear combination of the basis shapes $\mathbf{B}_d$ with the weights l_{id}, $d \in \{1,\ldots,k\}$. Since $\mathbf{W}_i = \mathbf{R}_i\mathbf{S}_i$ defines a 2D scene observed by an orthographic camera $\mathbf{R}_i$ for every frame, $\mathbf{W}_i$ can be written as

$$\mathbf{W}_i = \begin{bmatrix} l_{i1}\mathbf{R}_i & \ldots & l_{ik}\mathbf{R}_i \end{bmatrix} \begin{bmatrix} \mathbf{B}_1 & \ldots & \mathbf{B}_k \end{bmatrix}^\top, \tag{4.8}$$

and for the whole $\mathbf{W}$ as

$$\mathbf{W} = \mathbf{M}\mathbf{B} = \begin{bmatrix} \mathbf{M}_1 \\ \vdots \\ \mathbf{M}_F \end{bmatrix} \begin{bmatrix} \mathbf{B}_1 \\ \vdots \\ \mathbf{B}_k \end{bmatrix}. \tag{4.9}$$

Thus, MP factorises $\mathbf{W}$ into the motion matrix $\mathbf{M}$ (it contains camera poses $\mathbf{R}_i$ multiplied with weights l_{ik}) and a set of the basis shapes $\mathbf{B}$ using alternating least squares, *i.e.,* either $\mathbf{B}$ or $\mathbf{M}$ is fixed while for the other one is being optimised. Further, the motion matrix $\mathbf{M}$ is projected onto the motion manifold (*i.e.,* a Lie group with augmented scalar weights) which guarantees the correct block structure of the motion matrix and camera pose matrices contained in it. The projection is performed for each $l_{id}\mathbf{R}_i$ submatrix of $\mathbf{M}_i$ individually and can be written in a form of an optimisation problem as

$$\min_{\mathbf{R}_i, l_{i1},\ldots,l_{ik}} ||\mathbf{M}_i - [l_{i1}\mathbf{R}_i| \ldots |l_{ik}\mathbf{R}_i]||^2_{\mathscr{F}} \tag{4.10}$$

or, after reordering, writing out the squared Frobenius norm and eventually bringing the expression to a quadratic form:

$$\min_{\mathbf{R}_i} \mathbf{r}_i^\top \left[-\sum_{d=1}^{k} \mathbf{m}_{id}\mathbf{m}_{id}^\top \right] \mathbf{r}_i \text{ so that} \tag{4.11}$$

$$\mathbf{R}_i\mathbf{R}_i^\top = \mathbf{I}_{2\times 2}, \tag{4.12}$$

with $\mathbf{r}_i = \text{vec}(\mathbf{R}_i^\top)$ and $\mathbf{m}_{id} = \text{vec}(\mathbf{M}_{id}^\top)$. The quadratic form in Eq. (4.11) has the form as in Eq. (4.1) and is convex-relaxed as stated in Sec. 4.1.1, Eqs. (4.1)–(4.3). A detailed derivation of the convex relaxation is given in the recent work by Dodig *et al.* [83].

Solving Eq. (4.11) leads to a provably optimal $\mathbf{M}_i$ update. As proposed in [216], we use a warm-start strategy in combination with a Newton-like iterative optimisation approach [85], *i.e.,* we compute only $\mathbf{M}_0$ using the convex-relaxed Eq. (4.11) is each iteration, and the remaining $\mathbf{M}_i$ matrices are obtained based on the optimal $\mathbf{M}_0$ estimate. Despite this strategy is theoretically suboptimal, it was empirically shown to converge to a local minimum in multiple experiments while significantly reducing the overall runtime [216]. Once recovered, the $\mathbf{R}_i^\top$ are used to update the weights:

$$l_{id} = \frac{1}{2}\text{tr}\left[\mathbf{M}_{id}^\top\mathbf{R}_i\right]. \tag{4.13}$$

Given the weights l_{id}, projection of $\mathbf{M}$ onto the motion manifold is complete — which, in turn, allows to obtain a current estimate of $\mathbf{S}$ and an update of $\mathbf{M}$. MP is an iterative approach, and requires an initial estimate of **M**. Further details on the core MP method can be found in [216].

Secs. 4.1.3.1 and 4.1.3.2 describe how Eq. (4.1) can be convex-relaxed and brought to the general form as accepted by an efficient C++ SDP solver. The modification given in the following leads to the proposed AMP method.

4.1.3.1 Coefficient Splitting

A symmetric matrix $Z = [z_{ij}]$ equals to its own transpose, *i.e.*, $z_{ij} = z_{ji}$. Moreover, the following property for symmetric matrices holds:

$$\mathrm{tr}(AZ) = \sum_{j=1}^{n}\sum_{i=1}^{n} a_{ij} z_{ij}, \tag{4.14}$$

where $A = [a_{ij}]$ is a real square matrix. Suppose a constraint is given in the form

$$c_{11}z_{11} + c_{12}z_{12} + \ldots + c_{1n}z_{1n} + \ldots + c_{nn}z_{nn} = b. \tag{4.15}$$

The key observation is that the constraint in Eq. (4.15) can be written with a *coefficient splitting* as

$$\sum_{i=1}^{n} c_{ii}z_{ii} + \sum_{i \neq j} \frac{c_{ij} + c_{ji}}{2}(z_{ij} + z_{ji}) = b \tag{4.16}$$

(again, this holds for a symmetric matrix $Z = [z_{ij}]$). Considering Eq. (4.14), Eq. (4.16) can be written as $\mathrm{tr}(AZ) = b$, where

$$A = \begin{pmatrix} c_{11} & \frac{c_{12}+c_{21}}{2} & \frac{c_{13}+c_{31}}{2} & \cdots \\ \frac{c_{12}+c_{21}}{2} & c_{22} & \frac{c_{23}+c_{32}}{2} & \cdots \\ \frac{c_{13}+c_{31}}{2} & \frac{c_{23}+c_{32}}{2} & c_{33} & \cdots \\ \vdots & \vdots & \vdots & \ddots \end{pmatrix}. \tag{4.17}$$

In Eq. (4.17), only the divisor value 2 preserves matrix symmetry, *i.e.*, an arbitrary re-distribution of the coefficients is not possible. In particular, if $c_{ii} = 1$, then $a_{ii} = 1$; if $c_{ij} = 1$, then $a_{ij} = a_{ji} = 0.5$ $(i \neq j)$; if $c_{ij} = 0$, then $a_{ij} = a_{ji} = 0$ $(i \neq j)$. Note that a constraint can also be formulated in terms of an upper/lower triangular submatrix of X. In this case, if c_{ij} defines a coefficient, then c_{ji} always equals to 0 (no coefficient splitting is required).

4.1.3.2 Constraints in the Unified Form

Now, using the properties of the matrix trace (Sec. 4.1.3.1), we can advance the convex relaxation of the quadratic optimisation problem given in Eq. (4.2) into the

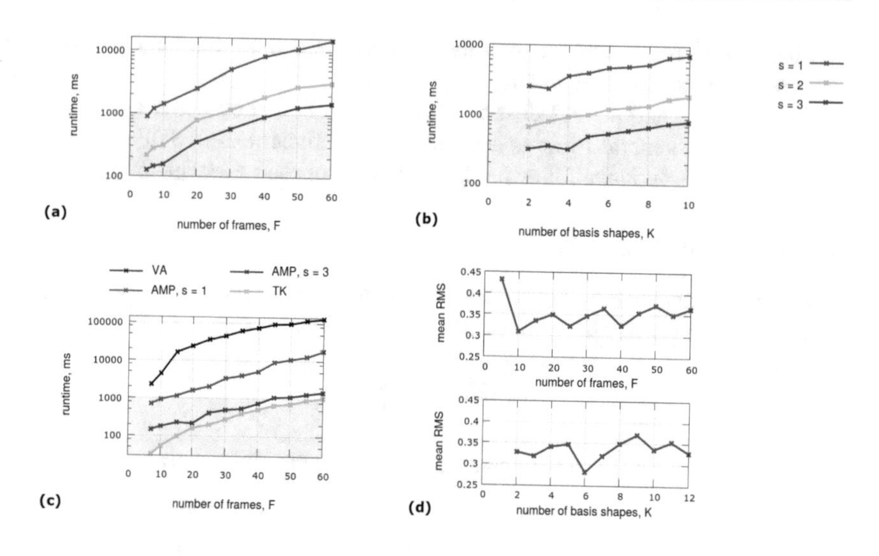

Figure 4.1: Results of the AMP runtime measurements on the synthetic flag sequence [105] for different subsampling factors (1 or w/o subsampling, 2 and 3) : (a) the number of basis shapes K is fixed to 3, the number of frames is varying; (b) the number of frames is fixed to 20, the number of basis shapes is varying. (c) runtime functions in the number of frames for VA [104], rigid factorisation [281] and AMP on the Xeon E5-1650 platform; AMP is an order of magnitude faster than VA; AMP achieves a comparable to rigid factorisation runtime if the number of points equals to approx. 10% compared to the TK case; (d) mean RMS error for AMP on the synthetic flag sequence as a function of the number of frames for $K = 4$ (top) and a function of the number of basis shapes for $F = 30$ (bottom). In (a-c), light yellow colour marks configurations which require less than a second to complete.

standard form of an SDP problem given by Expr. (4.4). The main matrices — on which elements (submatrices or particular values) constraints are imposed — are X and Y. In the following, we will use the proposition

Proposition 1. *If a matrix* $U = \begin{pmatrix} U_1 & 0 \\ 0 & U_2 \end{pmatrix}$ *has a block structure, then* U *is PSD if both* U_1 *and* U_2 *are PSD. Conversely, if* U *is PSD, then the principal submatrices*[1] U_1 *and* U_2 *are PSD.*

Proof. To show validness of Prop. 1, we use one of the necessary and sufficient conditions for positive semidefiniteness: a symmetric matrix U is PSD if $x^\top U x \succeq 0$ for all vectors x (this is also definition of PSD).
Consider symmetric matrix $\hat{U} = \begin{pmatrix} U_1 & 0 \\ 0 & 0 \end{pmatrix}$. Since U_1 is PSD, $y^\top U_1 y \succeq 0$ holds for all vectors y. Moreover, $\hat{y}^\top \hat{U} \hat{y} \succeq 0$, where $\hat{y}$ is an arbitrary vector. Hence, $\hat{U}$ is PSD. Analogously, consider symmetric matrix $\tilde{U} = \begin{pmatrix} 0 & 0 \\ 0 & U_2 \end{pmatrix}$. Since U_2 is PSD, $z^\top U_2 z \succeq 0$ holds for all arbitrary vectors z. Furthermore, $\tilde{z}^\top \tilde{U} \tilde{z} \succeq 0$, where $\tilde{z}$ is an arbitrary vector. Hence, $\tilde{U}$ is PSD. Now consider the sum $\hat{U} + \tilde{U}$:

$$x^\top \hat{U} x + x^\top \tilde{U} x \succeq 0, \text{ or} \tag{4.18}$$

$$x^\top (\hat{U} + \tilde{U}) x \succeq 0. \tag{4.19}$$

Thus, Eq. (4.19) implies that $U = \hat{U} + \tilde{U}$ is PSD.

To show positive semidefiniteness of U_1, we choose a vector $\tilde{x}$ with non-zero entries at the first $k = \dim(U_1)$ positions. Since U is PSD, $\tilde{x}^\top U \tilde{x} \succeq 0$. Hence, $x_{1\times k}^\top U_1^{k\times k} x_{k\times 1} \succeq 0$ for all vectors $x_{k\times 1}$, *i.e.*, U_1 is PSD. Analogously, we choose a vector $\tilde{x}$ with non-zero entries at the last $l = \dim(U_2)$ positions, and $x_{1\times l}^\top U_2^{l\times l} x_{l\times 1} \succeq 0$ for all vectors $x_{l\times 1}$, *i.e.*, U_2 is PSD. □

Using Prop. 1 we can join X and Y into a block matrix U and demand its positive semidefiniteness:

$$U = \begin{bmatrix} X_{6\times 6} & 0 \\ 0 & Y_{4\times 4} \end{bmatrix} = \begin{bmatrix} A_{3\times 3} & B & 0 \\ B^\top & C_{3\times 3} & 0 \\ 0 & 0 & Y_{4\times 4} \end{bmatrix} \overset{!}{\succeq} 0. \tag{4.20}$$

Thus, if the matrix U will be found, X and Y will be guaranteed to be PSD.

[1] a principal submatrix is a square submatrix obtained by removing k rows and columns with the same indexes (whenever i-th row is removed from U, the i-th column is also removed).

4.1.3.3 Constraints on the Traces

The SDP problem we are interested in, contains three constraints on the matrix trace. Constraint $\mathrm{tr}(A) = 1$ can be rewritten as $\sum_i^3 a_{ii} = 1$ or equivalently as

$$\mathrm{tr}(A_1 U) = \mathrm{tr}\left(\begin{bmatrix} \mathbf{I}_3 & 0 \\ 0 & \mathbf{0}_{7\times 7} \end{bmatrix} U\right) = 1. \tag{4.21}$$

In the same manner, we obtain constraint expressions for $\mathrm{tr}(C) = 1$:

$$\mathrm{tr}(A_2 U) = \mathrm{tr}\left(\begin{bmatrix} \mathbf{0}_{3\times 3} & 0 & 0 \\ 0 & \mathbf{I}_3 & 0 \\ 0 & 0 & \mathbf{0}_{4\times 4} \end{bmatrix} U\right) = 1 \tag{4.22}$$

and $\mathrm{tr}(B) = 0$:

$$\mathrm{tr}(A_3 U) = \mathrm{tr}\left(\begin{bmatrix} 0 & \frac{\mathbf{I}_3}{2} & 0 \\ \frac{\mathbf{I}_3}{2} & 0 & 0 \\ 0 & 0 & 0 \end{bmatrix} U\right) = 0. \tag{4.23}$$

In total, we obtain three constraints on the traces of the submatrices of X, and, accordingly, first three A_i.

4.1.3.4 Constraints on the Combined Matrix Y

To formulate constraints on Y, we consider all its unique elements separately. Since Y is symmetric, it has ten non-repeating elements. We denote a_{ij} and c_{ij} elements of A and C in Eq. (4.2) and (4.3), respectively. Consider matrix Y in Eq. (4.3) element-wise. We see that the element y_{11} is constrained as

$$y_{11} = 1 - a_{11} - c_{11}. \tag{4.24}$$

Therefore, A_4 must fetch correct elements and equate them to 1. We obtain a sparse matrix with the elements $[1;1]$, $[4;4]$ and $[7;7]$ being equal to 1, and zeros otherwise. In a similar way, A_5 and A_6 for the elements y_{22} and y_{33} can be obtained, with non-zero elements shifted along the diagonal by 1 and 2 positions, respectively. Accordingly, A_7 (element y_{44}) has only a single non-zero entry in the position $[10;10]$. We proceed further with the elements of matrix Y. Thus, $y_{12} = -a_{12} - c_{12}$. As these elements are not on the diagonal of U, coefficient splitting is required, *i.e.*,

$$\frac{u_{12}+u_{21}}{2} + \frac{u_{45}+u_{54}}{2} + \frac{u_{78}+u_{87}}{2} = 0, \tag{4.25}$$

and the constraint matrix A_8 reads as

$$A_8 = \begin{bmatrix} 0 & 0.5 & 0 & 0 & 0 & 0 & 0 & 0 & 0 & 0 \\ 0.5 & 0 & 0 & 0 & 0 & 0 & 0 & 0 & 0 & 0 \\ 0 & 0 & 0 & 0 & 0 & 0 & 0 & 0 & 0 & 0 \\ 0 & 0 & 0 & 0 & 0.5 & 0 & 0 & 0 & 0 & 0 \\ 0 & 0 & 0 & 0.5 & 0 & 0 & 0 & 0 & 0 & 0 \\ 0 & 0 & 0 & 0 & 0 & 0 & 0 & 0 & 0 & 0 \\ 0 & 0 & 0 & 0 & 0 & 0 & 0 & 0.5 & 0 & 0 \\ 0 & 0 & 0 & 0 & 0 & 0 & 0.5 & 0 & 0 & 0 \\ 0 & 0 & 0 & 0 & 0 & 0 & 0 & 0 & 0 & 0 \\ 0 & 0 & 0 & 0 & 0 & 0 & 0 & 0 & 0 & 0 \end{bmatrix}. \tag{4.26}$$

Note that since A_8 is symmetric, it expresses constraints on both elements y_{12} and y_{21}. For the elements y_{13}, y_{31}, y_{23}, y_{32}, y_{14}, y_{41}, y_{24}, y_{42} and y_{34}, y_{43} we proceed in a similar manner and obtain $A_9 - A_{13}$, respectively:

$$A_9 = \begin{bmatrix} 0 & 0 & 0.5 & 0 & 0 & 0 & 0 & 0 & 0 & 0 \\ 0 & 0 & 0 & 0 & 0 & 0 & 0 & 0 & 0 & 0 \\ 0.5 & 0 & 0 & 0 & 0 & 0 & 0 & 0 & 0 & 0 \\ 0 & 0 & 0 & 0 & 0 & 0.5 & 0 & 0 & 0 & 0 \\ 0 & 0 & 0 & 0 & 0 & 0 & 0 & 0 & 0 & 0 \\ 0 & 0 & 0 & 0.5 & 0 & 0 & 0 & 0 & 0 & 0 \\ 0 & 0 & 0 & 0 & 0 & 0 & 0 & 0 & 0.5 & 0 \\ 0 & 0 & 0 & 0 & 0 & 0 & 0 & 0 & 0 & 0 \\ 0 & 0 & 0 & 0 & 0 & 0 & 0.5 & 0 & 0 & 0 \\ 0 & 0 & 0 & 0 & 0 & 0 & 0 & 0 & 0 & 0 \end{bmatrix},$$

$$A_{10} = \begin{bmatrix} 0 & 0 & 0 & 0 & 0 & 0 & 0 & 0 & 0 & 0 \\ 0 & 0 & 0.5 & 0 & 0 & 0 & 0 & 0 & 0 & 0 \\ 0 & 0.5 & 0 & 0 & 0 & 0 & 0 & 0 & 0 & 0 \\ 0 & 0 & 0 & 0 & 0 & 0 & 0 & 0 & 0 & 0 \\ 0 & 0 & 0 & 0 & 0 & 0.5 & 0 & 0 & 0 & 0 \\ 0 & 0 & 0 & 0 & 0.5 & 0 & 0 & 0 & 0 & 0 \\ 0 & 0 & 0 & 0 & 0 & 0 & 0 & 0 & 0 & 0 \\ 0 & 0 & 0 & 0 & 0 & 0 & 0 & 0 & 0.5 & 0 \\ 0 & 0 & 0 & 0 & 0 & 0 & 0 & 0.5 & 0 & 0 \\ 0 & 0 & 0 & 0 & 0 & 0 & 0 & 0 & 0 & 0 \end{bmatrix},$$

$$A_{11} = \begin{bmatrix} 0 & 0 & 0 & 0 & 0 & 0 & 0 & 0 & 0 & 0 \\ 0 & 0 & 0 & 0 & 0 & -0.5 & 0 & 0 & 0 & 0 \\ 0 & 0 & 0 & 0 & 0.5 & 0 & 0 & 0 & 0 & 0 \\ 0 & 0 & 0 & 0 & 0 & 0 & 0 & 0 & 0 & 0 \\ 0 & 0 & 0.5 & 0 & 0 & 0 & 0 & 0 & 0 & 0 \\ 0 & -0.5 & 0 & 0 & 0 & 0 & 0 & 0 & 0 & 0 \\ 0 & 0 & 0 & 0 & 0 & 0 & 0 & 0 & 0 & 0.5 \\ 0 & 0 & 0 & 0 & 0 & 0 & 0 & 0 & 0 & 0 \\ 0 & 0 & 0 & 0 & 0 & 0 & 0 & 0 & 0 & 0 \\ 0 & 0 & 0 & 0 & 0 & 0 & 0.5 & 0 & 0 & 0 \end{bmatrix},$$

$$A_{12} = \begin{bmatrix} 0 & 0 & 0 & 0 & 0 & 0.5 & 0 & 0 & 0 & 0 \\ 0 & 0 & 0 & 0 & 0 & 0 & 0 & 0 & 0 & 0 \\ 0 & 0 & 0 & -0.5 & 0 & 0 & 0 & 0 & 0 & 0 \\ 0 & 0 & -0.5 & 0 & 0 & 0 & 0 & 0 & 0 & 0 \\ 0 & 0 & 0 & 0 & 0 & 0 & 0 & 0 & 0 & 0 \\ 0.5 & 0 & 0 & 0 & 0 & 0 & 0 & 0 & 0 & 0 \\ 0 & 0 & 0 & 0 & 0 & 0 & 0 & 0 & 0 & 0 \\ 0 & 0 & 0 & 0 & 0 & 0 & 0 & 0 & 0 & 0.5 \\ 0 & 0 & 0 & 0 & 0 & 0 & 0 & 0 & 0 & 0 \\ 0 & 0 & 0 & 0 & 0 & 0 & 0 & 0.5 & 0 & 0 \end{bmatrix},$$

$$A_{13} = \begin{bmatrix} 0 & 0 & 0 & 0 & -0.5 & 0 & 0 & 0 & 0 & 0 \\ 0 & 0 & 0 & 0.5 & 0 & 0 & 0 & 0 & 0 & 0 \\ 0 & 0 & 0 & 0 & 0 & 0 & 0 & 0 & 0 & 0 \\ 0 & 0.5 & 0 & 0 & 0 & 0 & 0 & 0 & 0 & 0 \\ -0.5 & 0 & 0 & 0 & 0 & 0 & 0 & 0 & 0 & 0 \\ 0 & 0 & 0 & 0 & 0 & 0 & 0 & 0 & 0 & 0 \\ 0 & 0 & 0 & 0 & 0 & 0 & 0 & 0 & 0 & 0 \\ 0 & 0 & 0 & 0 & 0 & 0 & 0 & 0 & 0 & 0 \\ 0 & 0 & 0 & 0 & 0 & 0 & 0 & 0 & 0 & 0.5 \\ 0 & 0 & 0 & 0 & 0 & 0 & 0 & 0 & 0.5 & 0 \end{bmatrix}.$$

In total, we obtain 13 constraints $\text{tr}(A_i U) = a_i$ with

$$a = (1\,1\,0\,1\,1\,1\,1\,0\,0\,0\,0\,0\,0)^\top. \tag{4.27}$$

So far, we have all components required for the approximation of Eq. (4.1)[2].

4.1.4 Implementation

In the AMP implementation, we use several lightweight libraries (*e.g., eigen3* [100] for linear algebra, *lapack* [25] especially for svd decomposition) as well as the CSDP library as an SDP solver. Our implementation is single core CPU only which enables compilation and execution on mobile platforms (though, a GPU can be used if available).

[2] matrix C' in Expr. (4.4) equals to identity in our case

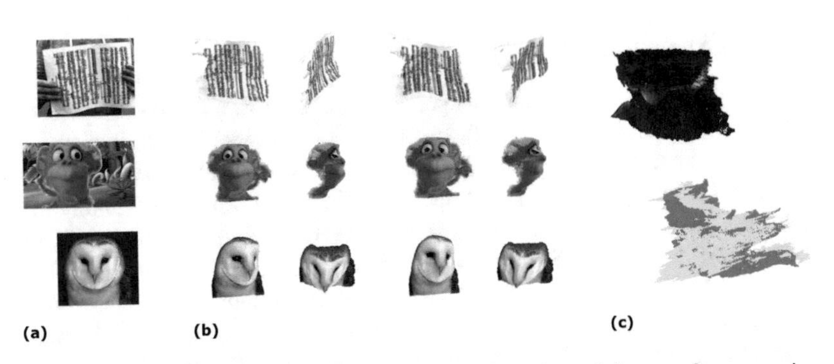

Figure 4.2: Qualitative evaluation of AMP: (a) selected frames from *music notes* (top), *monkaa* (middle) [191] and *barn owl* (bottom) [82] sequences; (b) reconstructions of the above-mentioned sequences; the surfaces are shown pairwise as frontal and side views; (c) a selected reference surface reconstruction from the *shaman2* sequence [61] obtained by VA (top) and an AMP reconstruction shown in cyan overlayed with the reference shown in purple (bottom). The mean RMS error for this 30-frame long sequence amounted to 0.42.

4.1.4.1 CSDP Solver

CSDP is a C/C++ library for solving SDP problems [47, 48]. It is built upon the *lapack* [25] library and provides an efficient implementation of the Primal-Dual Interior Point (PDIP) algorithm for SDP proposed by Helmberg *et al.* [140]. PDIP has polynomial complexity. The *Interior point* means that the method converges to an optimum through the interior of the polyhedron (which is the solution space), whereas the optimum always lies on its surface. *Primal-dual* refers to the property of the algorithm to solve both the primal and the dual SDP problems simultaneously. This property together with additional feasibility constraints on the solution space brings the advantage of the increased stability as well as a faster convergence. CSDP enables SDP programs to be solved in polynomial time and the constraints are provided in the form as stated above in Expr. (4.4). In computer vision, CSDP was used for camera calibration [8], image clustering [177] and dimensionality reduction of image data [306]. The constraints as derived in Sec. 4.1.3.2 are provided to the CSDP library as a configuration file containing non-zero elements of A_i matrices.

4.1.5 Experiments

MP was extensively evaluated on sparse datasets [216], and played a role of the baseline in several works [13, 104]. In this section, we describe runtime evaluation of AMP in the dense setting and show some qualitative results. We test AMP on a mobile platform with 6 GB RAM and Intel i5-2410M processor running at 2.30 GHz. We also report runtimes on a more powerful desktop machine with 32 GB RAM, Intel Xeon E5-1650 and NVIDIA Titan X GPU for comparing AMP with rigid factorisation approach [281] and a heterogeneous implementation of VA [104]. We choose several sequences for the experiments, *i.e., synthetic flag* [105], *music notes* [119], *monkaa* [191], *shaman2* [61], *barn owl* [82], *face* [104] and *heart surgery* [259].

4.1.5.1 Quantitative Evaluation

For quantitative evaluation, we use the ground truth optical flow of the synthetic flag sequence available from [105], and run AMP in different modes. In the first mode, the number of basis shapes K is fixed to 3 and the number of frames is varied. In the second mode, the number of frames F is fixed to 20 and the number of basis shapes K is varied. In both modes, we perform the experiment for three different values of the correspondence subsampling factor s. For instance, if $s = 2$, correspondences are decimated so that every second point in both image directions is included into the input measurement matrix $\mathbf{W}$. When undecimated, reconstructions contain $8.2 \cdot 10^4$ points. During the execution, the influence of the operating system workloads is minimised through launching the executable file without a graphics environment in a command line terminal. We report average runtimes for 10 runs for every mode and every value of the respective variable. Fig. 4.1-(a),(b) illustrates results of the experiment for both modes.

As can be observed in Fig. 4.1-(a), runtime grows as a superlinear function of the number of frames. Depending on the number of points, a shift along the time axis is observed, whereas the function graph preserves its form. A similar effect if observed in Fig. 4.1-(b) for the case of varying number of basis shapes. Allowing more complex deformations results in a higher computational complexity for a fixed number of frames. However, in certain cases ($K = 3$ for $s = 1$ or for $K = 4$ for $s = 3$), deviations from the monotonous growth can occur. The deviations are explained by a faster convergence of the CSDP solver, due to better initialisations, and are possibly related to an optimal number of basis shapes for a given image sequence. From Fig. 4.1-(a),(b) it can also be seen that for many configurations, the execution time is below one second. Thus, for $2 \cdot 10^4$ points and 10 frames, the execution time amounted to 311.7 milliseconds which allows building a window-based approach on top of AMP and save processing time for other workloads. An optimal window size depends on an application and can vary from 5 to 15 frames.

Next, we compare runtimes of AMP, VA [104] and rigid Tomasi-Kanade (TK) factorisation approach [281] on the Intel Xeon platform with a GPU. We use own heterogeneous C++/CUDA C implementation of VA and an optimised C++ version of the rigid factorisation. VA was shown to outperform MP in terms of reconstruction accuracy for dense cases due to the Total Variation (TV) term [104]. However, due to the same reason — the TV term as well as the proximal splitting — it is computationally expensive. The runtime of TK factorisation serves as a lower runtime bound for every NRSfM approach. The runtimes as a function of number of frames for the comparison are plotted in Fig. 4.1-(c) (runtime comparison as functions of number of basis shapes for VA is omitted, since the number of basis shapes is determined in VA automatically as proposed by Dai *et al.* [76]). For VA, we set 10 iterations with 10 primal-dual alternations each, unless the algorithm converges in early iterations. Despite AMP is a single core CPU implementation, it finishes in average one order of magnitude faster than VA. On the other hand, TK finishes approximately ten times faster than AMP. Using the decimation factor of 3, the runtimes of AMP are approaching the runtimes of TK. Thus, by decreasing the number of points by the factor of nine, we are able to achieve a runtime comparable to TK for the synthetic flag sequence.

In the next experiment on synthetic data, we perform measurement of mean RMS error for the dense case. The mean RMS error is defined as $e_{3D} = \frac{1}{F}\sum_{f=1}^{F}\frac{\left\|\mathbf{S}_f^{ref}-\mathbf{S}_f\right\|_{\mathscr{F}}}{\left\|\mathbf{S}_f^{ref}\right\|_{\mathscr{F}}}$, where $\mathbf{S}_f^{ref}$ are ground truth 3D shapes. We take the 3D motion capture data [309] and create measurement matrix by projecting it using an orthographic camera (which is identity $\mathbf{I}_{2\times3}$ — the third coordinate is omitted). Note that in the reference frame, the object is not observed frontally but is rotated by approx. 30°. This makes reconstruction more challenging. Due to the orthographic camera, ambiguities in the initial rotation can occur, and reconstructions are rigidly pre-aligned to the ground truth using a transformation estimation algorithm (Procrustes analysis). Fig. 4.1-(d) shows the measured mean RMS as a function of the number of frames (top) and basis shapes (bottom). Starting from 10 frames, the 3D error periodically increases with the number of frames. Since the synthetic flag sequence represents motion with strong deformations, AMP accumulates an error and becomes less accurate locally. This example shows that the 3D error can vary considerably with the number of basis shapes. An optimal parameter K depends on the type of the observed motion and deformations.

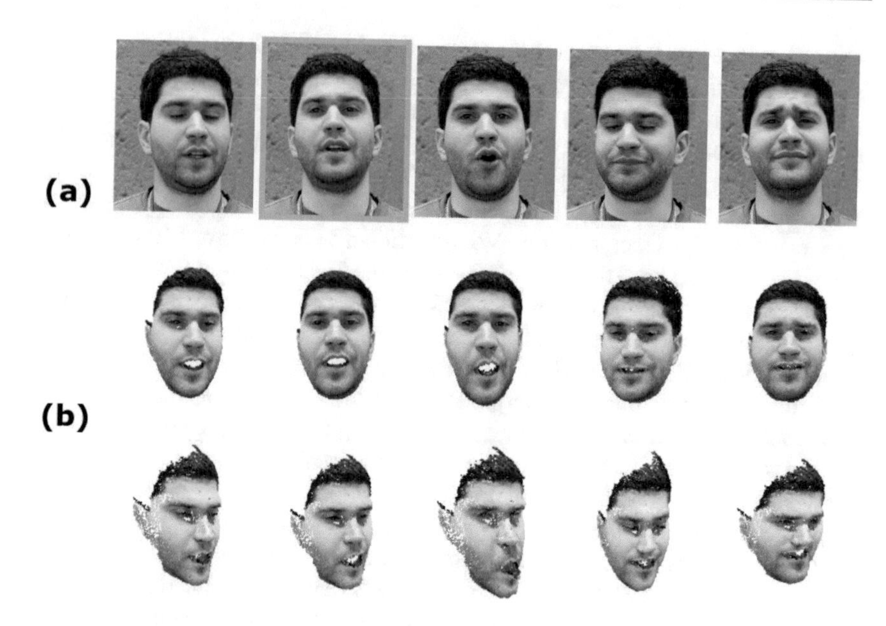

Figure 4.3: Results on the face sequence [104]: (a) exemplary sequence frames with the reference frame highlighted by yellow colour; (b) corresponding reconstructions (two different perspectives).

4.1.5.2 Qualitative Results on Real and Rendered Image Sequences

We show results of dense reconstructions for several real-world image sequences. The experiment shows that AMP is able to generate spatially smooth results for the dense cases if correspondences are estimated accurately and context-aware, *e.g.*, by multi-frame optical flow (MFOF) approaches such as [105, 267]. Some exemplary results of AMP on several image sequences are given in Fig. 4.2-(a), (b). The shown reconstructions are smooth and detailed. Bending of the music notes sheet is reasonably conveyed through non-rigid deformations. Head movements of the owl are realistically explained by combined rotational effects and non-rigid deformations. Fig. 4.2-(c) shows an example of reconstruction of the *shaman2* sequence by VA (top) as well as AMP, given point reprojections as an input. This experiment shows that AMP can reconstruct an overall appearance of the scene, despite a relatively high mean RMS error of 0.42. In many applications, this accuracy can be sufficient. An excerpt from the 120 frames long *face* sequence together with exemplary reconstructions (two different perspectives) are shown in Fig. 4.3. Every shape contains $\sim 2.8 \cdot 10^4$ points and the complete reconstruction

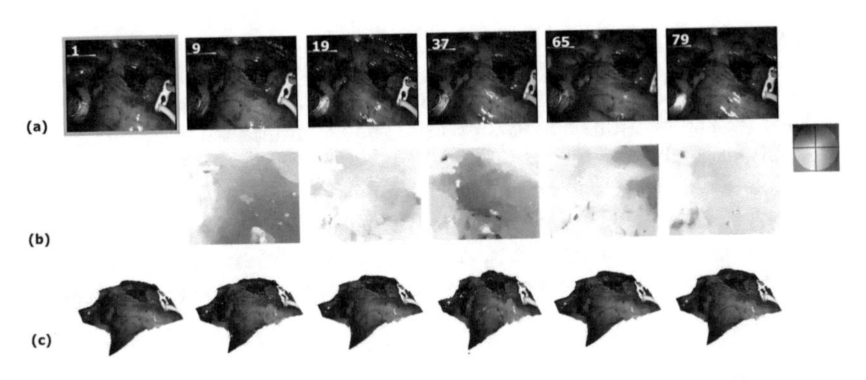

Figure 4.4: Results on the heart sequence [259]: (a) the input image sequence with the reference frame highlighted by yellow colour; (b) optical flow computed using TV-L^1 approach of Zach *et al.* [323], together with the colour key [29] on the right; (c) corresponding reconstructions recovered by AMP. Correspondence computation and non-rigid surface reconstruction with AMP are obtained within 33.5 seconds.

is accomplished within 10 seconds. Note how well the mouth expressions are reflected in the reconstructed point clouds.

The experiment on the *heart* bypass surgery sequence demonstrates the performance of AMP in the medical context. This sequence contains 80 frames and has several distinctive attributes. First, self-occlusion effects are almost not presented (except, perhaps, the areas of drifting specular effects). Second, point displacements from every frame to every other frame are comparably small due to the periodicity of heart beating and the fixed camera. These circumstances create conditions for application of a two-frame optical flow algorithm to compute correspondences from a key frame to every other frame in the sequence, instead of the computationally expensive MFOF. Thus, we compute correspondences using the real-time capable TV-L^1 optical flow approach of Zach *et al.* [323] in 21 seconds (computation is sequential). Based on the established dense point correspondences, AMP computes realistic reconstructions with $68 \cdot 10^4$ points each in 11.5 seconds. Only several frames exhibit insignificant surface fluctuations due to the highest point displacements (*e.g.*, see Fig. 4.4, frame 37). This experiment demonstrates that an accurate dense non-rigid reconstruction, 80 frames in length, can be obtained in combination with AMP within a half minute after the acquisition; adaptation for a real-time window-based operation is conceivable.

4.1.5.3 Discussion

Due to the runtimes achieved in the experiments, AMP is suitable for building interactive and real-time applications (*e.g.*, robotic or medical ones) on CPU only, possibly in combination with other methods. We believe that AMP, being suitable for mobile platforms, can at the same time provide initialisation for other, more accurate NRSfM methods. We also observe that AMP scales well with the number of points. Either 10^3 or 10^6 points, AMP is running reliably; we have not observed any side effects as the number of points increases unless there is not enough memory in the system.

4.1.6 Conclusion

The methodology proposed in this section enables convex relaxation problems formulated on a high level of abstraction to be solved with an efficient SDP solver such as CSDP. The feasibility of the proposed methodology is demonstrated on the example of MP algorithm, for which an efficient implementation or an implementation on a mobile platform was not previously possible. The proposed algorithm, *i.e.*, AMP allows obtaining accurate dense reconstructions with tens of thousands of points on a mobile platform in seconds; it outperforms the more accurate VA (own optimised heterogeneous implementation) in regard to the runtime by at least the factor of 10. We discovered that AMP scales well with the number of points, the number of frames as well as the number of basis shapes. Combined with suitable methods for correspondence computation, AMP can produce spatially and temporally accurate and realistic results in challenging real scenarios. These properties suggest that AMP can be advantageously placed in the context of modern real-time and interactive applications in such areas as robotics, medicine, and many others. Readers interested in real-time and interactive applications of NRSfM may wish to further investigate the temporal smoothness constraint, as AMP reconstructions are currently independent of the frame order.

4.2 Scalable NRSfM with Few Prior Assumptions

4.2.1 Introduction and an Overview of Contributions

In this section, we design an NRSfM approach which is easy to implement, computationally efficient, can potentially run interactively and can be used for a variety of realistic datasets. We call the new method Scalable Monocular Surface Reconstruction, or SMSR (Sec. 4.2.3). In our method, optimal basis selection is performed automatically which reduces the overall number of unknowns. Moreover, a non-convex formulation for camera pose estimation is solved iteratively using singular value thresholding (SVT) and proximal gradient. Surface estimation is formulated as an optimisation problem where guaranteed convergence to global minima is achieved using alternating direction method of multipliers (ADMM) [49, 179].

Besides, we propose a novel pre-processing step for input matrix. Thus, we reinitialised the measurement matrix with smooth trajectory constraints. This provides consistently accurate shape recovery for a wide variety of datasets, both sparse and dense (Sec. 4.2.3.2). Due to lack of dense datasets with ground truth for quantitative evaluation, we propose a two-step scheme. First, 3D shapes are estimated using an accurate and computationally expensive Variational Approach (VA) [104] followed by applying a rotation pattern in the second step (Sec. 4.2.4.1).Eventually, we found realistic datasets which allow to demonstrate strengths of the method compared to the state of the art. We perform an extensive evaluation of SMSR on a variety of sparse and dense datasets, both synthetic and real ones (Sec. 4.2.4). Across various evaluation scenarios, SMSR achieves *consistent state-of-the-art level accuracy* compared to the competing approaches.

4.2.2 Related Work

SMSR is an algorithmic mixture of several ideas found in several approaches. We believe that dense NRSfM methods should be well scalable with the number of points, *i.e.,* perform equally well and stable on a broad range of datasets, with small and large deformations. Thus, we adopt constraints in the metric and trajectory spaces and our formulation is most closely related to [173]. Compared to [125], we apply trajectory space constraints for smoothing of the measurement matrix and not throughout the whole optimisation. As a distinctive characteristic, we use proximal gradient method for optimisation [35]. To find the final shapes, we coarsely follow Block Matrix Method (BMM) [76]. Instead of using a fixed point continuation algorithm to relax rank minimisation to a nuclear norm, we use ADMM, similar to the patch reconstruction step of Lee *et al.* [173]. In contrast to the latter method, our approach is not piecewise. One major advantage of using ADMM is a guaranteed global minimum for a convex optimisation problem.

4.2.3 Scalable Monocular Surface Reconstruction (SMSR) Approach

SMSR consists of two main steps. First, the camera motion is estimated followed by 2D shape alignment. Next, a time-varying 3D surface is recovered. In the following, we describe the proposed algorithm.

4.2.3.1 Problem Formulation

We formulate NRSfM as a factorisation approach assuming every non-rigid shape follows LRSM. The measurement matrix $\mathbf{W} \in \mathbb{R}^{2F \times N}$ combining coordinates of N tracked points thoughout F frames is decomposed as a product of two matrices:

$$\underbrace{\begin{bmatrix} \mathbf{W}_1 \\ \mathbf{W}_2 \\ \vdots \\ \mathbf{W}_F \end{bmatrix}}_{\mathbf{W}} = \underbrace{\begin{bmatrix} \mathbf{R}_1 & & & \\ & \mathbf{R}_2 & & \\ & & \ddots & \\ & & & \mathbf{R}_F \end{bmatrix}}_{\mathbf{R}} \underbrace{\begin{bmatrix} \mathbf{S}_1 \\ \mathbf{S}_2 \\ \vdots \\ \mathbf{S}_F \end{bmatrix}}_{\mathbf{S}}, \tag{4.28}$$

where $\mathbf{R}$ is a quasi-block-diagonal matrix of camera poses, $\mathbf{S}$ is the block matrix with non-rigid shapes, $\mathbf{R}_f \in \mathbb{R}^{2\times 3}$ and $\mathbf{S}_f \in \mathbb{R}^{3\times N}$ are camera pose and shape for f^{th} frame, respectively. Every frame in $\mathbf{W}$ is registered to the centroid of the structure. Alternatively, the measurement matrix can be written as

$$\mathbf{W} = \mathbf{R}\underbrace{(\mathbf{C} \otimes \mathbf{I}_3)\mathbf{B}}_{\mathbf{S}} = \mathbf{M}\mathbf{B}, \tag{4.29}$$

where $\otimes$ denotes Kronecker product, $\mathbf{I}_3$ is a 3×3 identity matrix, $\mathbf{B} \in \mathbb{R}^{3K\times N}$ denotes set of basis shapes of cardinality K and $\mathbf{C} \in \mathbb{R}^{F\times K}$ is the corresponding *coefficient matrix*; $\mathbf{M} \in \mathbb{R}^{2F\times 3K}$ is a combined block diagonal shape-coefficient matrix (an element of the motion manifold). We utilise singular value decomposition to find a $3K$-rank approximation of $\mathbf{W}$ as $\mathbf{W} \cong \mathbf{M}'\mathbf{B}'$. An "implicit" non-unique solution can be obtained up to an invertible matrix $\mathbf{Q} \in \mathbb{R}^{3K\times 3K}$ (corrective transformation):

$$\mathbf{M}'\mathbf{Q} \cong \mathbf{M}, \quad \mathbf{Q}^{-1}\mathbf{B}' \cong \mathbf{B}. \tag{4.30}$$

It can be noticed that $\mathbf{M}'\mathbf{Q}$ is a scaled orthogonal matrix. Let $\mathbf{M}'_{2f-1:2f} \in \mathbb{R}^{2\times 3K}$ denote the $f^{th} \in \{1,\ldots,F\}$ pair of rows of $\mathbf{M}$ and $\mathbf{Q}_k \in \mathbb{R}^{3K\times 3}$ denote the $k^{th} \in \{1,\ldots,K\}$ column triplet of $\mathbf{Q}$. Then, for every $\mathbf{M}'_{2i-1:2i}$ submatrix, the following product holds:

$$\mathbf{M}'_{2f-1:2f}\mathbf{Q}_k = c_{fk}\mathbf{R}_f, \tag{4.31}$$

where c_{fk} is the element in f^{th} row and k^{th} column of $\mathbf{C}$. Let $\mathbf{F} = \mathbf{Q}\mathbf{Q}^\mathsf{T}$ be a positive semi-definite matrix. Using orthogonality constraint, two systems of linear equations can be obtained for each f:

$$\begin{cases} \mathbf{M}'_{2f-1}\mathbf{F}_k\mathbf{M}'^T_{2f-1} = \mathbf{M}'_{2f}\mathbf{F}_k\mathbf{M}'^T_{2f} = c^2_{fk}\mathbf{I}, \\ \mathbf{M}'_{2f-1}\mathbf{F}_k\mathbf{M}'^T_{2f} = 0. \end{cases} \tag{4.32}$$

The system of equations (4.32) can be rewritten as

$$\underbrace{\begin{bmatrix} \mathbf{M}'_{2f-1} \otimes \mathbf{M}'^T_{2f-1} - \mathbf{M}'_{2f} \otimes \mathbf{M}'^T_{2f} \\ \mathbf{M}'_{2f-1} \otimes \mathbf{M}'^T_{2f} \end{bmatrix}}_{\mathbf{A}_f} \operatorname{vec}(\mathbf{F}_k) = 0, \tag{4.33}$$

where $\operatorname{vec}(\cdot)$ is vectorisation operator defined as $\operatorname{vec}(\mathbb{R}^{m\times n}) \Rightarrow \mathbb{R}^{mn\times 1}$. Here, we use the property:

$$\operatorname{vec}(\boldsymbol{\Lambda}\boldsymbol{\Psi}\boldsymbol{\Upsilon}^T) = (\boldsymbol{\Upsilon} \otimes \boldsymbol{\Lambda}) \operatorname{vec}(\boldsymbol{\Psi}), \tag{4.34}$$

which holds for real matrices $\boldsymbol{\Lambda}$, $\boldsymbol{\Upsilon}$ and $\boldsymbol{\Psi}$. We denote the matrix on the left side of Eq. (4.33) with $\mathbf{A}_f$. Assembling equations for all $\mathbf{A}_f$ by stacking leads to a single equation

$$\mathbf{A} \operatorname{vec}(\mathbf{F}_k) = 0, \tag{4.35}$$

where $\mathbf{A} = [\mathbf{A}_1^T, \mathbf{A}_2^T, \ldots, \mathbf{A}_F^T]$. The optimisation problem in Eq. (4.35) — finding an optimal $\mathbf{F}_k$ — can be solved with linear least-squares by minimising

$$\|\mathbf{A} \operatorname{vec}(\mathbf{F}_k)\|^2 \tag{4.36}$$

[199]. The latter problem is non-convex due to the rank-3 constraint on $\mathbf{F}_k$, and can be efficiently solved using proximal gradient method [35] in an iterative manner. In each iteration, we solve the following subproblem:

$$\min_{\mathbf{F}_k} \|\mathbf{F}_k - \mathbf{F}_k^*\|^2, \text{ s.t } \operatorname{rank}(\mathbf{F}_k) = 3, \tag{4.37}$$

$$\operatorname{vec}(\mathbf{F}_k^*) = \operatorname{vec}(\mathbf{F}_k^0) - \frac{1}{L_{\mathbf{A}}} \mathbf{A}^\top \mathbf{A} \operatorname{vec}(\mathbf{F}_k^0). \tag{4.38}$$

In Eq. (4.37), $\mathbf{F}_k^0$ is an initial estimate of $\mathbf{F}_k$ and $L_{\mathbf{A}}$ denotes Lipschitz constant, *i.e.*, the largest eigenvalue of $\mathbf{A}^\top\mathbf{A}$. Once an optimal $\mathbf{F}$ is found, the corrective transformation $\mathbf{Q}$ is recovered by Cholesky decomposition. Further, $\mathbf{R}$ is computed using Eqs. (4.31)–(4.32) with a known $\mathbf{Q}$.

4.2.3.2 Smooth Shape Trajectory

For an enhanced accuracy, we add smoothness constraint on point trajectories. It enforces the non-rigid shapes to change gradually over time, *i.e.*, a trajectory represented by a K-dimensional $f^{th} \in \{1,..,F\}$ vector $c_f|_{1:K} = c(f)$ must lie in a linear subspace which constraints a point $c_{f,k}$ to vary smoothly over time. Thus, we represent $\mathbf{C}$ by K compact cosine series:

$$\mathbf{C} = \Omega_d\,[x_1, \quad \ldots \quad , x_K] = \Omega_d\mathbf{X} \quad \text{with} \quad x_k \in \mathbb{R}^d, \tag{4.39}$$

where $k \in \{1,..,K\}$ and d are numbers of low frequency DCT coefficients in the shape trajectory $\mathbf{X} \in \mathbb{R}^{d \times K}$ which represents $\mathbf{C}$ compactly over a truncated DCT domain $\Omega_d \in \mathbb{R}^{F \times d}$. The advantage of the constraints in the trajectory space is a known DCT basis which reduces considerably the number of unknowns. This leads to a faster overall convergence. Using the representation in trajectory space, $\mathbf{M}$ can be written as

$$\mathbf{M} = \mathbf{R}(\mathbf{C} \otimes \mathbf{I}_3) = \mathbf{R}(\Omega_d \mathbf{X} \otimes \mathbf{I}_3). \tag{4.40}$$

Using the pre-computed camera motion $\mathbf{R}$ (see **Sec.** 4.2.3.1), $\mathbf{M}$ is treated as a function of $\mathbf{X}$ only. Further, $\mathbf{B}$ can be computed as

$$\mathbf{B} = \mathbf{M}^{\dagger}\mathbf{W}, \tag{4.41}$$

whereby $\dagger$ denotes Moore-Penrose pseudo-inverse operator [114]. Estimating DCT shape trajectory $\mathbf{X}$ is formulated as a problem of minimising the squared reprojection error in the Frobenius norm:

$$\min_{\mathbf{M}} ||\mathbf{W} - \mathbf{W}^*||^2_{\mathscr{F}}, \quad \text{and} \quad \mathbf{W}^* = \mathbf{MB} = \mathbf{MM}^{\dagger}\mathbf{W}. \tag{4.42}$$

$\mathbf{X}$ is initialised to $\mathbf{X}_0 = [\mathbf{I}_K 0]$ which is $K \times K$ identity matrix with padding of additional zeros. The higher frequency DCT coefficients can be then estimated by iterative Gauss-Newton minimisation of Eq. (4.42). Shape basis matrix $\mathbf{B}$ is not directly used in the next part of the pipeline. Instead, only a 2D projection obtained from the newly computed matrix $\mathbf{S} = (\Omega_d \mathbf{X} \otimes \mathbf{I}_3)\mathbf{B}$ is used as the new measurement matrix $\mathbf{W}$. The described pre-processing step reinitialises the input matrix after imposing the smooth trajectory constraint on the recovered shapes.

4.2.3.3 Non-Rigid Shape Recovery

We use the new $\mathbf{W}$ to recover the final shapes and follow the formulation proposed in [76]. A rearranged shape matrix

$$\mathbf{S}^{\#} = \begin{bmatrix} X_{11} \dots X_{1N} & Y_{11} \dots Y_{1N} & Z_{11} \dots Z_{1N} \\ \vdots \quad\quad \vdots & \vdots \quad\quad \vdots & \vdots \quad\quad \vdots \\ X_{F1} \dots X_{FN} & Y_{F1} \dots Y_{FN} & Z_{F1} \dots Z_{FN} \end{bmatrix} \tag{4.43}$$

with an additional constraint $\text{rank}(\mathbf{S}^{\#}) < K$ compactly represents the optimal non-rigid structure:

$$\min_{\mathbf{S}} ||\mathbf{S}^{\#}||_*, \quad \text{s.t.} \quad \mathbf{W} = \mathbf{RS}, \tag{4.44}$$

where $||.||_*$ denotes the nuclear norm. We assume the mean 3D component is dominant in $\mathbf{S}^{\#}$ and can be removed in temporal dimension. Eq. (4.44) can be modified as

$$\min_{\mathbf{S}} ||\mathbf{S}^{\#}\mathbf{P}||_*, \quad \text{s.t.} \quad \mathbf{W} = \mathbf{R}\mathbf{S}, \tag{4.45}$$

where $\mathbf{P} = (\mathbf{I} - \frac{1}{F}\mathbf{1}\mathbf{1}^T)$ is the orthogonal projection and $\mathbf{1}$ is a vector of ones being its own null space. Eq. (4.45) is an optimisation problem over a convex function. We use for optimisation ADMM which can be considered as the inner-loop version of augmented Lagrangian method [179]:

$$L(\mathbf{S}, \mathbf{Y}, \mu) = ||\mathbf{S}^{\#}\mathbf{P}||_* + \mathbf{Y}^T(\mathbf{W} - \mathbf{R}\mathbf{S}) + \frac{\mu}{2}||\mathbf{W} - \mathbf{R}\mathbf{S}||^2_{\mathscr{F}}, \tag{4.46}$$

where $\mathbf{Y}$ denotes the vector of Lagrange multipliers and $\mu > 0$ is the positive step size. Eq. (4.45) is solved by the optimisation

$$\max_{\mathbf{Y}} \min_{\mathbf{S}} L(\mathbf{S}, \mathbf{Y}, \mu). \tag{4.47}$$

$\mathbf{S}$ is initialised as a planar structure and later updated iteratively by SVT [179] using coordinate descent. Lagrange multipliers are updated using sub-gradients:

$$\mathbf{S}^{t+1} = \arg\min_{\mathbf{S}} L(\mathbf{S}, \mathbf{Y}^t, \mu^t), \tag{4.48}$$

$$\mathbf{Y}^{t+1} = \mathbf{Y}^t + \mu^t(\mathbf{W} - \mathbf{R}\mathbf{S}^{t+1}), \tag{4.49}$$

$$\mu^{t+1} = \rho\,\mu^t, \tag{4.50}$$

where μ grows geometrically in every iteration t with the growth rate ρ. One advantage of ADMM is a guaranteed global minimum for convex problems and we use the above modified formulation to estimate the final shapes.

4.2.4 Experimental Results

We perform experiments on both synthetic as well as real data and compare the proposed approach with several existing approaches, *i.e.*, Metric Projection (MP) [215] (for the dense datasets we also use accelerated MP [121]), Point Trajectory Approach (PTA) [21], smooth time trajectories approach by Gotardo *et al.* (CSF1) [125], complementary rank-3 spaces approach CSF2 [127], Dai *et al.* (BMM) [76], VA [104] and Lee *et al.* [173]. We run experiments on a system with 32 GB RAM and Intel Core-i5200U processor. For SMSR implementation we used *Matlab* programming tool [277]. For quantitative performance evaluation of the methods, we compute normalised mean 3D error e_{3D} given by

$$e_{3D} = \frac{1}{\sigma FN} \sum_{f=1}^{F} \sum_{j=1}^{n} e_{fj}, \; \sigma = \frac{1}{3F} \sum_{f=1}^{F} (\sigma_{fx} + \sigma_{fy} + \sigma_{fz}), \tag{4.51}$$

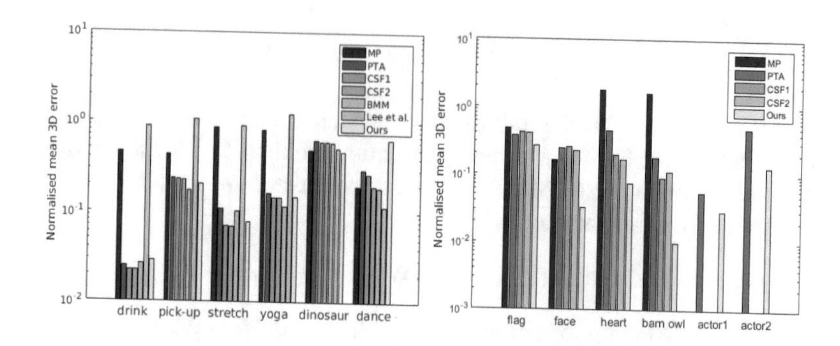

Figure 4.5: Comparison of the log scale normalised mean 3D error of SMSR (proposed) and several other methods over sparse and dense benchmark datasets. Missing bars in the figure for some approaches corresponds to N.A. in Table 4.2.

where σ is the normalised variance and e_{fj} is the reconstruction error (*i.e.*, Euclidean distance for the jth 3D point of fth frame) between ground truth $\mathbf{G}$ and recovered shape $\mathbf{X}$ given by $e_{fj} = ||\mathbf{G}_j^f - \mathbf{X}_j^f||_{\mathscr{F}}^2$, where $||.||_{\mathscr{F}}^2$ is the squared Frobenius norm. Before computing the error metrics, Procrustes analysis is performed to align recovered shapes with the ground truth [128].

4.2.4.1 Dense Datasets with Ground Truth

To generate more datasets for quantitative evaluation, we propose to employ the available dynamic 3D reconstructions. For a given input, we firstly reconstruct non-rigid shapes with VA [104]. Next, the 3D shapes are transformed to the fronto-parallel position relative to the image plane and projected onto it with a virtual orthographic camera. Thus, ground truth point correspondences are obtained. The 3D shapes and point correspondences can now be utilised as a dataset for NRSfM evaluation with a known ground truth. We obtained three datasets in this way — *face* [104], *heart* [259] and *barn owl* [82]. While generating 2D measurement matrices, for all sequences we apply the recovered camera poses to the dynamic shapes.

Additionally, we use the actor dataset of Beeler *et al.* with given dynamic 3D shapes [36]. The sequence was acquired using motion capture techniques and interpolation of key frame shapes; it contains $\approx 1.1 \times 10^6$ points and over 300 frames with subtle details of an actor's face with different facial expressions and

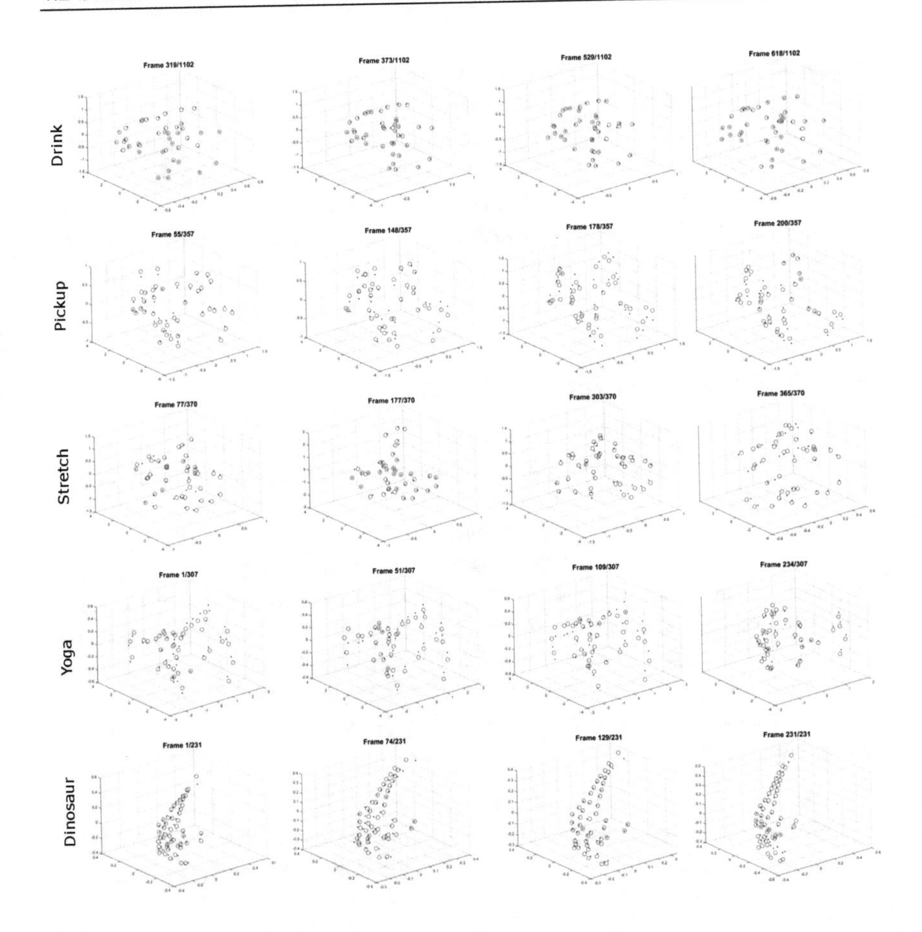

Figure 4.6: Qualitative evaluation of SMSR (proposed approach) for sparse synthetic and real benchmark datasets [21, 282]. Red circles denote ground truth and blue dots denote recovered 3D points.

emotions. We use every odd frame from the first 101 frames in the experiment and create two modified datasets, namely *Actor1* — with the introduced left-right head movement — and *Actor2* — with the introduced left-right up-down head movement. Similar to the *face, heart* and *barn owl* sequences, we project the 3D shapes by an orthographic camera and obtain 2D correspondences. In the similar manner — using the new sequences *Actor1* and *Actor2* — we created modified datasets *Actor1 Sparse* and *Actor2 Sparse*, as some implementations were only able to execute with less points ($\approx 3.8 \times 10^4$).

4.2.4.2 Evaluation on Sparse Datasets

We evaluate SMSR on challenging synthetic datasets *Drink, Pickup, Stretch, Yoga* with added per frame angular displacement of 5 degrees [22]. They contain 41 tracked points and up to 1102 frames. Further, SMSR is evaluated on real sparse benchmark *Dinosaur* and *Dance* sequences introduced in [22].

In Table 4.1, we show that the performance of the proposed method attains very close to the respective best performing method for nearly all cases (for *Drink* and *Yoga* sequences, SMSR matches the best e_{3D} up to the second decimal digit). Moreover, CSF1 achieves e_{3D} of 0.023 and 0.602 for *Drink* sequence and *Dinosaur* sequence and our method achieves 0.028 and 0.467 for the respective sequences. Hence, SMSR shows robust behaviour and steady performance on all sparse benchmark datasets.

Table 4.1: Normalised mean 3D error e_{3D} for benchmark datasets [21,126,282]. The number of bases K which has led to the best result is given in the brackets.

Method	*Drink*	*Pick-up*	*Stretch*	*Yoga*	*Dinosaur*	*Dance*
MP [215]	0.46	0.433	0.855	0.804	**0.489**	0.195
PTA [21]	0.025(13)	0.237(12)	0.109(12)	0.163(11)	0.627(2)	0.296(5)
CSF1 [125]	**0.022**(6)	0.23(6)	**0.071**(8)	**0.147**(7)	0.602(2)	0.269(2)
CSF2 [127]	**0.022**(6)	0.228(3)	**0.068**(8)	**0.147**(7)	0.6(2)	0.196(7)
BMM [76]	0.027(12)	**0.173**(12)	0.103(11)	**0.115**(10)	0.587	0.186(10)
Lee *et al.* [173]	0.875	1.069	0.9	1.228	0.508	**0.116**
SMSR (*ours*)	**0.029**	**0.202**	**0.078**	**0.149**	**0.468**	0.641

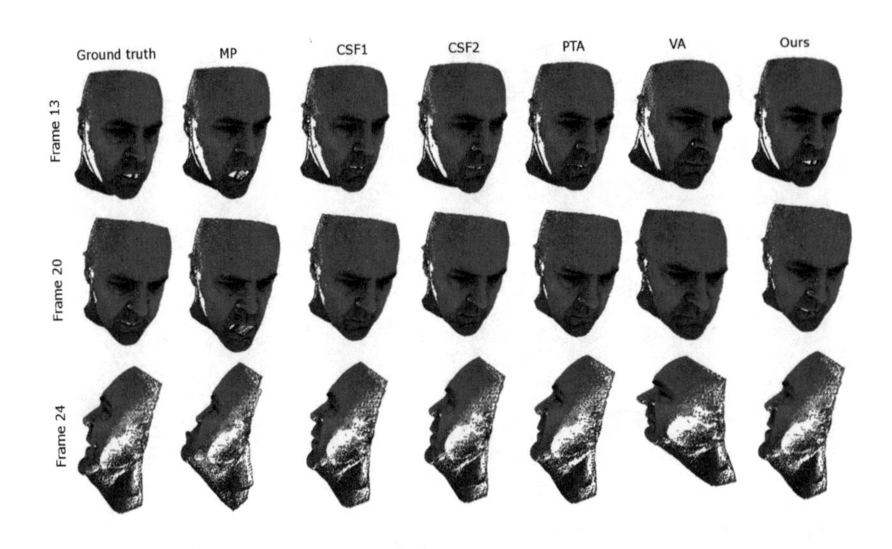

Figure 4.7: Qualitative comparison of MP [215], CSF1 [125], CSF2 [127], PTA [21], VA [104] and our SMSR approach for the *Actor1 Sparse* sequence.

4.2.4.3 Evaluation on Dense Datasets

We use in the experiments the challenging dense synthetic mocap *flag* dataset with ground truth surfaces [105, 309] with $\approx 10^4$ points and 60 frames. To the best of our knowledge, no real dense dataset with the ground truth is available for quantitative evaluation of NRSfM yet. Aiming to fill this gap, we create further synthetic dense datasets with ground truth using publicly available sequences (see Sec. 4.2.4.1), *i.e.,* the *face* sequence with $\approx 28 \times 10^3$ points and 120 frames [104] (it contains sudden emotion changes between frames) and the *heart* bypass surgery sequence [259] with $\approx 68 \times 10^3$ points and 30 frames. We used multi-frame optical flow [267] to compute dense correspondences. To demonstrate the scalable nature of SMSR , we have tested it on a dense real *barn owl* video acquired outdoors [82] with a higher number of points and frames ($\approx 2 \times 10^5$ points, 202 frames).

Additionally, we use four synthetic face sequences introduced in [104] denoted by *Face Seq1 – Face Seq4*. Each sequence contains 28887 points per frame. Note that the initial shape alignment by imposing smooth trajectories constraints (described in Sec. 4.2.3.2) takes longer time to converge where the total number of point tracks is high. As a result, for these sequences, the initial shape alignment is disabled.

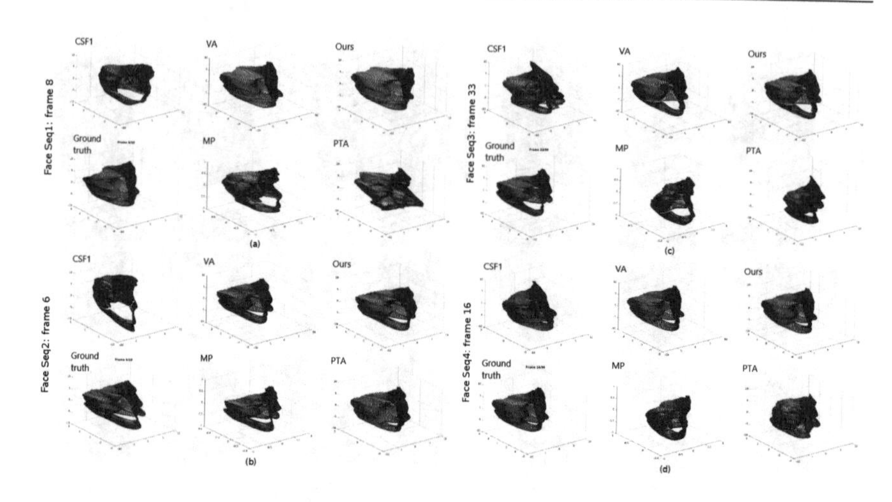

Figure 4.8: (a)-(d)Qualitative performance evaluation of MP [215], PTA [21], CSF1 [125], VA [104] and proposed approach for synthetic face sequences [104].

Table 4.2: Normalised mean 3D error e_{3D} for benchmark datasets [36, 82, 104, 309]. The number of bases K which has led to the best result is given in the brackets. N.A. is used where evaluation is not possible because of very high memory requirement or infeasible runtime. Likewise, BMM [76] and Lee *et al.* [173] could not be evaluated. The newly generated ground truth (see Sec. 4.2.4.1) does not contain connectivity data structure, thus running VA was not possible.

Method	*flag*	*face*	*heart*	*barn owl*	*Actor1*	*Actor2*
MP [215]	0.4756(8)	0.1604(2)	1.8231(4)	1.6277(4)	N.A.	N.A.
PTA [21]	0.3755(2)	0.2410(3)	0.4509(2)	0.1805(2)	0.0559(2)	0.4941(3)
CSF1 [125]	0.4202(4)	0.2581(2)	0.1941(2)	0.0896(2)	N.A.	N.A.
CSF2 [127]	0.4021(4)	0.2226(2)	0.1667(2)	0.1101(2)	N.A.	N.A.
SMSR (*ours*)	**0.2631**	**0.0321**	**0.075**	**0.0099**	**0.0287**	**0.1318**

Table 4.2 shows performance for mocap synthetic flag sequence [309] and real sequences [36, 82, 104]. The proposed method achieves the best performance and outperforms other approaches by a considerable margin. From Table 4.2 we can infer that other approaches exhibit high variation in performance across tested datasets and scale poorly for different deformations, number of points and frames.

Table 4.3: Normalised mean 3D error e_{3D} for the modified benchmark dataset [36, 104]. The number of bases K which has led to the best result in given in the brackets. BMM [76] and Lee *et al.* [173] could not be evaluated for *Actor1 Sparse* and *Actor2 Sparse* sequences because of infeasible memory requirement and very high runtime. The results of VA are included as presented in [104]. N.A. is used for the newly generated datasets (see Sec. 4.2.4.1) which does not contain connectivity data structure, thus running VA was not possible.

Method	MP [215]	PTA [21]	CSF1 [125]	CSF2 [127]	VA [104]	SMSR
Actor1 Sparse	0.5226(3)	**0.0418**(2)	0.3711(2)	0.3708(2)	N.A.	**0.0352**
Actor2 Sparse	0.2737(2)	**0.0532**(2)	0.2275(3)	0.2279(3)	N.A.	**0.0334**
Face Seq1	0.7251(2)	0.3933(2)	0.5325(3)	0.4677(3)	**0.1058**	**0.1893**
Face Seq2	0.6633(2)	**0.1871**(2)	0.9266(3)	0.7909(3)	**0.1014**	0.2133
Face Seq3	0.5676(4)	0.1706(4)	0.5274(3)	0.5474(3)	**0.0811**	**0.1345**
Face Seq4	0.5038(4)	0.2216(4)	0.5392(4)	0.5292(3)	**0.0806**	**0.0984**

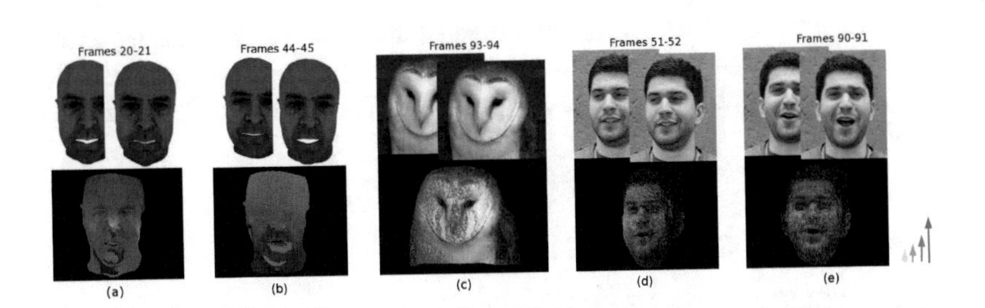

Figure 4.9: Visualisation of the 3D motion fields in the sense of the NRSfM-Flow framework [119] from the 3D shapes recovered by our SMSR method, for the *Actor* [36] (a, b), *barn owl* [82] (c) and *face* [104] (d,e) dense datasets. The scheme on the right shows relative vector lengths and corresponding colours.

In contrast, SMSR performs robustly and is well scalable for sparse and dense data and different types of deformations.

In Table 4.3, we provide a quantitative comparison for several NRSfM methods on the newly created datasets *Actor1 Sparse*, *Actor2 Sparse* and synthetic face sequences *Face Seq1 – Face Seq4*. The e_{3D} obtained by our approach remains lower than 0.04 for the modified actor datasets evidencing the consistent performance of our method. The qualitative performance evaluation for *Actor1 Sparse* sequence is shown in Fig. 4.7. SMSR accurately recovers actor's face with distinct fine details while most of the other approaches fail. According to Table 4.3, SMSR achieves second best accuracy with e_{3D} being less than 0.21 while VA achieves best results for all synthetic face sequences, it is found that these sequences contain overexaggerated facial deformations (*cf.* results of VA on sequences *Actor1 Sparse* and *Actor2 Sparse* with smaller and realistic deformations). We visualise 3D motion fields on the surface following the methodology proposed in [119]. Accordingly, in Fig. 4.9 we show two frames per dataset in the top row and the colour-coded magnitudes of the recovered 3D motion fields overlayed with the recovered shapes in the bottom row. Finally, Fig. 4.10 shows reconstructions obtained using various approaches — the third and fourth rows show that the SMSR reconstructions of *Face Seq1 – Face Seq4* remain very close to those of VA and ground truth.

SMSR requires only a few user-specified settings — the step size μ and the growth rate ρ. In all experiments, we set $\mu = 1$ and $\rho = 1.02$. For all sparse benchmark datasets, SMSR takes less than 5 *sec* for the reconstructions, and between 40 and 300 *sec* for the dense datasets (*flag, heart, face*). Other sequences, *i.e., barn owl* and *Actor* sequences can be reconstructed offline in feasible time (≤ 2500 *sec*) because of the high number of points. Note that our current implementation is not optimised. With hardware acceleration SMSR can be suitable for interactive applications.

4.2.4.4 NRSfM Challenge 2017

We tested SMSR on the NRSfM Challenge 2017 benchmark[3] [57]. This benchmark represents a set of five measurement matrices corresponding to five different monocular scene observations with small frame-to-frame changes — *Articulated Joints, Balloon Deflation, Paper Bending, Rubber Stretching* and *Paper Tearing* — and reports average root mean square (RMS) error in *mm* for seventeen NRSfM methods with publicly available source code and methods submitted for evaluation. RMS is defined as

$$\frac{1}{F}\sum_{f=1}^{F}\frac{\left\|\mathbf{X}^f-\mathbf{G}^f\right\|_{\mathscr{F}}}{\mathbf{G}^f}, \tag{4.52}$$

with reconstructions $\mathbf{X}^f$ and ground truth shapes $\mathbf{G}^f$. Measurement matrices for both perspective and orthographic cameras are provided. We reconstructed all sequences with the proposed SMSR and submitted to the portal (note that the

[3] see `http://nrsfm2017.compute.dtu.dk/benchmark` for more details

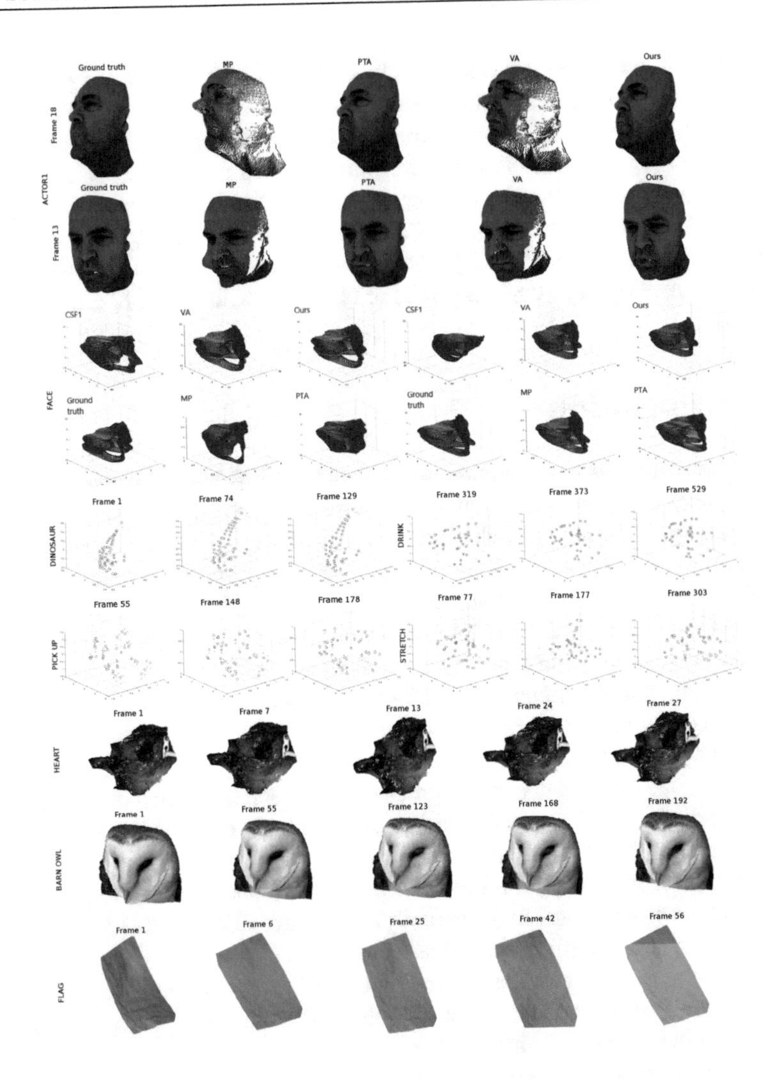

Figure 4.10: (Top 2 rows): Qualitative results on the *Actor1* [36] dataset; (rows 3–4): qualitative results on dense datasets *Face1* (left) and *Face2* (right) [104]; (rows 5–6): qualitative results on sparse datasets *Dinosaur, Pickup, Drink, Stretch* [21, 282]; (last three rows): qualitative results of SMSR on dense benchmark datasets *Heart* [259], *Barn Owl* [82] and *Flag* [309].

ground truth 3D shapes are not publicly available). We use default parameters across all test sequences. In the mode without self-occlusions, SMSR achieves the average RMS error of $41.84\,mm$ (ranging from $31.71\,mm$ for *Balloon Deflation* to $58.12\,mm$ for *Articulated Joints*) with the standard deviation of $9.2\,mm$ — an intermediate overall accuracy. The low value of the standard deviation accentuates the scalable nature of our approach. SMRS outperforms Lee *et al.* [173] and Dai *et al.* [76], though [22, 125, 127, 215] achieve lower errors on the new benchmark. The evaluation methodology and the results of NRSfM Challenge 2017 are published in [155].

4.3 Conclusion

This section presents a new method for scalable NRSfM which involves two main steps — camera motion estimation and 3D surface update. The primary advantage of the proposed SMSR approach is superior scalability and consistent performance across different datasets which are two core synergic properties which distinguish our SMSR method from the competing methods. SMSR imposes constraints in metric and trajectory spaces and involves various widely-used high-level operations in different steps including proximal gradient method, iterative Gauss-Newton optimisation, singular value thresholding, coordinate descent, Cholesky decomposition as well as multiple matrix-vector and matrix multiplications. We often observed a quick convergence of our method. The reconstruction error obtained in the extensive experiments by the proposed approach is low and consistent over a large variety of publicly available real and synthetic datasets including sparse and dense ones. To facilitate and enrich quantitative evaluation, we evaluated the performance of several approaches on ground truth tracks obtained as projections of trail VA reconstructions. In the majority of the cases, SMSR achieves the lowest normalised mean 3D error among all tested methods or reaches close to the respective best performing method. One reason for this gain is the new pre-processing step, *i.e.*, smoothing of the shape trajectory. Another possible reason is the convergence property of the ADMM minimisation.

A current limitation of the proposed technique lies in handling large deformations such as those occurring during tissue bending and twisting. Extending SMSR for the handling of large deformations is an interesting direction for future work. Another possible direction could be adopting the method for online applications and embedded devices.

5 Shape Priors in Dense Non-Rigid Structure from Motion

THIS chapter is devoted to two NRSfM methods with shape priors obtained on-the-fly. Both static and dynamic shape priors are investigated. *Static shape prior* refers to a single prior 3D state, and *dynamic shape prior* refers to a series of states. In both cases, we assume that a shape prior is obtained on-the-fly from an image subsequence. Both proposed methods are robust to occlusions and inaccuracies in dense point tracks.

Sec. 5.1 introduces an accurate solution to dense orthographic NRSfM in scenarios with severe occlusions or, likewise, inaccurate correspondences. We integrate a shape prior term into a variational optimisation framework. It allows penalising irregularities of the time-varying structure on the per-pixel level if correspondence quality indicator such as an occlusion tensor is available. We assume that several non-occluded views of the scene are available and sufficient to estimate an initial shape prior, though the entire observed scene can exhibit non-rigid deformations. Experiments on synthetic and real image data show that the proposed framework significantly outperforms state-of-the-art methods for correspondence establishment in combination with the state-of-the-art NRSfM methods. Together with the profound insights into optimisation methods, implementation details for heterogeneous platforms are provided.

In Sec. 5.2, we propose a hybrid approach that extracts prior knowledge directly from the input sequence and uses it as a dynamic shape prior for sequential surface recovery in scenarios with recurrence. The energy functional of our Dynamic Shape Prior Reconstruction (DSPR) method is optimised with multi-start gradient descent at real-time rates for new incoming dense point tracks. The proposed versatile multi-purpose framework with a new core NRSfM approach outperforms several other methods in the ability to handle inaccurate and noisy point tracks, provided we have access to a representative (in terms of the deformation variety) image subsequence. Comprehensive experiments evaluate different aspects of DSPR including convergence properties and accuracy under different types of disturbing effects. We also perform a joint study of tracking methods and reconstruction and

V. Golyanik, *Robust Methods for Dense Monocular Non-Rigid 3D Reconstruction and Alignment of Point Clouds*,

show applications of DSPR to shape compression and heart reconstruction under occlusions. We achieve state-of-the-art metrics (accuracy and compression ratios) in different evaluation scenarios.

5.1 Static Shape Prior for Explicit Occlusion Handling

5.1.1 Motivation and Contributions

Nevertheless, support for real-world image sequences is still limited due to the systematic violation of assumptions on the degree as well as the type of motion and deformation presented in a scene. Moreover, severe self- and external occlusions occur frequently, which results in noisy and erroneous correspondences. Since methods for computing correspondences are limited in compensating for occlusions, NRSfM methods should be able to cope with missing data and the associated disturbing effects robustly.

In this chapter, a novel dense orthographic NRSfM approach is proposed which can cope with severe occlusions — Shape Prior based Variational Approach (SPVA) — along with a scheme for obtaining a shape prior from several non-occluded frames. The latter relies on a realistic assumption that a scene is not-occluded in a reference frame and there are some non-occluded views. Influence of the shape prior can be controlled by series of occlusion maps — an occlusion tensor — obtained from a measurement matrix and an input image sequence. In contrast to template-based reconstruction, the shape prior is computed automatically in our framework, and we do not rely on the rigidity assumption. The proposed methods are combined into a joint correspondence computation, occlusion detection, shape prior estimation and surface recovery *framework*, and evaluated against different state-of-the-art non-rigid recovery pipeline configurations. SPVA surpasses state of the art in real scenarios with large occlusions or noisy correspondences, both in terms of the reconstruction accuracy and processing time. To the best of our knowledge, our method is the first to stably handle severe external occlusions in dense scenarios without requiring an expensive correspondence correction step.

5.1.2 Related Work

The proposed method is based on factorising the measurement matrix into shapes and camera motion and operates on an image batch. The idea of factorisation was initially proposed for the rigid case [281] and adopted for the non-rigid case in [53] where every shape is represented by a linear combination of basis shapes. This statistical constraint can be interpreted as a basic form of a shape prior, and reflects the assumption on the linearity of deformations. This setting is known to perform well for moderate deformations and many successor methods built upon the idea of metric space constraints [216, 228, 238, 282]. In contrast, SPVA determines optimal basis shapes implicitly by penalising nuclear norm of the shape matrix as proposed in [76].

For robustness to occlusions and missing data, several policies were proposed so far. One is to compensate for disturbing effects in the preprocessing step. Associating image points with their entire trajectories over an image sequence, Multi-Frame Optical Flow (MFOF) methods allow to detect occlusions and robustly estimate correspondences in occluded regions [103, 231, 232, 267]. These methods perform well if occlusions are rather small or of a short duration. Support of longer occlusions is, however, limited which results in reduced accuracy of NRSfM methods.

Another policy is to account for missing data and incorrect correspondences during surface recovery. In [282], Gaussian noise in measurements is explicitly modelled in the motion model. Authors report accurate results on perturbed inputs with an additive normally distributed noise. The shape manifold learning approach of [270] is withstandable against Gaussian noise (levels up to 12% result in reconstructions of a decent accuracy). A method based on the recently introduced low-rank force prior includes a term accounting for a Gaussian noise in the measurements and was shown to handle 11.5% of missing data caused by short-time occlusions [13]. Due to a variational formulation, the approach of Garg *et al.* [104] can compensate for small amount of erroneous correspondences, provided an appropriate solution initialisation is given. Due to a mode shape interpretation, the method of Agudo *et al.* [12] can perform accurately when 40% of points are randomly removed from the input. Some other methods can also handle noisy and missing correspondences [14, 216], but in scenarios limited to short and local occlusions. In contrast, our method can cope with large and long occlusions.

Some NRSfM approaches allow integration of an explicit shape prior into the surface recovery procedure. Del Bue [56] proposed to jointly factorize measurement matrix and a pre-defined shape prior. The method showed enhanced performance under degenerate non-rigid deformations. The shape prior represented a single pre-defined static shape acquired by an external procedure or pre-computed basis shapes. Tao *et al.* [270] proposed to adopt a graph-based manifold learning technique based on diffusion maps where the shapes are constrained to lie on the pre-computed non-linear shape prior manifold. In this scheme, the basis shapes can be different for every frame and hence the method can reconstruct strong deformations. However, the approach requires a representative training set with a computationally expensive procedure (especially for the case of dense reconstructions) for embedding of new

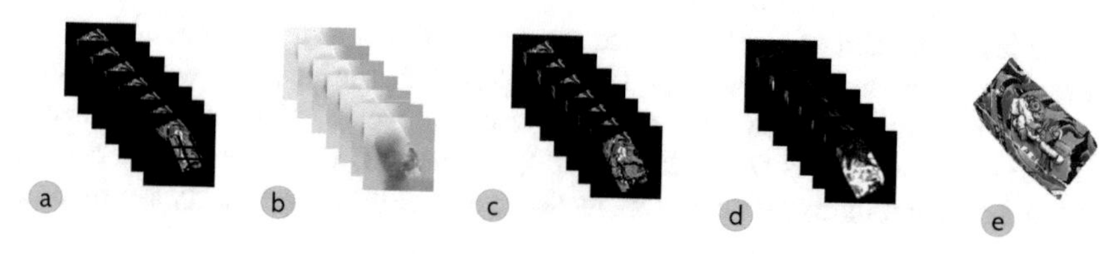

Figure 5.1: An overview of the considered pipeline: (a): the input to the pipeline is an image sequence of a non-rigidly deforming scene; (b): first stage of the pipeline is point tracking with multi-frame optical flow [103]; (c): the second stage is occlusion tensor estimation (shown in (d); brighter values indicate higher occlusion probability of the pixel). Next, a shape prior is estimated relying on the total intensity criterion. The correspondences, occlusion tensor and estimated shape prior are inputs for the Shape Prior based Variational Approach (SPVA); (e): example of a shape prior.

shapes not presented in the training set. Recent template-based reconstruction methods employ a similar principle as us [180,319]. Thus, Yu *et al.* proposed to estimate a template shape at rest from several first frames provided sufficient cues for a multi-view reconstruction [319]. This estimate is based on rigidity assumption and the accuracy of the method depends on this step; an external pre-aligned template can also be used. Similarly, our approach estimates a shape prior from several initial frames. We also assume the initial views to be occlusion-free, but our method neither assumes rigidity nor requires a known template.

In our core approach, an estimated shape prior is integrated into a joint variational framework. It is most closely related to [76] due to the nuclear norm, and Variational Approach (VA) [104] due to the spatial integrity, *i.e.,* Total Variation (TV). Additionally, our energy functional includes a soft shape prior term. Camera poses are recovered in a closed-form through the projection of affine approximations on the $\mathfrak{so}(3)$ manifold (which is up to two orders of magnitude faster than non-linear optimisation). To detect occlusions, we propose a novel lightweight scheme relying on [103]. This approach differs from Taetz *et al.* [267] which corrects measurements in the pre-processing step but requires multiples of the computational time compared to [103].

5.1.3 Variational Approach with a Shape Prior (SPVA)

Suppose N points are tracked throughout an image sequence with F frames and the input is assembled in the measurement matrix $\mathbf{W} \in \mathbb{R}^{2F \times N}$ so that every pair of rows contains x and y coordinates of a single frame, respectively. During scene acquisition, an orthographic camera observes a non-rigidly deforming 3D scene $\mathbf{S} \in \mathbb{R}^{3F \times N}$. Similarly, every treble of rows contains x, y and z coordinates of an instantaneous scene. $\mathbf{W}$ depends on a scene, relative camera poses as well as a camera model as

$$\mathbf{W} = \mathbf{P}\mathbf{R}_{3D}\mathbf{S}, \tag{5.1}$$

where $\mathbf{R}_{3D} \in \mathbb{R}^{3F \times 3F}$ is a block-diagonal matrix with camera poses for every frame and $\mathbf{P} \in \mathbb{R}^{2F \times 3F}$ is a combined camera projection matrix with entries $\begin{pmatrix} 1 & 0 & 0 \\ 0 & 1 & 0 \end{pmatrix}$. Here, we additionally assume that the measurements are registered to the origin of the coordinate system and translation is resolved. The objective is to reconstruct a time varying shape $\mathbf{S}$ and relative camera poses $\mathbf{R}_{3D}$. In other words, we seek a realistic factorisation of $\mathbf{W}$. Since the third dimension is lost during the projection, $\mathbf{W}$ will be factorised in $\mathbf{P}\mathbf{R}_{3D} = \mathbf{R} \in \mathbb{R}^{2F \times 3F}$ and $\mathbf{S}$. In a post-processing step, $\mathbf{R}_{3D}$ can be estimated by imposing orthonormality constraints on rotation matrices, *i.e.*, entries of $\mathbf{R}$.

If additional information about shape of a scene is available, it can be used to constrain the solution space. We formulate NRSfM as a variational energy minimisation problem, and the most natural form of the shape prior is $\mathbf{S}_{\text{prior}} \in \mathbb{R}^{3F \times N}$, *i.e.*, a matrix containing prior shapes for every frame. In SPVA, $\mathbf{S}_{\text{prior}}$ influences the optimisation procedure in a flexible manner according to the required per frame and per pixel control. Next, depending on the control granularity, several energies are proposed, and for each energy, an optimisation method is derived.

5.1.3.1 Per Sequence Shape Prior

A per-sequence shape prior is the strongest prior, *i.e.*, it allows to constrain the solution space for the whole sequence at once. Minimiser has the simplest form among all types, and the shape prior term has only a single weight parameter γ. The energy takes on the following form:

$$\underset{\mathbf{R},\mathbf{S}}{\operatorname{argmin}} \frac{\lambda}{2} \|\mathbf{W} - \mathbf{R}\mathbf{S}\|_{\mathcal{F}}^2 + \frac{\gamma}{2} \|\mathbf{S} - \mathbf{S}_{\text{prior}}\|_{\mathcal{F}}^2 + \sum_{f,i,p} \|\nabla \mathbf{S}_f^i(p)\| + \tau \|\mathrm{P}(\mathbf{S})\|_*, \tag{5.2}$$

where $\sum_{f,i,p} \|\nabla \mathbf{S}_f^i(p)\|$ denotes TV with the gradient $\nabla \mathbf{S}_f^i(p)$ of the shape $\mathbf{S}_f$, $f \in \{1, \ldots, F\}$ at the point $p \in \{1, \ldots, N\}$ in the direction i; $\|\cdot\|_*$ and $\|\cdot\|_{\mathcal{F}}$ denote nuclear and Frobenius norms, respectively, and the operator $\mathrm{P}(\cdot)$ permutes $\mathbf{S}$ into the matrix of the dimensions $F \times 3N$ (the point coordinates are rearranged framewise into single rows). The energy in Eq. (5.2) contains data, shape prior, smoothness and linear subspace model terms, respectively.

If **R** or **S** is fixed, the energy is convex in **S** and **R** variables, respectively. Such kind of energies, also called biconvex, can be optimised by Alternating Convex Search (ACS). In ACS, optimisation is performed for **R** and **S** while **S** or **R** is respectively fixed. Suppose **S** is fixed. In this case, the only term which depends on **R** is the data term. We seek a solution to the problem

$$\underset{\mathbf{R}}{\operatorname{argmin}} \frac{\lambda}{2} \|\mathbf{W} - \mathbf{RS}\|_{\mathscr{F}}^2. \tag{5.3}$$

The idea is to find an unconstrained solution **A** minimising Eq. (5.3) and to project it blockwise into the $\mathfrak{so}(3)$ group in a closed-form. The projection will yield an optimal rotation matrix **R** [202]. First, we consider the sum of the separate data terms for every frame f in the transposed form:

$$\sum_f \|\mathbf{W}_f^\top - \mathbf{S}_f^\top \mathbf{R}_f^\top\|_{\mathscr{F}}^2. \tag{5.4}$$

Here, the property of invariance of Frobenius norm under transposition is used. Now an optimal matrix $\mathbf{A}_f$ can be found which minimises the data term in Eq. (5.4) by projecting $\mathbf{W}_f^\top$ onto the column space of $\mathbf{S}_f^\top$ in a closed form:

$$\mathbf{A}_f = (\mathbf{S}_f \mathbf{S}_f^\top)^{-1} \mathbf{S}_f^\top \mathbf{W}_f^\top. \tag{5.5}$$

Note that the matrix $\mathbf{S}_f \mathbf{S}_f^\top$ has dimensions 3×3 which supports a low memory complexity of the optimisation. Next, we decompose $\mathbf{A}_f^\top$ with singular value decomposition (svd) and find $\mathbf{R}_f$ as follows:

$$\operatorname{svd}(\mathbf{A}_f^\top) = \operatorname{svd}(\mathbf{W}_f \mathbf{S}_f (\mathbf{S}_f \mathbf{S}_f^\top)^{-1}) = \mathbf{U}\Sigma\mathbf{V}^\top \tag{5.6}$$

$$\mathbf{R}_f = \mathbf{U}\mathbf{C}\mathbf{V}^\top, \tag{5.7}$$

where $\mathbf{C} = \operatorname{diag}(1, 1, \ldots, 1, \operatorname{sign}(\det(\mathbf{U}\mathbf{V}^\top)))$. We favour the least squares solution for the sake of computational efficiency (see Sec. 5.1.5 for implementation details).

Next, we consider the energy functional in Eq. (5.2) with a fixed **R**. We seek a solution to the problem

$$\underset{\mathbf{S}}{\operatorname{argmin}} \frac{\lambda}{2} \|\mathbf{W} - \mathbf{RS}\|_{\mathscr{F}}^2 + \frac{\gamma}{2} \|\mathbf{S} - \mathbf{S}_{\text{prior}}\|_{\mathscr{F}}^2 + \sum_{f,i,p} \|\nabla \mathbf{S}_f^i(p)\| + \tau \|\mathrm{P}(\mathbf{S})\|_*. \tag{5.8}$$

This minimisation problem is convex, but it involves different norms and therefore cannot be solved in the standard way. After applying proximal splitting, we obtain two subproblems with an auxiliary variable $\bar{\mathbf{S}}$:

$$\underset{\mathbf{S}}{\operatorname{argmin}} \frac{1}{2\theta}\|\mathbf{S}-\bar{\mathbf{S}}\|_{\mathscr{F}}^2+\frac{\gamma}{2}\|\mathbf{S}-\mathbf{S}_{\text{prior}}\|_{\mathscr{F}}^2+ \frac{\lambda}{2}\|\mathbf{W}-\mathbf{R}\mathbf{S}\|_{\mathscr{F}}^2+\sum_{f,i,p}\|\nabla\mathbf{S}_f^i(p)\| \tag{5.9}$$

$$\underset{\bar{\mathbf{S}}}{\operatorname{argmin}} \frac{1}{2\theta}\|\mathbf{S}-\bar{\mathbf{S}}\|_{\mathscr{F}}^2+\tau\|\mathrm{P}(\bar{\mathbf{S}})\|_*\,. \tag{5.10}$$

The minimisation problem in Eq. (5.10) involves a squared Frobenius norm and the nuclear norm. It is of the form

$$\underset{\mathbf{Z}}{\operatorname{argmin}} \frac{1}{2}\|\mathbf{B}-\mathbf{Z}\|_{\mathscr{F}}^2+\eta\,\|\mathbf{Z}\|_* \tag{5.11}$$

and can be solved by a soft-impute algorithm (in our case, $\eta=\theta\tau$). We rewrite the nuclear norm as

$$\|\mathbf{Z}\|_* := \min_{\mathbf{U},\mathbf{V}:\,\mathbf{Z}=\mathbf{U}\mathbf{V}} \frac{1}{2}\left(\|\mathbf{U}\|_{\mathscr{F}}^2+\|\mathbf{V}\|_{\mathscr{F}}^2\right). \tag{5.12}$$

The solution to this problem is given by $\mathbf{Z}=\mathbf{U}\mathbf{D}_\eta\mathbf{V}$, where

$$\operatorname{svd}(\mathbf{Z})=\mathbf{U}\mathbf{D}\mathbf{V}, \mathbf{D}=\operatorname{diag}(\sigma_1,\ldots,\sigma_r), \text{and} \tag{5.13}$$

$$\mathbf{D}_\eta=\left(\max(\sigma_1-\eta,0),\ldots,\max(\sigma_r-\eta,0)\right). \tag{5.14}$$

The energy in Eq. (5.9) is convex, but — because of the TV regulariser — not differentiable. Nevertheless, the problem can be dualised with Legendre-Fenchel transform. The primal-dual form is then given by

$$\underset{\mathbf{S}}{\operatorname{argmin}}\max_{q} \frac{1}{2\theta}\|\mathbf{S}-\bar{\mathbf{S}}\|_{\mathscr{F}}^2+\frac{\lambda}{2}\|\mathbf{W}-\mathbf{R}\mathbf{S}\|_{\mathscr{F}}^2+ \frac{\gamma}{2}\|\mathbf{S}-\mathbf{S}_{\text{prior}}\|_{\mathscr{F}}^2+\sum_{f,i,p}\left(\mathbf{S}_{fi}(p)\nabla^* q_f^i(p)-\delta\left(q_f^i(p)\right)\right), \tag{5.15}$$

where q is the dual variable that contains the 2-dimensional vectors $q_f^i(p)$ for each frame f, coordinate i and pixel p. $\nabla^*=-\operatorname{div}(\cdot)$ is the adjoint of the discrete gradient operator ∇, and δ is the indicator of the unit ball. In the primal-dual algorithm used to solve the problem, firstly the differential $\mathbf{D}_q$ of the dual part is initialised. Next, the gradient w.r.t. $\mathbf{S}$ is computed and set to zero to obtain a temporal minimiser $\bar{\mathbf{S}}$. Next, $\mathbf{D}_q$ is updated. The algorithm alternates between finding $\bar{\mathbf{S}}$ and updating $\mathbf{D}_q$ until convergence. The gradient operator $\nabla_{\mathbf{S}}$ applied to

Algorithm 1 SPVA: Variational NRSfM with a Shape Prior

Input: measurements $\mathbf{W}$, $\mathbf{S}_{prior}$, parameters λ, γ, τ, θ, $\eta = \theta\tau$
Output: non-rigid shape $\mathbf{S}$, camera poses $\mathbf{R}$

1: **Initialisation:** $\mathbf{S}$ and $\mathbf{R}$ under rigidity assumption [281]
2: **STEP 1. Fix S, find an optimal R *framewise*:**
3: $\text{svd}(\mathbf{WS}(\mathbf{SS}^\mathsf{T})^{-1}) = \mathbf{U\Sigma V}^\mathsf{T}$
4: $\mathbf{R} = \mathbf{UCV}^\mathsf{T}$, where
$\mathbf{C} = \text{diag}(1, 1, \ldots, 1, \text{sign}(\det(\mathbf{UV}^\mathsf{T})))$
5: **STEP 2. Fix R; find an optimal S:**
6: **while** not converge **do**
7: **Primal-Dual:** fix $\bar{\mathbf{S}}$; *find an intermediate* $\mathbf{S}$ *(Eq. (5.9))*
8: **Initialisation:** $q_f^i(p) = \mathbf{0}$
9: **while** not converge **do**
10: $$\mathbf{D}_q = \begin{pmatrix} \nabla^* q_1^1(1) & \cdots & \nabla^* q_1^1(N) \\ \vdots & \ddots & \vdots \\ \nabla^* q_F^3(1) & \cdots & \nabla^* q_F^3(N) \end{pmatrix}$$
11: $\mathbf{S} = \left(\lambda \mathbf{R}^\mathsf{T}\mathbf{R} + \gamma + \frac{1}{\theta}\mathbf{I}\right)^{-1}$
12: $(\lambda \mathbf{R}^\mathsf{T}\mathbf{W} + \frac{1}{\theta}\bar{\mathbf{S}} + \gamma\mathbf{S}_{\text{prior}} - \mathbf{D}_q)$
13: **for** $f = 1, ..., F$; $i = 1, ..., 3$; $p = 1, ..., N$ **do**
14: $q_f^i(p) = \frac{q_f^i(p) + \sigma\nabla\mathbf{S}_f^i(p)}{\max(1, \|q_f^i(p)+)\sigma\nabla\mathbf{S}_f^i(p)\|)}$
15: **end while**
16: **Soft-Impute:** fix $\mathbf{S}$; *find an intermediate* $\bar{\mathbf{S}}$ *(Eq. (5.10))*
17: $\text{svd}(\text{P}(\mathbf{S})) = \mathbf{UDV}^\mathsf{T}$, where $\mathbf{D} = \text{diag}(\sigma_1, ..., \sigma_r)$
18: $\bar{\mathbf{S}} = \mathbf{UD}_\eta\mathbf{V}^\mathsf{T}$, where
$\mathbf{D}_\eta = \text{diag}\left(\max(\sigma_1 - \eta, 0), ..., \max(\sigma_r - \eta, 0)\right)$
19: **end while**

the energy in Eq. (5.15) yields

$$(\lambda \mathbf{R}^\mathsf{T}\mathbf{R} + \gamma + \frac{1}{\theta})\mathbf{S} - (\lambda \mathbf{R}^\mathsf{T}\mathbf{W} + \frac{1}{\theta}\bar{\mathbf{S}} + \gamma\mathbf{S}_{\text{prior}} - \mathbf{D}_q). \tag{5.16}$$

The minimiser $\bar{\mathbf{S}}$ is obtained by imposing $\nabla_\mathbf{S}(\cdot) \stackrel{!}{=} 0$ as

$$\left(\lambda \mathbf{R}^\mathsf{T}\mathbf{R} + \gamma + \frac{1}{\theta}\mathbf{I}\right)^{-1}\left(\lambda \mathbf{R}^\mathsf{T}\mathbf{W} + \frac{1}{\theta}\bar{\mathbf{S}} + \gamma\mathbf{S}_{\text{prior}} - \mathbf{D}_q\right). \tag{5.17}$$

An overview of the entire algorithm is given in Alg. 1. STEP 1 and STEP 2 are repeated until convergence.

5.1.3.2 Per Frame Shape Prior

In the case of a shape prior which is different for every frame, the data term reads

$$\mathbf{E}_{\text{data}} = \frac{\gamma}{2} \|\Gamma(\mathbf{S} - \mathbf{S}_{\text{prior}})\|_{\mathcal{F}}^2, \tag{5.18}$$

where Γ is a diagonal matrix controlling the influence of the shape prior for individual frames. Following the same principles as in Sec. 5.1.3.1, we derive the minimiser $\bar{\mathbf{S}}$ of the primal-dual formulation of Eq. (5.9) as

$$\left(\lambda \mathbf{R}^{\mathsf{T}}\mathbf{R} + \gamma \Gamma^{\mathsf{T}}\Gamma + \frac{1}{\theta}\mathbf{I}\right)^{-1}\left(\lambda \mathbf{R}^{\mathsf{T}}\mathbf{W} + \frac{1}{\theta}\bar{\mathbf{S}} + \gamma \Gamma^{\mathsf{T}}\Gamma \mathbf{S}_{\text{prior}} - \mathbf{D}_q\right). \tag{5.19}$$

Entries of Γ adjust the parameter γ framewise. If Γ contains zero and non-zero values, it can be interpreted as a binary shape prior indicator for every frame[1].

5.1.3.3 Per Pixel Per Frame Shape Prior

Per pixel per frame shape prior is the most general form of the proposed constraint; integration of it is more challenging. Firstly, we obtain the matrices $\tilde{\mathbf{S}} \in \mathbb{R}^{3FN}$, $\tilde{\bar{\mathbf{S}}} \in \mathbb{R}^{3FN}$, $\tilde{\mathbf{S}}_{\text{prior}} \in \mathbb{R}^{3FN}$ and $\tilde{\mathbf{D}}_q \in \mathbb{R}^{3FN}$ from $\mathbf{S}$, $\bar{\mathbf{S}}$, $\mathbf{S}_{\text{prior}}$ and $\mathbf{D}_q$, respectively, by applying the permutation operator $\mathrm{P}(\cdot)$ and stacking point coordinates of all frames into a vector (*e.g.*, $\tilde{\mathbf{S}} = \mathrm{vec}(\mathrm{P}(\mathbf{S}))$, and analogously for the remaining matrices). Similarly, we obtain matrix $\tilde{\mathbf{W}} \in \mathbb{R}^{2FN}$ as

$$(\underbrace{\mathbf{W}_{11}\mathbf{W}_{21}\cdots\mathbf{W}_{1N}\mathbf{W}_{2N}}_{\text{all points of frame 1}}\cdots\underbrace{\mathbf{W}_{(2F-1)1}\mathbf{W}_{(2F)1}\cdots}_{\text{all points of frame } F})^{\mathsf{T}}. \tag{5.20}$$

Accordingly, the rotation matrix is adjusted. The resulting matrix $\tilde{\mathbf{R}} \in \mathbb{R}^{2FN \times 3FN}$ is a quasi-block diagonal. It contains FN blocks of size 2×3, *i.e.*,

$$\tilde{\mathbf{R}} = \mathrm{diag}\{\boxed{\mathbf{R}_1}\ldots\boxed{\mathbf{R}_F}\ldots\}. \tag{5.21}$$

We introduce a diagonal matrix $\tilde{\Gamma} \in \mathbb{R}^{3FN \times 3FN}$ containing weights per frame per point coordinate. After applying proximal splitting, $\mathrm{P}(\cdot)$ and $\mathrm{vec}(\cdot)$ operators, the minimisation problem in Eq. (5.9) alters to

$$\underset{\tilde{\mathbf{S}}}{\operatorname{argmin}} \frac{\lambda}{2}\|\tilde{\mathbf{W}} - \tilde{\mathbf{R}}\tilde{\mathbf{S}}\|_{\mathcal{F}}^2 + \frac{1}{2\theta}\|\tilde{\mathbf{S}} - \tilde{\bar{\mathbf{S}}}\|_{\mathcal{F}}^2 + \frac{\gamma}{2}\|\tilde{\Gamma}(\tilde{\mathbf{S}} - \tilde{\mathbf{S}}_{\text{prior}})\|_{\mathcal{F}}^2 + \sum_{f,i,p}\|\nabla \mathbf{S}_f^i(p)\|. \tag{5.22}$$

[1] in the case if Γ contains only zeroes and ones, $\Gamma^{\mathsf{T}}\Gamma = \Gamma$.

The gradient of the function in Eq. (5.22) reads

$$\nabla_{\tilde{\mathbf{S}}} = (\lambda\tilde{\mathbf{R}}^{\mathsf{T}}\tilde{\mathbf{R}} + \frac{1}{\theta}\mathbf{I}_{3FN} + \gamma\tilde{\Gamma}^{\mathsf{T}}\tilde{\Gamma})\tilde{\mathbf{S}} - (\lambda\tilde{\mathbf{R}}^{\mathsf{T}}\tilde{\mathbf{W}} + \frac{1}{\theta}\tilde{\tilde{\mathbf{S}}} + \gamma\tilde{\Gamma}^{\mathsf{T}}\tilde{\Gamma}\tilde{\mathbf{S}}_{\text{prior}} - \tilde{\mathbf{D}}_q) \stackrel{!}{=} 0. \tag{5.23}$$

Finally, the minimiser of Eq. (5.23) is obtained as

$$\tilde{\mathbf{S}} = (\underbrace{\lambda\tilde{\mathbf{R}}^{\mathsf{T}}\tilde{\mathbf{R}}}_{\text{block-diagonal}} + \underbrace{\frac{1}{\theta}\mathbf{I}_{3FN}}_{\text{diagonal}} + \underbrace{\gamma\tilde{\Gamma}^{\mathsf{T}}\tilde{\Gamma}}_{\text{diagonal}})^{-1}(\lambda\tilde{\mathbf{R}}^{\mathsf{T}}\tilde{\mathbf{W}} + \frac{1}{\theta}\tilde{\tilde{\mathbf{S}}} + \gamma\tilde{\Gamma}^{\mathsf{T}}\tilde{\Gamma}\tilde{\mathbf{S}}_{\text{prior}} - \tilde{\mathbf{D}}_q). \tag{5.24}$$

Note that the factor on the left side of Eq. (5.24) represents a block-diagonal matrix. Its inverse can be found by separately inverting FN blocks of size 3×3. After $\tilde{\mathbf{S}}$ is computed, we obtain $\mathbf{S}$ by an inverse permutation.

5.1.4 Obtaining Shape Prior

In this section, we revise the method for occlusion tensor estimation, and formulate a criterion for a set of views to be suitable for the shape prior estima tion.

5.1.4.1 Occlusion Tensor Estimation

An occlusion tensor is a probabilistic space-time occlusion indicator. We refer to occlusion maps as slices of the occlusion tensor corresponding to individual frames. If occlusion tensor is available, it is possible to control a shape prior with the per pixel per frame granularity (see Sec. 5.1.3.3).

Occlusion tensor is computed from $\mathbf{W}$ and a reference image. For every frame, a corresponding occlusion map equals to a Gaussian-weighted difference between a backprojection of the frame to the reference frame and the reference frame itself. Thus, the occlusion indicator triggers a higher response for areas which cannot be backprojected accurately due to occlusions, specularities, illumination inconstancy, large displacements, highly non-rigid deformations, or a combination of those. As a result, the occlusion tensor accounts for multiple reasons for inaccuracies in correspondences. A similar scheme was applied in [231, 267].

Consider dense flow fields computed by an MFSF, or displacements of pixels visible in the reference frame throughout the whole image sequence:

$$\mathbf{u}(x;\mathbf{n}) = \begin{bmatrix} u(x,\mathbf{n}) \\ v(x,\mathbf{n}) \end{bmatrix} : \Omega \times \{1,...,F\} \to \mathbb{R}^2, \tag{5.25}$$

Algorithm 2 Estimation of Occlusion Tensor $\mathbf{E}(x)$

Input: dense flow fields $\mathbf{u}(x;\mathbf{n})$, a reference frame $\mathbf{I}(x,\mathbf{r})$, Gaussian kernel $G_{k\times k}$
Output: occlusion maps $\mathbf{E}(x,\mathbf{n})$
1: **for every frame** $\mathbf{n} \in \{2,\dots,F\}$ **do**
2: $\quad \mathbf{w}(\mathbf{n},\mathbf{r}) = \mathbf{I}(x,\mathbf{n}) - \mathbf{u}(x;\mathbf{n})$ (backprojection to $\mathbf{I}(x,\mathbf{r})$)
3: $\quad$ image difference $\mathbf{B}(x) = \mathbf{w}(\mathbf{n},r) - \mathbf{I}(x,\mathbf{r}) \hat{=}$
4: $\quad$ **for every pixel** x **do**
5: $\quad\quad \mathbf{B}(x) = \left\| (x_r^{\mathbf{w}} - x_r^{\mathbf{I}})^2 + (x_g^{\mathbf{w}} - x_g^{\mathbf{I}})^2 + (x_b^{\mathbf{w}} - x_b^{\mathbf{I}})^2 \right\|_2$
6: $\quad$ **end for**
7: $\quad \mathbf{E}(x,\mathbf{n}) = \mathbf{B}(x) * G$
8: $\quad$ postprocess $\mathbf{E}(x)$
9: **end for**

where $\Omega \in \mathbb{R}^2$, $\mathbf{n}$ denotes a frame index, $\mathbf{u}(x,\cdot)$ denotes a 2D displacement of a point x through the image sequence ($u(x,\mathbf{n})$ and $v(x,\mathbf{n})$ denote displacements in u and v directions, respectively). Let $\mathbf{r}$ be the index of the reference frame and $\mathbf{I}(x,\mathbf{r})$ a reference frame. Occlusion maps $\mathbf{E}(x,\mathbf{n}) : \Omega \times \{2,...,F\} \to \mathbb{R}$ can be obtained from the dense correspondences and the reference frame according to the algorithm summarised in Alg. 2. Firstly, a backprojection $\mathbf{B}(\mathbf{n},\mathbf{r})$ of every frame $\mathbf{I}(x,\mathbf{n})$ to the reference frame is performed. In this step, reverse point displacements are applied to every frame $\in \{2,\dots,F\}$ with an optional interpolation for missing parts. Secondly, image differences between the warped images and the reference image are computed. Therefore, we take L_2-norms of sums of channel-wise differences (RGB) for every pixel convolved with the Gaussian kernel $G_{k\times k}$ of an odd width k (see Alg. 2, rows 3–7). The result is postprocessed (normalised and discretised) so that the estimated values lie in the interval $[0;255]$. The resulting image series represents occlusion tensor with per-pixel occlusion probabilities for every frame. If required, occlusion maps can be binarised. The algorithm exhibits data parallelism (on the frame and pixel levels) and is well suitable for implementation on a GPU.

The complexity of the occlusion tensor estimation is $\mathscr{O}(F\mathbf{whJ}^2)$ with $\mathbf{w}$, $\mathbf{h}$ and $\mathbf{J}$ being the width and height of a frame and size of a square Gaussian kernel, respectively. For a few dozens of frames of common resolutions as they occur in NRSfM problems, the whole computation can be performed on a GPU in less than a second. Examples of occlusion maps are given in Figs. 5.2, 5.8.

5.1.4.2 Total Intensity Criterion

Given an occlusion tensor, we determine the set of frames suitable for the shape prior estimation using the accumulative *total intensity criterion*:

$$\sum_{f=1}^{F_{sp}} \left\| \int_{\Omega} du\,dv \right\|_2 \leq \varepsilon. \tag{5.26}$$

In Eq. (5.26), Ω denotes an image domain of a single frame, F_{sp} denotes the length of the sequence suitable for shape prior estimation, and ε is a non-negative scalar value. In other words, as far as the frames are not significantly occluded (regardless of in which image region occlusions happen), they can be used for the estimate. The obtained shape prior is rigidly aligned with an initialisation obtained with [281] since a different number of frames significantly affect the initial alignment of the reconstructions. Therefore, we employ Procrustes analysis on 3D points corresponding to unoccluded image pixels ($15-20$ points are uniformly selected). An example of the aligned initialisation and the shape prior for the flag sequence with a hash occlusion pattern is shown in Fig. 5.6-(a).

In Fig. 5.3, the results of the algorithm operating based on the total intensity (TI) criterion are given. TI accumulates non-zero pixel values for all frames until the frame f [2]. As can be seen in the comparison of the images and corresponding TI plots, sudden increases in TI happen if an occlusion begins. Low occlusion probabilities cause gradual increases of TI. When comparing the frame indexes with the starting occlusions and the responses of the TI indicators, the correlation is clearly seen. The sensitivity threshold ε, depends on the size and duration of occlusion. Suppose occlusions do not happen, and the sequence is infinitely long. In this case, the threshold ε will still be reached after a finite number of frames. However, this property is still desirable, because the length of the sequence for the shape prior estimation should be narrowed down. TI can be augmented with or in some cases replaced by a differential TI criterion such as:

$$\frac{TI(F_{sp}+1) - TI(F_{sp}-1)}{2} \leq \varepsilon', \tag{5.27}$$

where ε' is a threshold on the derivative and F_{sp} is the last frame suitable for the shape prior estimation.

5.1.5 Experiments

The proposed approach is implemented in C++/CUDA C [210] for a heterogeneous platform with a multi-core CPU and a single GPU. We run experiments on a

[2] in this sense, the TI criterion is similar to a cumulative distribution function.

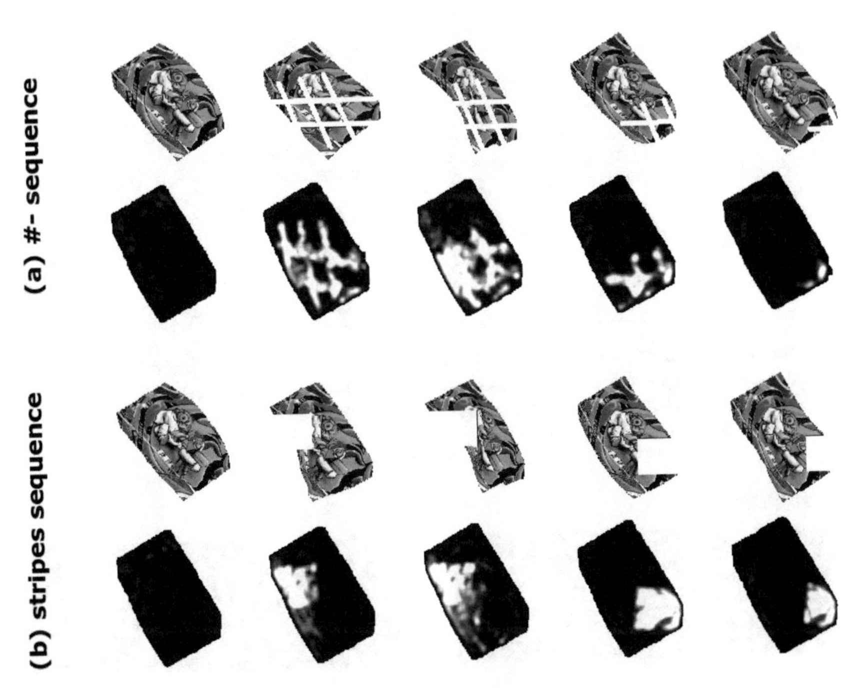

Figure 5.2: Exemplary frames from the modified flag sequences [103] with the computed occlusion maps: (a) #-sequence; (b) *stripes* sequence.

machine with Intel Xeon E5-1650 CPU, NVIDIA GK110 GPU and 32 GB RAM. While finding an optimal $\mathbf{R}$ (Eqs. (5.4)–(5.6)), the most computationally expensive operation is $\mathbf{S}\mathbf{S}^{\mathsf{T}}$. This operation can be accomplished by six vector dot products and only $\mathbf{S}$ needs to be stored in memory. It is implemented as a dedicated GPU function, together with the computation of $\mathbf{D}_q$ (Alg. 1, rows 10, 13, 14). Compared to the C++ version, $12-15$x speedup is achieved.

To compute dense correspondences, Multi-Frame Subspace Flow (MFSF) [103] is used, and to estimate a shape prior, we run [104] on several initial frames of the sequence as described in Sec. 5.1.4. If not available for a respective sequence, segmentations of the reference frames are computed with [234].

As our objective is to jointly evaluate correspondence establishment under severe occlusions and non-rigid reconstruction, we perform joint evaluation of different pipeline configurations. We compare occlusion-aware MFOF [267] + VA [104], MFSF [103] + AMP [121], MFSF + VA, MFSF + SPVA (the proposed method). For every configuration, we report mean Root-Mean-Square (RMS) error metric

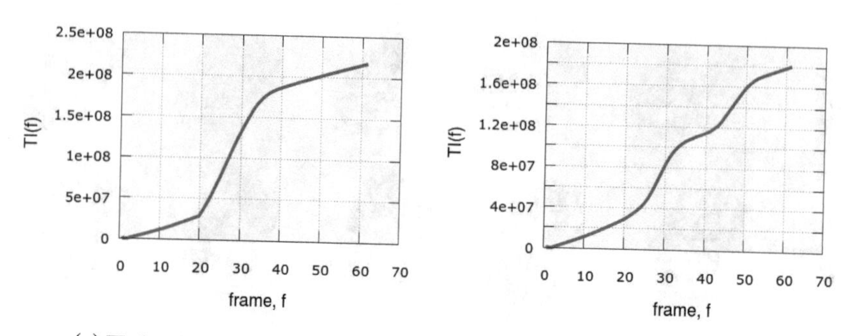

(a) TI plots for #-sequence (left) and stripes sequence (right). Deep yellow colour marks occluded, and light yellow colour marks unoccluded frames. Note the difference in function slopes for occluded and unoccluded regions, as well as the correlation of the change in slopes with the beginning of occlusions in (b).

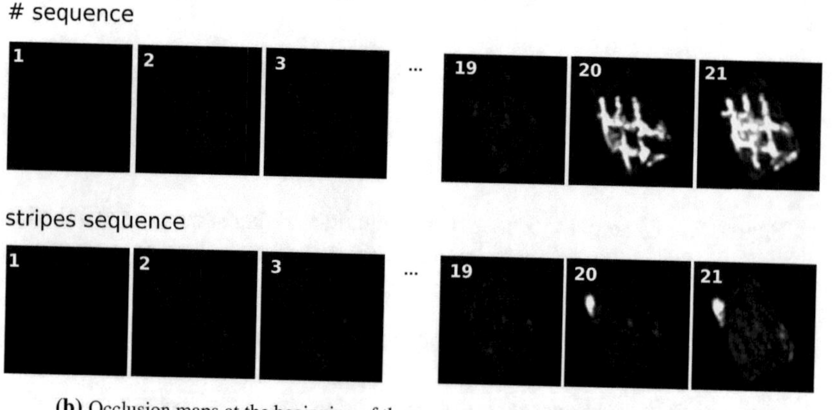

(b) Occlusion maps at the beginning of the sequences and when occlusions start.

Figure 5.3: Plots of TI function of the number of frames for #- and stripes sequences (a), excerpts when occlusions begin (b).

defined as $e_{3D} = \frac{1}{F}\sum_{f=1}^{F} \frac{\left\|\mathbf{S}_f^{ref} - \mathbf{S}_f\right\|_{\mathscr{F}}}{\left\|\mathbf{S}_f^{ref}\right\|_{\mathscr{F}}}$, where $\mathbf{S}_f^{ref}$ is a ground truth surface. Finally, we show results on real image sequences and compare results qualitatively.

5.1.5.1 Evaluation Methodology

For the joint evaluation, a dataset with a ground truth geometry and corresponding images is required. There is one dataset known to the authors which partially fulfils

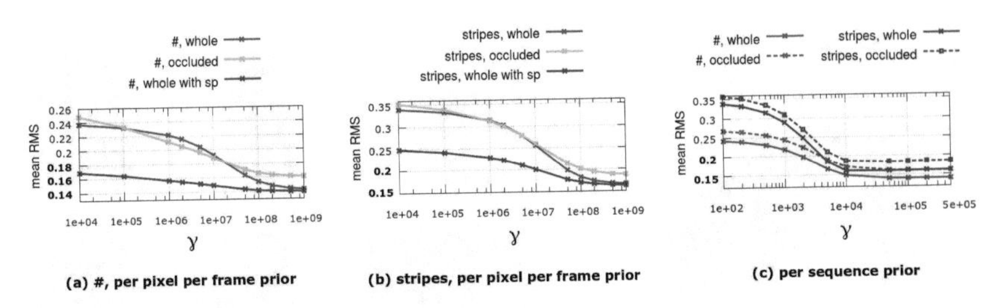

Figure 5.4: Results of the quantitative evaluation of the proposed method in the configuration MFSF [103] + SPVA: (a) per pixel per frame mode on the #-sequence; (b) per pixel per frame mode on the *stripes* sequence; (c) per sequence mode on both sequences. "whole": mean RMS is computed on all frames of the respective sequence, "occluded": mean RMS is computed only on the occluded frames; "whole with sp": the algorithm is initialised with the shape prior in the non-occluded frames. Bold font (mean RMS) highlights parameter values which outperform occlusion-aware MFOF [267] + VA [104].

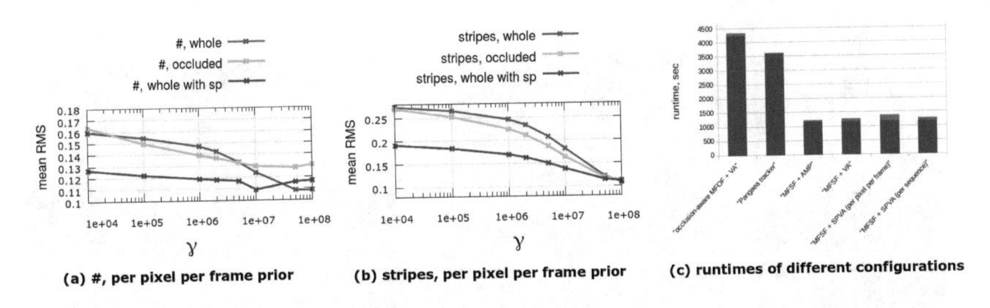

Figure 5.5: Results of the quantitative evaluation on the flag sequence with the dense segmentation mask. In (a) and (b), the notation is the same as in Fig. 5.4. Reconstructions obtained on the unoccluded ground truth optical flow are used as a reference for comparison; (c) runtimes of different pipeline configurations on the dense flag dataset (blue colour marks correspondence computation, orange marks NRSfM, except for Pangaea which is a template-based method). The fastest configuration MFSF [103] + AMP [121] is only ca. 4% faster than the proposed configuration with SPVA which is the most accurate.

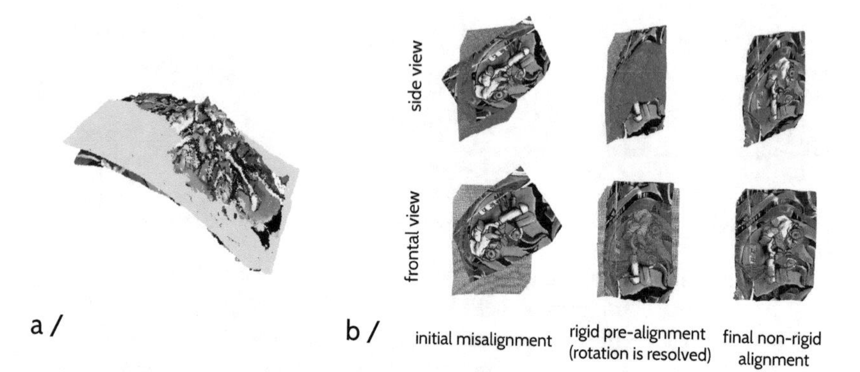

Figure 5.6: (a): Initialisation obtained under rigidity assumption overlayed with a shape prior (cyan); (b): Non-rigid alignment of the ground truth geometry with an exemplary reconstruction for correspondence establishment. Here, we use Extended CPD (see Chapter 8). The point correspondences are eventually used in the quantitative evaluation (3D error metric).

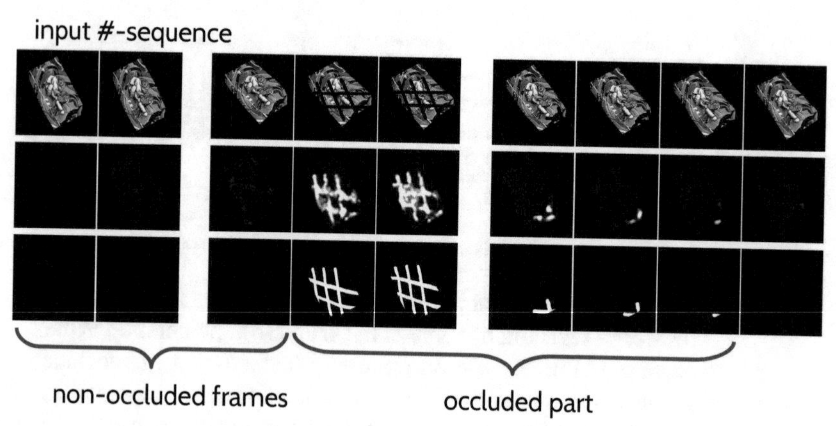

Figure 5.7: Exemplary frames of the hashtag sequence. On the first several unoccluded frames the shape prior is obtained. It is then used to stabilise the reconstructions of the frames with external occlusions.

the requirements — the synthetic flag sequence initially introduced in [103]. This dataset originates from mocap measurements and contains images of a waving flag rendered by a virtual orthographic camera. The flag dataset was already used for evaluation of NRSfM [9, 12] and MFOF algorithms [103, 267], but not for a joint evaluation, to the best of our knowledge. To generate orthographic views, the mocap flag data was projected onto an image plane (with an angle of approx. 30° around the x axis) and a texture was applied on it (here the texture does not reflect distortion effects associated with the view which is different from the frontal one). More details on the dataset can be found in [235]. Using the rendered images, we evaluate MFOF and NRSfM jointly.

First, we extend the flag dataset with several data structures. The ground truth surfaces contain 9622 points, whereas the rendered images are of the resolution 500×500. If the corresponding segmentation mask for the reference frame is applied, $8.2 \cdot 10^4$ points are fetched. To overcome this circumstance, we create a segmentation mask which fetches the required number of points as in the ground truth. Therefore, we project the ground truth surface corresponding to the reference frame onto the image plane and obtain a sparse segmentation mask. When applied to the dense $\mathbf{W}$, the sparse mask fetches 9622 points. To establish point correspondences between the ground truth and reconstructions, we apply non-rigid point set registration with correspondence priors [122]. This procedure needs to be preformed only once on a single ground truth surface and a single flag reconstruction with 9622 points, and the correspondence index table is used during computation of the mean RMS. Non-rigid registration does not alter any reconstruction which is evaluated for mean RMS.

Second, we introduce severe occlusions into the flag image sequence which go beyond those added for evaluation in [267] in terms of the occlusion duration and size of the occluded regions. We overlay two different patterns with the clean flag sequence — a grid # and stripes patterns. The resulting sequences contain 20 and 29 occluded frames, respectively. See Figs. 5.7–5.2 for exemplary frames and the corresponding occlusion maps.

5.1.5.2 Experiments on Synthetic Data

We compare several framework configurations on the synthetic flag sequence and report mean RMS and runtimes. We also evaluate the influence of the shape prior term by varying γ in several shape prior modes.

Results of the experiment are summarised in Fig. 5.4. Occlusion-aware MFOF+VA achieves the mean RMS error of 0.18(0.219) (in brackets, mean RMS only on occluded frames is reported) for the #-sequence and 0.195(0.209) on the *stripes*.

MFSF+VA achieves 0.239(0.256) and 0.341(0.355) for the #- and *stripes* sequences, respectively. MFSF+SPVA achieves 0.143 (0.161) and 0.167(0.187) for the #- and *stripes* sequence, respectively, in the per pixel per frame mode, and 0.140(0.160) and 0.160(0.183) in the per sequence mode. At the same time, runtime of the MFSF+SPVA in the per frame mode is almost equal to the runtime of MFSF+VA — the difference is less than 1% — whereas the configuration MFSF+SPVA in the per pixel per frame mode takes only 3% more time.

The configuration with the fastest MFSF and the proposed SPVA achieves the lowest mean RMS; it is comparable in the runtime to the fastest configuration with AMP. We are 3.4 times faster than the second best NRSfM based configuration with the computationally expensive occlusion-aware MFOF+VA. As can be seen in Fig. 5.4, performance of SPVA depends on γ. As expected, mean RMS is the lowest for a particular finite γ value and grows as γ increases. The drop in the accuracy happens because the shape prior term becomes so dominant that even less probably occluded areas are regularised out. If γ is infinite, all frames (all pixels with non-zero occlusion map) are set to the shape prior which leads to a suboptimal solution. In the per sequence mode (Fig. 5.4-(c)) $\gamma \in [0; 10^6]$, whereas in per pixel per frame mode $\gamma \in [0; 10^9]$ (γ is split between all pixels weighted with the occlusion map values). Experiment shows that the transitions are gradual with the gradual changes of γ.

Besides, we perform a comparison of our method with the recent template-based method of Yu *et al.* [319] — Pangaea tracker. SPVA can be classified as a hybrid method for monocular non-rigid reconstruction. Firstly, the assumption of a template-based technique — an exact 3D shape is known for at least a single frame — is not fulfilled in our case (a shape prior is not an exact reconstruction). Secondly, we obtain shape prior automatically, whereas in template-based methods [34,223,245,280,319], a template is assumed to be known in advance. Nevertheless, the comparison with such a method as [319] is valuable. Ultimately, research in the area of template-based reconstruction shifts in the direction of hybrid methods, *i.e.,* there is an endeavour to find a way to obtain a template automatically and under non-rigid deformations. The SPVA framework is perhaps the first attempt in this direction, and in this experiment, we demonstrate that a template-based method can work with a shape prior obtained with the proposed approach (see Sec. 5.1.4) and produce accurate results. Pangaea tracker achieves almost equal mean RMS of 0.172 (0.191) for both #- and *stripes* sequences. We discovered that this template-based method is stable against textureless occlusions, but an error can accumulate if occlusions are permanent and large. Still, Pangaea tracker achieves the second best result after the combination MFSF+SPVA and outperforms the more recent occlusion-aware MFOF+VA pipeline on the sparse flag sequence. Table 5.2 summarizes the lowest achieved mean RMS errors for all tested combinations.

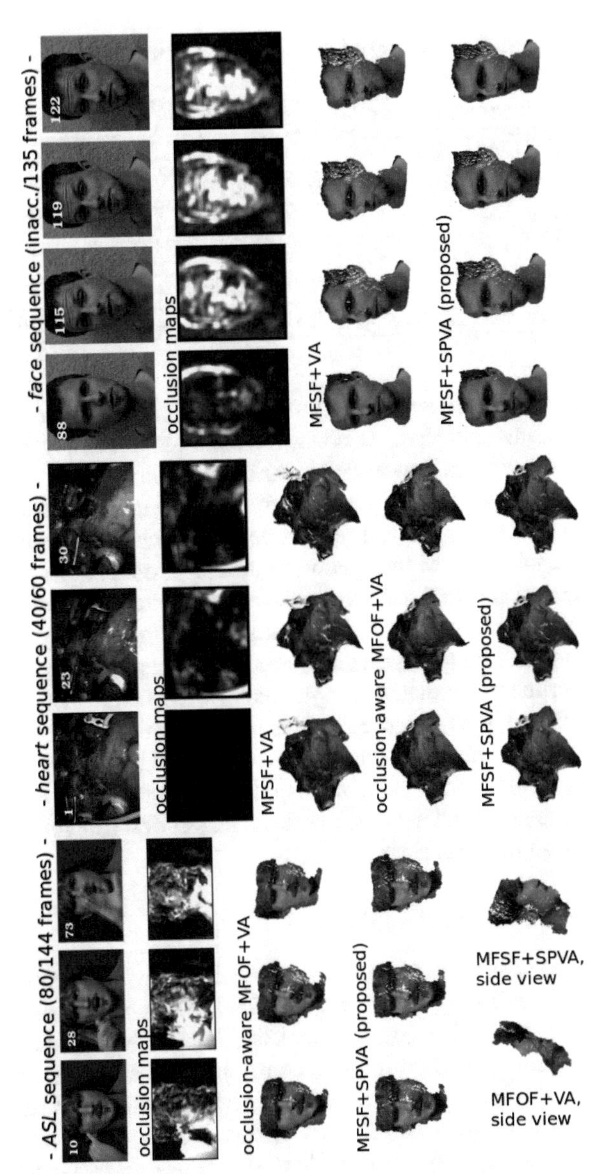

Figure 5.8: Qualitative results of the proposed SPVA framework and other pipeline combinations on several challenging real image sequences, *i.e.,* ASL [38], *heart* [259] and new *face*.

Table 5.1: Mean RMS errors of different algorithmic combinations for the #- and *stripes* sequences.

algorithmic combination	mean RMS, #	mean RMS, *stripes*
o.a. MFOF [267]+VA [104]	0.181 (0.219)	0.195 (0.209)
Pangaea tracker [319]	**0.172 (0.191)**	**0.172 (0.191)**
MFSF [103]+AMP [121]	0.297 (0.381)	0.460 (0.523)
MFSF [103]+VA [104]	0.239 (0.252)	0.341 (0.355)
MFSF [103]+SPVA, p. pix.	**0.143 (0.161)**	**0.167 (0.189)**
MFSF [103]+SPVA, p. seq.	**0.140 (0.160)**	**0.160 (0.184)**

The experiment with the varying γ is also repeated on the flag sequence with the dense segmentation mask. Here, we obtain reference reconstructions for comparison on the ground truth optical flow available for the unoccluded views. In this manner, it is possible to see how good the proposed pipeline alleviates side effects associated with occlusions and how close the reconstructions reach the reference. Moreover, the TV term is enabled, since the measurements are dense. Results are summarised in Fig. 5.5. The mean RMS relative to the reference reconstruction follows the similar pattern as in the case of the comparison with the sparse ground truth. In Fig. 5.5-(c), runtimes for all tested pipeline configurations are summarised.

In both experiments, a relatively high mean RMS is explained by two effects. As above mentioned, the reference frame is not a frontal projection of the ground truth, and no frontal views are occurring in the image sequence. Moreover, the flag sequence exhibits rather large non-rigid deformations. All evaluated methods including the proposed approach perform best if deformations are moderate deviations from some mean shape.

5.1.5.3 Experiments on Real Data

We tested SPVA on several challenging real-world image sequences: *American Sign Language (ASL)* [38], *heart surgery* [259], and a new *face*. Results are visualised in Fig. 5.8. *ASL* sequence depicts a face with permanent occlusions due to hand gesticulation. Only sparse reconstructions were previously shown on it [126, 137]. On this sequence, occlusion-aware MFOF performs poorly and marks whole frames starting from frame 20 as occluded. Consequently, the combination MFOF [267]+VA fails to reflect realistic head shapes, and it is seen distinctly in the side view. The proposed approach, using the shape prior obtained on first 17 frames provides realistic reconstructions. The *heart* sequence is a recording of a heart bypass surgery. 40 out of 60 frames are significantly occluded by a robotic arm. For the first time, dense reconstructions on this sequence were shown in [267]. The

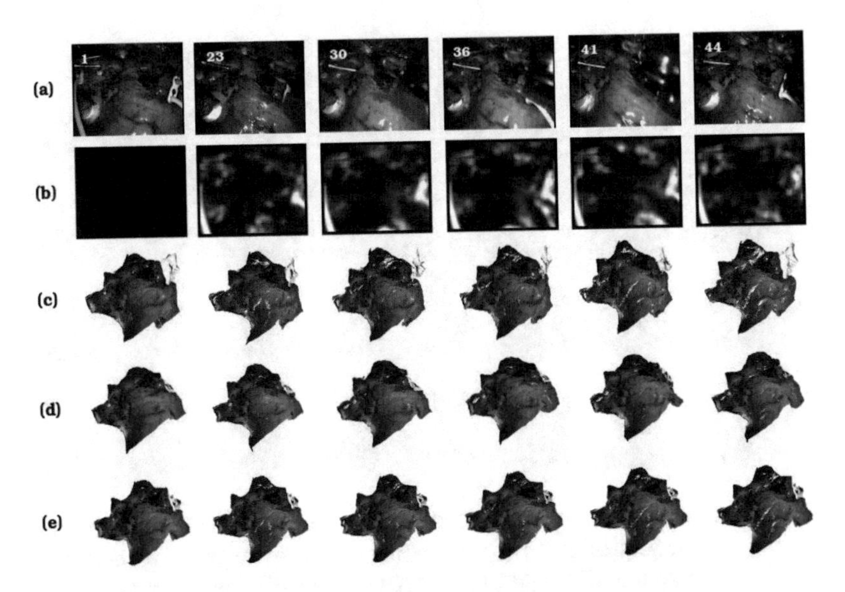

Figure 5.9: Experimental results on the heart sequence [259]: (a) the reference frame 1 and several occluded frames; (b) per-pixel occlusion maps for the frames shown above; (c) reconstructions with the MFSF [103]+VA [104]; (d): reconstructions with the occlusion-aware MFOF [267]+VA [104]; (e) reconstructions with the MFSF [103]+SPVA (per frame).

proposed SPVA achieves similar appearance, but the runtime is 30% lower. The new *face* sequence depicts a speaking person. No external occlusions are happening, but MFSF produces noisy correspondences due to large head movements. Thus, MFSF+VA outputs reconstructions with a bent structure in the nose area, whereas the shape prior in SPVA suppresses unrealistic twisting.

The Heart Surgery Sequence

The heart sequence originates from [259] and shows a patient's heart during bypass surgery naturally non-rigidly deforming. The sequence contains 60 frames; at frame 20, a robotic arm enters the scene and occludes large regions of the scene over multiple frames. The shape prior is estimated on 18 initial frames. We use an average occlusion map for every frame since some of the regions disappear or are occluded in most of the frames. The results of the experiment are shown in

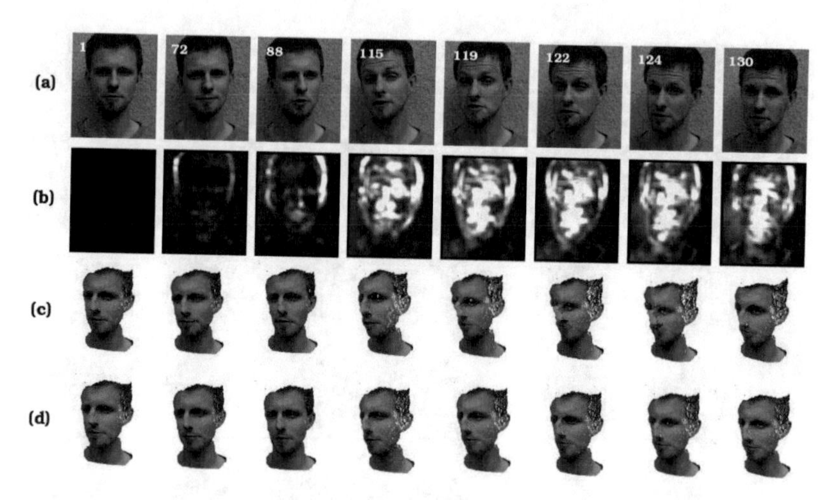

Figure 5.10: Experimental results on the face sequence: (a) several frames of the sequence; (b) corresponding occlusion maps; (c) reconstructions with the MFSF [103]+VA [104]; (d) reconstructions with the MFSF [103] + SPVA (per frame per pixel).

Fig. 5.9. Due to noisy initialisation under rigidity assumption, the combination MFSF+VA [104] produces reconstructions with severe inaccuracies and discontinuities (Fig. 5.9-(c)). Occlusion-aware MFOF+VA [104] generates visually consistent and smooth reconstructions, but we notice that natural non-rigid deformations of the heart are attenuated (due to oversmoothing). MFSF+SPVA produces visually consistent and accurate reconstructions which better reflect heart contractions.

The Human Face Sequence

The new human face sequence depicting a speaking person was recorded with a monocular RGB camera. It contains 135 frames. Due to large displacements and deformations, the MFSF method [103] gives inaccurate correspondences, especially around frame 120. The direct method [104] relying on data term corrupts the structure and does not preserve realistic point topology. Using the per pixel per frame shape prior, we are able to preserve the point topology in the corrupted regions while relying on the data term where correspondences are accurate. Fig. 5.10 illustrates several problematic frames, corresponding occlusion maps and reconstructions with combinations MFSF+VA [104] and MFOF+SPVA. The experiment shows that even in the absence of occlusions, a method for correspondence computation can

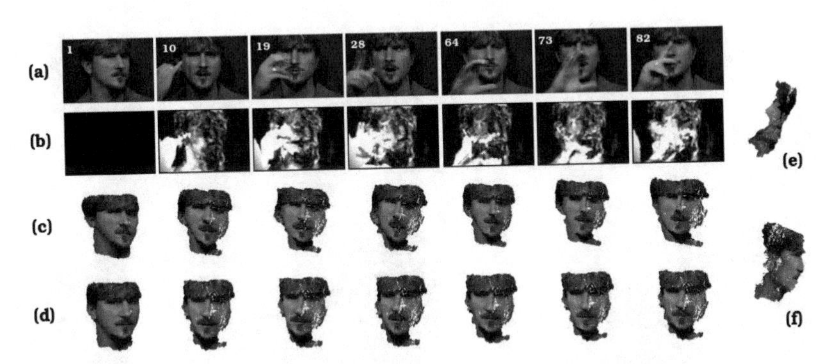

Figure 5.11: Experimental results on the ASL sequence [38]: (a) reference frame 1 and several frames with occlusions; (b) corresponding occlusion maps; (c) reconstructions with MFSF [103]+VA [104]; (d) reconstructions with MFSF [103]+SPVA (per frame per pixel); (e) side view of frame 28 for the configuration used in (c); (f) side view of frame 28 for the configuration used in (d).

perform poorly. In this case, reconstructions can exhibit such artefacts as broken point topology leading to corrupted reconstructions. Especially if it is not possible to recompute correspondences or available methods for obtaining correspondences do not improve the situation, our framework can be advantageous. For non-occluded regions (or regions with accurate correspondences), our approach strongly relies on the data term, whereas otherwise it strongly relies on the regularisation and shape prior terms.

The Sign Language Sequence

The American Sign Language (ASL) sequence F5_10_A_H17 is taken from [38]. It shows a communicating person and contains severe occlusions due to hand gesticulation. Out of 114 frames, 80 frames have occlusions. To compute a shape prior, the first twelve frames are used. The processing results are shown in Fig. 5.11. The combination MFSF [103]+VA [104] suffers from the complexity of the seemingly simple scene (Fig. 5.11-(c),(e)). In this experiment, the occlusion-aware MFOF could not improve the reconstruction accuracy. The reason is a mixture of a suboptimal reference view, occlusions due to significant head rotations and severe external occlusions. Our approach paired with MFSF is the only one which achieves a meaningful reconstruction on the F5_10_A_H17 sequence, perhaps

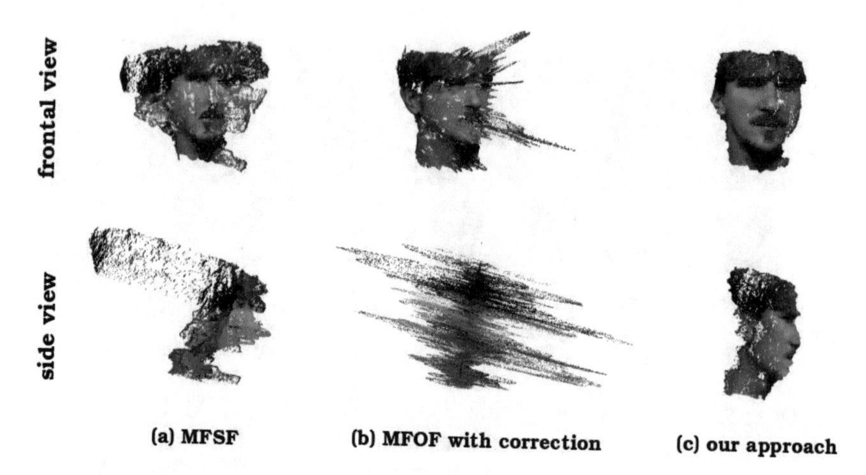

Figure 5.12: Results on the ASL sequence [38] with correspondence correction; in this example, method by Taetz *et al.* [267] does not improve reconstruction accuracy; (a) reconstruction example of MFSF [103]+VA [104]; (b) reconstruction example of MFOF [267]+VA [104]; (c) reconstruction example of MFSF [103]+SPVA (the proposed approach).

Table 5.2: Runtimes of different algorithm combinations for the sequences involved in the experiments, in seconds.

configuration	heart surgery [259] 360 × 288, 50 fr.	face (new) 241 × 285, 136 fr.	F5_10_A_H17 [38] 720 × 480, 114 fr.
MFSF [103] + VA [104]	481.0 + 119.3	728.9 + 35.7	3114.0 + 400.0
MFSF [103] + AMP [121]	481.0 + 20.4	728.9 + 26.4	3114.0 + 98.0
MFOF [267] + VA [104]	1592.8 + 119.2	2693.6 + 35.7	11995.3 + 300.5
MFSF [103] + SPVA	481.0 + 846.2	728.9 + 122.9	3114.0 + 1011.0

Table 5.3: Parameters of the proposed approach for different sequences.

sequence	λ	θ	τ	γ
heart surgery	10^4	10^{-5}	10^4	$5 \cdot 10^4$
human face	$5 \cdot 10^3$	10^{-5}	$5 \cdot 10^3$	10^3
ASL	$5 \cdot 10^4$	10^{-5}	$4.2 \cdot 10^3$	10^5

for the first time in the dense case (Fig. 5.11-(d), (f)). In Fig. 5.12, additional results are shown. We tested correspondence correction [267] and found out that it did not produce expected results. In comparison to reconstructions achieved by the combination MFSF [103]+VA [104] (see Fig. 5.12-(a)), the combination MFOF [267]+VA [104] (see Fig. 5.12-(b)) deteriorates the results. Large and long occlusions are mainly responsible for that. MFOF [267] can compensate for occlusions of small durations. If occlusions are large and permanent, the built-in occlusion indicator of the method can fail. As a side effect, the measurements are often oversmoothed in these cases. Another reason is the high default sensitivity of the occlusion indicator. Note that we have not tuned parameters of the occlusion indicators, because they are supposed to be universal. In this example, however, large regions of the scene are spuriously detected as occlusions, and the correction step relies on erroneous data.

Table 5.2 contains a summary of the performed experiments, including parameters of the sequences, types of the applied shape priors and runtimes for all combinations and sequences. Parameters of the proposed SPVA approach are summarised in Table. 5.3. Recall that in an energy-based formulation, relative weights of the different terms are important, and an absolute value of the energy does not have a direct interpretation. In all experiments, σ was set to 1.0, and per frame per pixel shape prior was used.

5.2 Intrinsic Dynamic Shape Prior for Dense Non-Rigid Structure from Motion

The main contribution of this section is a new technique for monocular non-rigid reconstruction with a dynamic shape prior (DSP), *i.e.*, a sequence-specific set of states obtained on a representative image subsequence. In the vast majority of real-world cases, not deformations but rather different angles of view (camera poses) cause different 2D measurements. It is assumed that the representative subsequence provides sufficient variety of deformations as they are likely to occur in a given scene, whereas the poses have to be just sufficient for an accurate reconstruction. While the DSP generation is offline, the reconstruction of new measurements with DSP is light-weight and well parallelisable. It implicitly assumes temporally-disjoint rigidity, *i.e.*, when a newly observed state is reoccurring relative to DSP.

For every new incoming measurement, the proposed shape-from-DSP or *Dynamic Shape Prior Reconstruction* (DSPR) approach finds a globally optimal 3D state corresponding to the 2D measurements and rigidly transforms it to the pose as observed in the measurements by alternating between *multi-start gradient descent* (MSGD) and camera pose estimation. Note that the pose in the incoming frames can be arbitrary and differ significantly from poses observed during the DSP generation in the generative subsequence, due to the decoupling property of the shape and pose in NRSfM. As a further contribution, we propose a new light-weight energy-based *Consolidating Monocular Dynamic Reconstruction* (CMDR) NRSfM approach as a step in DSP recovery from a representative sequence, even though any accurate existing dense NRSfM method can be employed for this task. Apart from real-time monocular reconstruction from noisy data, our main idea can also be applied to several related problems. Since DSP represents a compact footprint of the geometry carrying a learned sequence-specific deformation model, it suggests suitability of DSPR for geometry compression.

In the following, we describe DSPR in detail and evaluate its convergence properties and precision with different types and portions of noise in the measurements. A joint study of dense point tracking and reconstruction is also conducted. Next, we apply DSPR to shape compression and heart reconstruction. Among all tested methods, DSPR achieves either lowest reconstruction errors or exhibits fewer compression artefacts at equal compression ratios.

5.2.1 Related Work

Some recent works on NRSfM focus on dense [104] and scalable methods [27, 170] as well as methods which can reconstruct complex non-linear deformations [166, 332]. A distinct tendency is investigating new, often simple and, at the same time, overlooked ideas [75, 176] and models for NRSfM [13, 116]. More and more attention is paid to hybrid methods which make stronger assumptions as classic NRSfM but fewer assumptions than template-based methods [223] or domain-specific approaches which expect a known object class [43]. One example of hybrid methods is an approach for handling large occlusions with a static shape prior obtained on several unoccluded frames of a sequence proposed in Sec. 5.1 [116]. Another example is a method with a trained deformation model [224] which relies on a representative dataset for training. Our method has thrived on the ideas proposed in the works mentioned above. The most closely related methods to DSPR is SPVA proposed in Sec. 5.1 and the method of Li *et al.* [176].

Li *et al.* propose to exploit state recurrency in sparse NRSfM. While the local rigidity assumption has been previously investigated in NRSfM [228], a temporally-disjoint rigid clustering provides more accurate approximations for intense deformations. While reconstructing such deformations, a local rigidity method rapidly reaches its lower bound on the number of views necessary for the rigid reconstruction to produce meaningful results, eventually leading to the overall accuracy deterioration. Temporally-disjoint rigidity, in contrast, does not rely on connected temporal windows and is agnostic to the deformation intensity over a short period. Instead, other states related to a given state by rigidity are found in arbitrary regions of the sequence so that the rigidity assumption can be more consistently satisfied.

Nevertheless, the number of rigid clusters has to be set in advance. Besides, if some states are unique or degenerate (are not observed in other poses), they are assigned to some non-empty clusters and treated as noise. Thus, non-reoccurring states are reconstructed less accurately. Moreover, the method of Li *et al.* [176] requires computationally costly graph clustering and works for few sparse points. In contrast, DSPR fits an instance from DSP which is related to a given dense 2D measurement by rigidity. It does not explicitly cluster dense point tracks into bins corresponding to similar 3D states. Instead, we solve an opposite problem of finding a subsequence providing as diverse deformations as possible in as few views as possible, for the DSP recovery.

Compared to NRSfM with a static shape prior [116], we extract multiple states (reflecting the whole deformation model) from a representative sequence, instead of a single state. While Golyanik *et al.* aim at stability under large occlusions, their method also tends to overconstrain reconstructions. Our primary goal is a light-weight sequential scheme with recurrent state identification, and still, it is remarkably robust under large occlusions (see Sec. 5.2.3.6).

In the proposed CMDR for reconstruction of the representative sequence, we optimise up to several millions of parameters simultaneously with non-linear least squares (NLLS). The most closely related approach to CMDR is the template-based method of Yu *et al.* [319], with several differences: 1) instead of using a multiview reconstruction to obtain a template, we initialise shapes and camera poses with the rigid factorisation [281] directly; 2) we use trajectory regularisation instead of as-rigid-as-possible regulariser [257], and 3) the fitting term operates on point tracks and not on images. One of the NRSfM techniques which simultaneously formulate constraints in metric and trajectory spaces is Column Space Fitting (CSF2) method of Gotardo and Martinez [126]. Our trajectory smoothness term was rarely used in energy-based NRSfM so far. It allows integration of subspace constraints on point trajectories and originates from [22]. We demand smoothness of neighbouring trajectories by optimising the total variation of trajectory coefficients. A similar regulariser was previously applied in multi-frame optical flow (MFOF) [105], see Sec. 2.2.8 for more details.

5.2.2 The Proposed Approach with Dynamic Shape Prior

Our objective is 3D reconstruction of a current state $\mathbf{S}_f$ given incoming measurements $\mathbf{W}_f$, $f \in \{1,\ldots,F\}$ and a DSP $\mathbf{D} = \{\mathbf{D}_i\}$, $i \in \{1,\ldots,Q\}$ with Q temporal rigidity bases. We formulate dense sequential NRSfM as an energy minimisation problem of finding $\mathbf{D}_i$ related to $\mathbf{S}_f$ by a rigid transformation and pose $\mathbf{R}_f$ so that the product $\mathbf{R}_f\mathbf{D}_i$ explains the current observation $\mathbf{W}_f$:

$$\begin{aligned}\mathbf{E}(\mathbf{S}_f = \mathbf{D}_i, \mathbf{R}_f) = \alpha \left\|\mathbf{W}_f - \mathbf{I}_{2\times3}\mathbf{R}_f\mathbf{D}_{i:\lambda_i=1}\right\|_{\mathscr{F}} + \\ \beta \left\|\mathbf{D}_{i:\lambda_i=1} - \mathbf{S}_{f-1}\right\|_{\mathscr{F}} + \gamma(\|\boldsymbol{\lambda}\|_0 - 1)^2,\end{aligned} \tag{5.28}$$

where $\|\cdot\|_0$ denotes a zero-norm of a vector and $\mathbf{I}_{2\times3}$ models orthographic projection. The energy functional (5.28) contains a data term, temporal smoothness term and a DSP regularisation term. The data term ensures that the factorisation $\mathbf{R}_f\mathbf{D}_i$ is accurately projected to $\mathbf{W}_f$. The temporal smoothness term expresses the assumption on the gradual character of changes in the states as well as helps to converge faster. The DSP regularisation term ensures that a single $\mathbf{D}_i$ is required to explain the observations upon our model. This practice contrasts to some other approaches, where every reconstruction is encoded as a linear combination of some basis shapes (recovered during reconstruction or known in advance) [216]. In our model, DSP is assumed to provide a sufficient variety to cover the entire space of reoccurring deformations.

The energy functional (5.28) can be minimised iteratively, by alternatingly fixing $\mathbf{R}_f$ and releasing $\mathbf{D}_{i:\lambda_i=1} = \mathbf{S}_f$, and vice versa, in every iteration. When $\mathbf{S}_f$ is fixed, the only term dependent from $\mathbf{R}_f$ is the data term. $\mathbf{R}_f$ can be updated in the closed-form by projecting its affine update to the $SO(3)$ group or by linear least squares with quaternion parametrisation. When $\mathbf{R}_f$ is fixed, an optimal $\mathbf{D}_i$ can be found by taking the partial derivative of the energy subspace with the fixed $\mathbf{R}_f$, denoted by $\mathbf{E}_{\mathbf{R}_f}$, w.r.t. $\boldsymbol{\lambda}$ and equating it to zero:

$$\frac{\partial \mathbf{E}_{\mathbf{R}_f}(\mathbf{S})}{\partial \mathbf{S}} \frac{\partial \mathbf{S}}{\partial \boldsymbol{\lambda}} = 0. \tag{5.29}$$

The optimality criterion in Eq. (5.29) defines a state when a small change in the shape caused by a small change in the prior state does not change the energy. We minimise the energy functional (5.28) when $\mathbf{R}_f$ is fixed by multi-start gradient descent (MSGD) approach. Starting from multiple regularly sampled values of $\boldsymbol{\lambda}$, we compute differences in $\mathbf{E}_{\mathbf{R}_f}$ and update $\boldsymbol{\lambda}$ in the direction of the energy decrease. Multiple starting points are required to obtain a globally optimal solution since $\mathbf{E}$ is non-convex. The global minimum is obtained by comparing locally minimal energy values. MSGD is well parallelisable as every thread can converge or finish upon a boundary condition (*e.g.*, when leaving the assigned range of values) independently from other threads. Thanks to MSGD, DSPR executes with three-five frames per second on our hardware without parallelisation (see Sec. 8.3.3).

5.2.2.1 Obtaining Dynamic Shape Prior (DSP)

DSP generation includes an accurate 3D reconstruction of a representative image sequence with a general-purpose NRSfM method. In principle, we are free to choose any accurate and scalable NRSfM technique for the initial reconstruction. In the experiments, we use two accurate existing methods, *i.e.*, Garg *et al.* [104] and Ansari *et al.* [27]. Additionally, we propose a new energy-based NRSfM method which outperforms the approaches mentioned above in a subset of evaluation scenarios.

Our Core NRSfM Approach for DSP Acquisition

For the notational consistency, in this section, we denote the measurements of the representative sequence and the corresponding 3D shapes by $\mathbf{W}_{2F\times N} = \{\mathbf{W}_f\}$ and $\mathbf{S}_{3F\times N} = \{\mathbf{S}_f\}$, respectively, with N denoting the number of points in every frame. The new method minimises the following energy functional with the Gauss-Newton approach:

$$\mathbf{E}_{\text{CMDR}}(\mathbf{R},\mathbf{S},\mathbf{A}) = \alpha\,\mathbf{E}_{\text{fit}}(\mathbf{R},\mathbf{S}) + \beta\,\mathbf{E}_{\text{temp}}(\mathbf{S}) + \lambda\,\mathbf{E}_{\text{linking}}(\mathbf{S},\mathbf{A}) + \rho\,\mathbf{E}_{\text{reg}}(\mathbf{A}), \tag{5.30}$$

where $\mathbf{A}$ is a matrix with trajectory coefficients explained below. The data term constrains projections of the recovered shapes to agree with the 2D measurements:

$$\mathbf{E}_{\text{fit}}(\mathbf{R},\mathbf{S}) = \sum_f \left\|\mathbf{W}_f - \mathbf{I}_{2\times 3}\mathbf{R}_f\mathbf{S}_f\right\|_\varepsilon^2, \tag{5.31}$$

where $\|\cdot\|_\varepsilon$ is Huber norm. The temporal smoothness term imposes similarity on reconstructions of adjacent frames:

$$\mathbf{E}_{\text{temp}}(\mathbf{S}) = \sum_{f=2}^{F} \left\|\mathbf{S}_f - \mathbf{S}_{f-1}\right\|_2^2. \tag{5.32}$$

The linking term expresses our assumptions about the complexity of deformations. Here, we rely on K known basis trajectories Θ sampled from discrete cosine transform (DCT) at regular intervals:

$$\mathbf{E}_{\text{linking}}(\mathbf{S},\mathbf{A}) = \|\mathbf{S} - (\Theta\otimes\mathbf{I}_3)_{3F\times 3K}\,\mathbf{A}_{3K\times N}\|_2^2,\ \text{where} \tag{5.33}$$

$$\begin{aligned}
\Theta &= \left([\theta_{11} \ \dots \ \theta_{1K}] \ \dots \ [\theta_{F1} \ \dots \ \theta_{FK}]\right)^{\mathsf{T}}, \\
\theta_{tk} &= \frac{\sigma_k}{\sqrt{2}} \cos\left(\frac{\pi}{2F}(2t-1)(k-1)\right) \text{ and} \\
\sigma_k &= \begin{cases} 1 & \text{if } k = 1, \\ \sqrt{2} & \text{otherwise.} \end{cases}
\end{aligned} \tag{5.34}$$

In Eq. (5.33), $\mathbf{A}$ holds coefficients of linear combinations which approximate trajectories of recovered 3D points. The linking term connects or *links* the recovered trajectories to unknown though valid combinations of basis trajectories. Depending on the linking strength, recovered trajectories will eventually more or less accurate resemble valid combinations of basis trajectories.

Finally, the regularisation term imposes a temporal coherence constraint on 3D trajectories of adjacent points. Since the recovered 3D trajectories are parameterised by $\mathbf{A}_k$, the regularisation term can be expressed as

$$\mathbf{E}_{\text{reg}}(\mathbf{A}) = \sum_{n=1}^{N} \sum_{k=1}^{K} \|\nabla \mathbf{A}_{k,n}\|_{\varepsilon}^{2}. \tag{5.35}$$

To compute gradients of trajectory coefficients, Eq. (5.35) requires knowledge of points adjacencies. We compute point adjacency lookup table from the spatial arrangement of the points in the reference frame, if the latter is available.

Our core NRSfM approach is called Consolidating Monocular Dynamic Reconstruction (CMDR), as it unifies constraints in the metric and trajectory space into a single energy functional. At the beginning, $\mathbf{S}$ and $\mathbf{R}$ are initialised under rigidity assumption with [281] on the unaltered point tracks $\mathbf{W}$. α, λ and ρ are usually equivalued, while β is set an order of magnitude lower.

Postprocessing of DSP

After a reconstruction of the representative sequence is accomplished, we obtain L shapes $\mathbf{S}_l^{\sharp}$, $l \in \{1,\dots,L\}$. Next, we build a map of pairs $\chi = (||\mathbf{S}_l^{\sharp}||_{\mathscr{F}}, \mathbf{S}_l^{\sharp})$ where the shapes are arranged in the increasing order of $||\mathbf{S}_l^{\sharp}||_{\mathscr{F}}$. Starting from $\mathbf{S}_1^{\sharp}$, we iteratively include $\mathbf{S}_l^{\sharp}$ into DSP if the norm difference between the current $\mathbf{S}_l^{\sharp}$ and the latest included $\mathbf{D}_i$ exceeds some μ. By varying μ, we can control the cardinality of $\mathbf{D}$. Experimentally, we have observed a strong correlation between $||\mathbf{S}_l^{\sharp}||_{\mathscr{F}}$ values and the corresponding shapes. If two shapes are close to each other, their Frobenius norms are close likewise.

5.2.3 Experimental Evaluation

This section outlines the evaluation methodology and summarises the experimental results. The DSP is implemented in C++ for a single thread. All values are reported for a system with 32 Gb RAM and Intel Core i7-6700K processor with cores running at 4.00GHz on Ubuntu 16.04.3 LTS operating system.

5.2.3.1 Evaluation Methodology

We develop several tests with synthetic and real data for the evaluation of the convergence, accuracy and runtime aspects of DSPR. For DSP reconstruction, we use several NRSfM methods based on different principles , *i.e.*, Variational Approach (VA) [104], Scalable Monocular Surface Reconstruction (SMSR) [27] and the proposed CMDR. VA relies on the framework of variational optimisation, SMSR performs a series of closed-form updates and CMDR is based on non-linear least squares.

Depending on the evaluation scenario, we report different metrics characterising the accuracy of geometry and camera pose estimation. Let $\mathbf{S}'_f$ and $\mathbf{R}'$, $f \in \{1,\dots,F\}$, be the ground truth geometries and camera poses, respectively. As a shape fidelity metric, we report mean root-mean-square error (m. RMSE) for a set of views:

$$e_{3D} = \frac{1}{F}\sum_f \frac{||\mathbf{S}'_f - \mathbf{S}_f||_{\mathcal{F}}}{||\mathbf{S}'_f||_{\mathcal{F}}}, \tag{5.36}$$

where $||\cdot||_{\mathcal{F}}$ is Frobenius norm. Since the camera poses are recovered up to an arbitrary rotation, we find a single optimal corrective rotation $\mathbf{R}^{\sharp}$ aligning the recovered poses and the ground truth camera poses. Thus, for the evaluation purposes we solve the following energy minimisation problem:

$$\min_{\mathbf{R}^{\sharp}} \sum_f \left\| \mathbf{R}'_f - \mathbf{R}^{\sharp}\mathbf{R}_f \right\|_{\varepsilon}, \tag{5.37}$$

with $||\cdot||_{\varepsilon}$ denoting Huber norm with the threshold value $\varepsilon = 1.0$. After applying $\mathbf{R}^{\sharp}$ to all $\mathbf{R}_f$, we compute mean quaternionic error (m. QE) defined as

$$e_q = \frac{1}{F}\sum_f |\mathbf{q}'_f - \mathbf{q}_f|, \tag{5.38}$$

with $|\cdot|$ standing for the quaternion norm. The $\mathbf{q}'_f$ and $\mathbf{q}_f$ are the quaternions corresponding to $\mathbf{R}'_f$ and $[\mathbf{R}^{\sharp}\mathbf{R}_f]$, respectively.

Next, we evaluate the core CMDR approach individually (Sec. 5.2.3.2) and jointly with DSPR (Secs. 5.2.3.3-5.2.3.4). We perform self- and cross-convergence

Table 5.4: M. RMSE of several methods on the synthetic face sequences.

	TB [22]	MP [216]	VA [104]	DSTA [75]	CDF [116]	SMSR [27]	GM [170]	CMDR
seq. A	0.1252	0.0611	0.0346	0.0374	0.0886	0.0304	0.0294	0.0324
seq. B	0.1348	0.0762	0.0379	0.0428	0.0905	0.0319	0.0309	0.0369

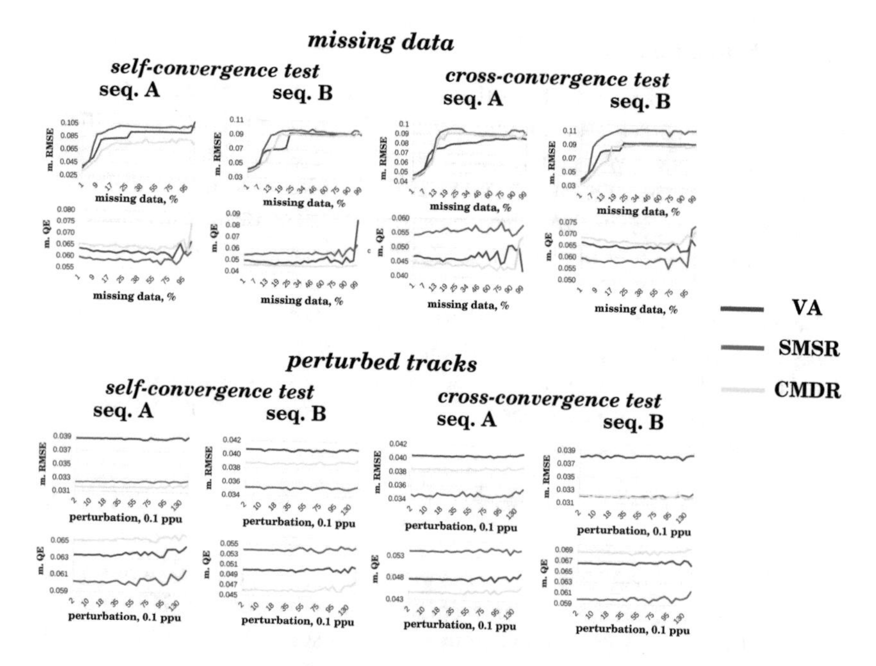

Figure 5.13: Results of the *self-convergence* and *cross-convergence* tests with missing and perturbed data. Three core approaches are tested, *i.e.*, VA [104], SMSR [27] and CMDR (ours). In all experiments, we report m. RMSE and m. QE as the functions of missing data ratio (in %) and perturbation (measured in 0.1 pixels per unit or *ppu*). Missing data is varied in the range $[0.0; 99\%]$ and the perturbation is varied in the range $[0; 15]$ pixels.

tests of DSPR with perturbed and missing data (Sec. 5.2.3.3), MSGD convergence tests (Sec. 5.2.3.4) and joint evaluation of flow and DSRP (Sec. 5.2.3.5). In the *self-convergence test,* DSP is reconstructed on ground truth point tracks, and the same tracks are used for the evaluation, whereas in the *cross-convergence test*, the

Table 5.5: M. RMSE and m. QE for SMSR [27], CMDR (our method for DSP reconstruction) and DSPR on perturbed tracks and tracks with missing entries.

METHOD		MISSING DATA				
		1%	3%	11%	17%	23%
SMSR [27]	m. RMSE	0.1001	0.1778	0.3365	0.4143	0.4849
	m. QE	0.2972	0.2973	0.2968	0.2968	0.2975
plain CMDR	m. RMSE	0.1001	0.1777	0.3365	0.4143	0.4849
	m. QE	0.0663	0.0663	0.0662	0.0663	0.0663
DSPR	m. RMSE	0.0327	0.03578	0.0754	0.0962	0.0994
	m. QE	0.0602	0.05984	0.0584	0.0585	0.0581

METHOD		PERTURBED DATA						
		0.4 px	1.2 px	1.6 px	2.0 px	3.0 px	4.0 px	5.0 px
SMSR [27]	m. RMSE	0.0455	0.0962	0.1243	0.1536	0.2232	0.2956	0.3885
	m. QE	0.2434	0.2999	0.3287	0.2450	0.3068	0.2280	0.3510
plain CMDR	m. RMSE	0.0646	0.1918	0.2541	0.2867	0.3571	0.4056	0.4522
	m. QE	0.0689	0.1077	0.1514	0.1711	0.4617	0.4578	0.4506
DSPR	m. RMSE	0.0324	0.0324	0.0324	0.0324	0.0324	0.0324	0.0325
	m. QE	0.0602	0.0601	0.0600	0.0601	0.0603	0.0600	0.0603

point tracks for the reconstruction of the shape prior and the DSP are different. In Secs. 5.2.3.2–5.2.3.4, we use two 99 frames long synthetic face sequences with known geometry and dense point tracks from [104]. Both A and B sequences originate from the same set of facial expressions. The difference lies in the series of camera poses applied to the interpolated expressions. Due to the different camera pose patterns, the sequences are of different difficulties for NRSfM. As a result, both the geometry and camera poses are always reconstructed differently (in many cases close to each other but not exactly in the same way). Thus, the synthetic face sequences offer an optimal testbed for the cross-convergence test. Finally, we run DSPR on real data and report shape compression ratios (Sec. 5.2.3.6).

5.2.3.2 Evaluation of CMDR Disjointly from DSPR

Although CMDR is evaluated jointly with DSPR in the next sections, we report m. RMSE of CMDR on synthetic face sequences, in comparison to several other methods. Table 5.4 summarises the results. The errors for Trajectory Basis (TB) [22], Metric Projections (MP) [216], VA [104] and Dense Spatio-Temporal Approach (DSTA) [75] are replicated from [75], and the numbers for Coherent Depth Fields (CDF) [116] and Grassmannian Manifold (GM) [170] are taken from the original papers. We recompute m. RMSE for SMSR [27] since the authors report another metric in their paper.

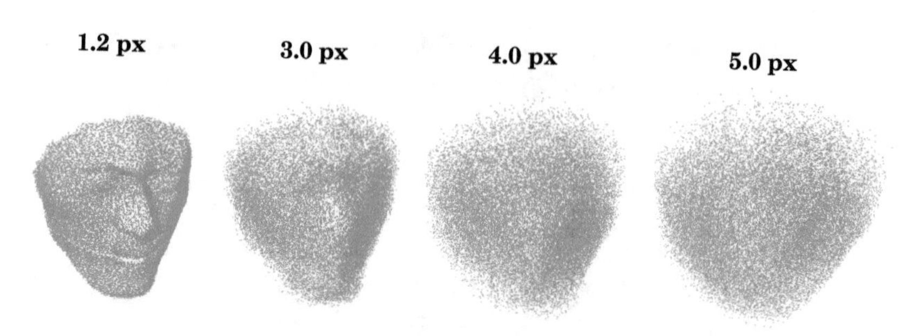

Figure 5.14: Reconstruction results of SMSR [27] on the perturbed point tracks (four different perturbation magnitudes), for frame 11 of *seq. A*.

Our approach is ranked third after GM [170] and SMSR [27], and the gap between the three best methods is narrow. Our m. RMSE is remarkably close to the currently most accurate GM method, even though we designed CMDR based on well-known principles, with a new unique blend and consolidation in the framework of NLLS.

5.2.3.3 Self- and Cross-Convergence Tests

The results of self- and cross-convergence tests with missing data and perturbed tracks are summarised in Fig. 5.13. Next, we analyse each case and visualise convergence patterns.

Missing Data

The amount of missing data is varied in the range $[1;99]\%$. We observe that at 30%, m. RMSE largely stabilises, and m. QE is very stable even with up to 75% of missing data for three cases out of four. This shows that much fewer points are often sufficient to recover the camera pose. In the cross-convergence test for sequence *A*, a 50% threshold is identifiable for two DSP generation methods (VA and SMSR). After surpassing the threshold, standard deviation of m. QE gradually increases, with the exception of CMDR. In the latter case, m. QE is stable across the range of missing data patterns up to 90%.

Perturbed Tracks

In the case of the perturbed data, DSPR is stable and accurate in the whole tested range of [0.1; 15] pixels of uniform perturbance per pixel. Across all experiments and test cases, m. RMSE is kept on the same level of accuracy and is nearly uninfluenced by the perturbations. On the contrary, m. QE is slightly affected by the increasing perturbation amplitude. Still, there is no observable qualitative difference in the estimation of the camera poses.

Altogether, this is an especially notable outcome in light of the results achieved by other methods. Fig. 5.14 congregates selected outcomes of SMSR executed on perturbed point tracks of *seq. A* arranged in ascending order of deterioration. As the perturbation magnitude increases, the point scattering effects are more and more distinct. Already at 3 pixels, the structure becomes barely recognisable. On the other hand, the appearance obtained on the tracks with missing data is reasonable but contains missing entries. Suddenly, with more $23-25\%$ of the missing data, no meaningful structure can be reconstructed by SMSR. Table 5.5 summarises the errors for SMSR and CMDR. The results of the plain CMDR follow the result pattern of SMSR.

In contrast, DSPR operates on the tracks with high amounts of missing data exceeding 25%. Even though the accuracy drops by the factor of $2-3$, the structure remains recognisable. Note that at the same time, the accuracy of the camera pose estimation is only marginally affected.

Convergence Patterns

A *convergence pattern* in DSPR refers to the sequence of states chosen from DSP for every new incoming measurement. In the self- and cross-convergence tests, it is possible to analyse convergence patterns quantitatively because the ground truth state identifiers are known. As a quantitative metric, we use an absolute distance from the chosen DSP state and the ground truth state for inducing the measurements denoted by η. Recall that two face sequences from [104] were used in the self- and cross-convergence tests. They contain 99 frames obtained from ten basis facial expressions by interpolation and differ in the applied to them series of rotations. The sequences are originally referred to as *sequence* 3 and *sequence* 4. Note that convergence pattern is an auxiliary metric as it does not take into account that DSP states can be similar or both well explain 2D observations. Consequently, even if η is large, e_{3D} can be small, and, conversely, large η can imply incorrect convergence and a high e_{3D}.

In Fig. 5.16, convergence patterns for all self- and cross-convergence experiments are visualised. We observe that the convergence pattern of the self-convergence test

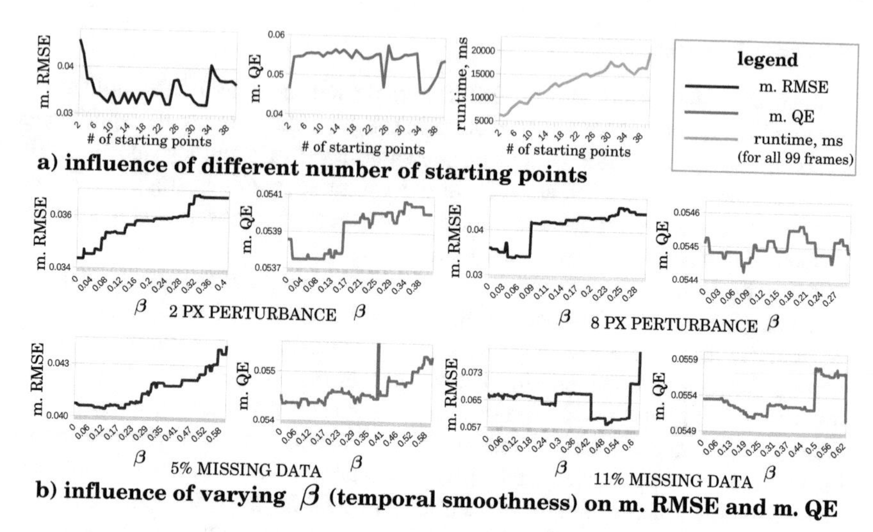

Figure 5.15: Results of the experiments with MSGD parameters: (a) the influence of the different number of starting points is evaluated on the measurements without noise; (b) the influence of β is evaluated on perturbed tracks and tracks with missing data with 20 starting points. In both cases, the reconstructions of seq. *B* was taken as a DSP and the clean tracks of seq. *A* are taken as the incoming dense point tracks.

mostly contains small η, except for the last ten shapes. The reason is that the last ten shapes are similar to other shapes of the interpolated sequence (*e.g.*, shape 94 is similar to shape 71), DSPR chooses a state with a higher η as a solution, and both states lead to an equally small e_{3D}.

Convergence pattern of the cross-convergence test is slightly different, as the structure is observed in different poses. Larger η at frame 15 and in the vicinity show that the convergence of DSPR is dependent on the accuracy of camera pose estimation. Moreover, note the differences in the convergence patterns due to either increasing perturbation magnitude or the missing data ratio. Some shapes are more sensitive to the disturbing effects compared to the others which can be explained by a decaying resolvability, *i.e.*, the method's ability to distinguish between the shapes.

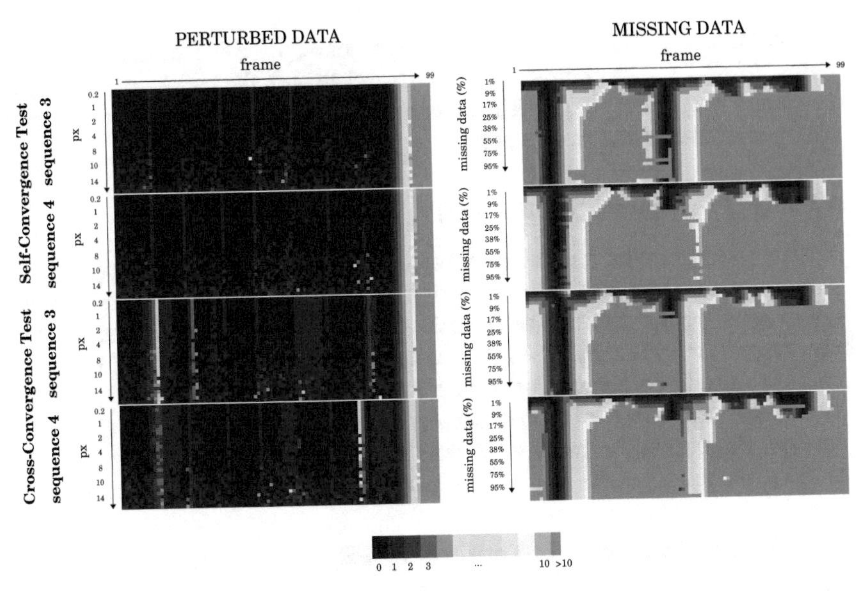

Figure 5.16: Convergence patterns observed in the self- and cross-convergence tests. Every square stands for the η shape distance for a given frame (x-axis) and type of the noise (y-axis, columnwise depending on the type of noise). The colour coding scheme for η is provided beneath.

5.2.3.4 Influence of the MSGD Parameters

We evaluate the influence of the different number of starting points in MSGD and perform an exemplary study for finding optimal weights for the data and the temporal smoothness terms. Therefore, we fix α and vary β (in the range $[0.0; 0.63]$ with the step $2 \cdot 10^{-3}$) under different number of MSGD starting points (in the range $[2; 40]$). The results of the experiment are summarised in Fig. 5.15 and elaborated in Secs. 5.2.3.4–5.2.3.4.

Varying Number of MSGD Starting Points

As expected, the runtime increases with the increasing number of starting points, and the dependency is close to a linear, see Fig. 5.15-(a). Starting from 6 sec. for two points, the runtime increases to 20 sec. for 40 points (for all 99 frames). In the region between 10 and 25 starting points, m. RMSE reaches the smallest value. In this region, we observe oscillations of the growing period and amplitude caused by regular shifts of the starting points and different convergence due to different camera poses. M. QE, on the contrary, does not correlate with the pattern of m. RMSE much and keeps at ca. 0.055. The latter phenomenon stems from the decoupled nature of geometry and camera poses.

Varying β (Temporal Smoothness)

Next, we vary β under four different types of noise — 2 and 8 pixels of uniform perturbances and 5% and 11% of missing data (Fig. 5.15-(b)). Under slight disturbances (2-pixels perturbance and 5% of missing data), m. RMSE and m. QE vary only slightly. The lowest errors are reached with a small β. By and large, the errors are smaller for the case of smaller disturbances. For 8 pixel perturbance and 11% of missing data, the optimal metrics are achieved with a larger β ($\beta \approx 0.05$ for 8-pixels perturbance and $\beta \approx 0.5$ for 11% of missing data) suggesting that the temporal smoothness term is gaining effectiveness for more noisy point tracks.

5.2.3.5 Joint Evaluation of Flow and DSPR

We perform a joint evaluation of DSP generation, the influence of optical flow and DSPR on the adapted *actor mocap* sequence of one hundred frames with $3.5 \cdot 10^4$ points in each shape. The actor mocap sequence is a synthetic dataset originating from real facial motion and deformation capture data of Valgaerts *et al.* [289]. It contains ground truth geometry, camera poses, corresponding rendered images, a reference image with the face segmentation mask and ground truth multi-frame optical flow (MFOF), *i.e.,* a series of optical flows between the reference frame and every other frame in the sequence. In our modification, we rotate the ground truth surfaces and project them into an image plane by ray tracing to render the images and the mask. The ground truth MFOF is obtained as the distances between the projections of the corresponding points in the image plane.

In addition to the ground truth MFOF, we compute dense correspondences by the optical flow method of Sun *et al.* [263] in the pairwise manner, as well as global MFOF with point trajectory regularisation over the whole batch [105]. The average endpoint error (AEPE) of two-frame optical flow (TFOF) and MFOF amount to 1.218 and 1.123, respectively. Next, we evaluate DSPR with ground truth flow, TFOF and MFOF measurements while using as the basis for DSP either ground truth geometry or reconstructions obtained by CDF [116] on the MFOF point tracks. In both cases, DSP contains 65 states after compression. Fig. 5.17 shows exemplary input frames and different types of flow fields; the associated table summarises the results. We see that the errors achieved with MFOF are slightly and consistently

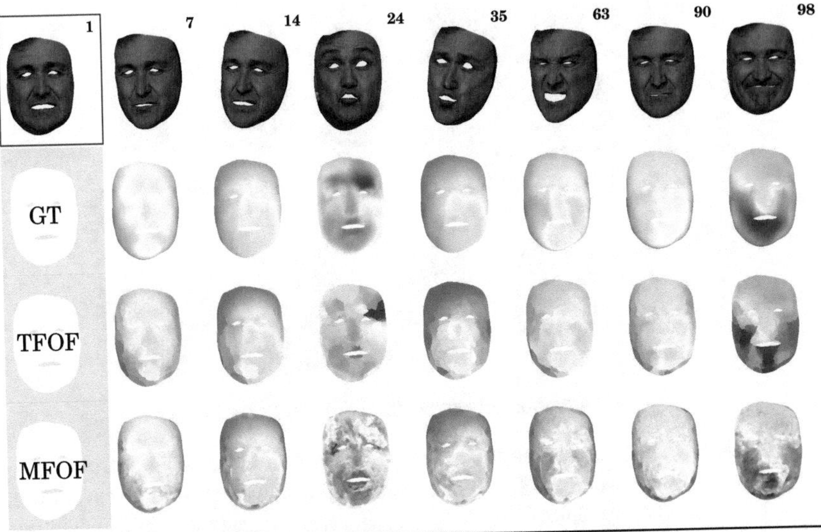

	GT shapes as DSP			CDF shapes as DSP		
	GT flow	TFOF	MFOF	GT flow	TFOF	MFOF
m. RMSE	0.00027	0.00042	0.00041	0.00525	0.00527	0.00527
m. QE	0.09136	0.09382	0.09271	0.12977	0.15017	0.13508

Figure 5.17: Exemplary frames from the adapted *actor mocap* sequence (first row), corresponding ground truth dense flow (second row), flow obtained by the method of Sun *et al.* [263] (third row) and MFOF [104] (fourth row). The table underneath lists errors for all evaluated combinations.

more accurate than those obtained with TFOF. Still, the TFOF errors do not worsen much attesting that DSPR tolerates less accurate and noisy point tracks.

5.2.3.6 Experiments with Real Data and Applications

We perform several tests with real data on the *face* [104], *back* [237], *heart bypass surgery* [259] and some newly recorded face sequences.

Apart from the monocular non-rigid reconstruction, several other modes of operation are conceivable for DSPR. First, if we rerun DSPR on the point tracks which are used to compute DSP, we obtain a compressed version of the reconstruction.

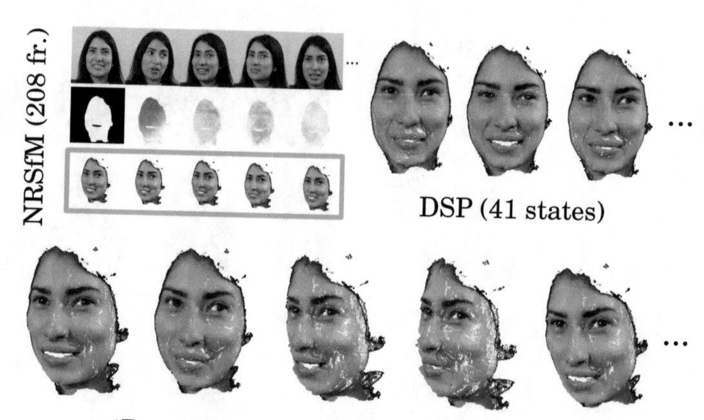

Figure 5.18: The input image sequence, dense point tracks and unrotated reconstructions for DSP estimation (top left), several states from the DSP (top right) and the reconstructions with DSPR of the displayed images (the bottom row).

With the increasing density and the number of views, the space required for storage of a dynamic reconstruction grows fast. Especially in embedded systems and mobile devices, the limits on the data transmission bandwidth can become noticeable. Hence, compression of dynamic reconstructions is of high practical relevance. In the compression mode, we need to save a DSP, a shape prior identifier and camera pose for every frame. This adds up to 12 bytes for camera pose in the axis-angle representation and one-two bytes for the shape prior identifier. Second, we are free to mix the sources of the DSP and incoming measurements. By computing correspondences between a reference frame of one sequence and frames observing a similar scene from another sequence, we can reproduce or reenact 3D deformation states as if they were another scene. Blending of different types of scenes is also possible if an explicit (in many cases, a complex) set of rules is defined allowing to relate both scenes.

Fig. 5.18 shows DSPR results on a new face sequence recorded by the CANON EOS 500D camera. The sequence contains 208 frames of the resolution 600×600 pixels in the representative sequence. From the initial 208 reconstructions obtained on MFOF [105] point tracks, we extract 42 DSP states with $\mu = 10.1$. DSPR runs on the representative subsequence and further frames with three frames per second.

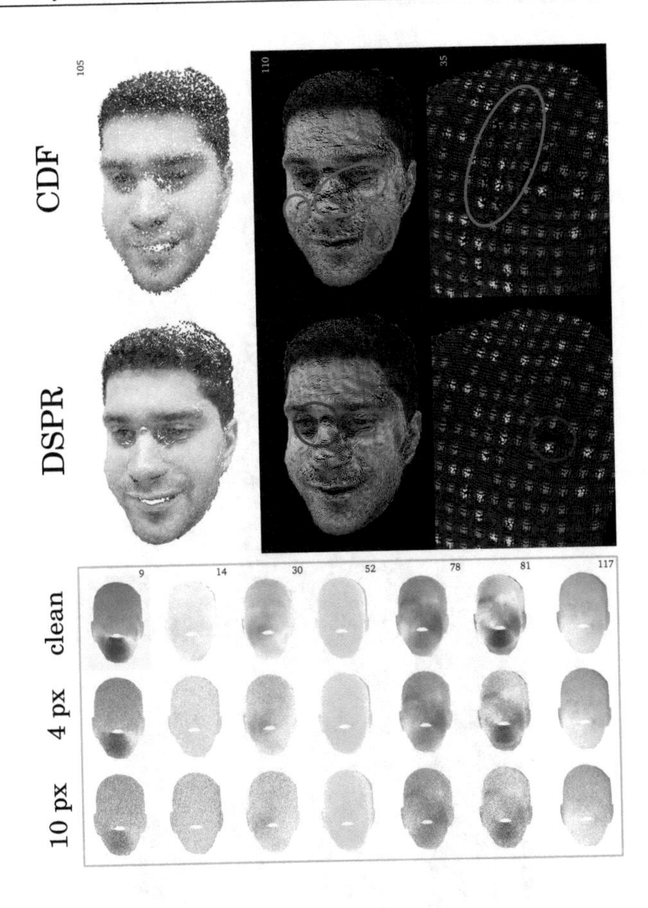

Figure 5.19: Exemplary reconstructions of CDF [116] and DSPR on noisy point tracks with 10.0 pixels of perturbation magnitude (top left) and the comparison of compressed reconstructions (top right). Compression artefacts highlighted in red are more pronounced for CDF, even though it achieves a $2.34 - 2.65$ times smaller compression ratio. The examples of clean and noisy point tracks with 4 and 10-pixels perturbance magnitudes are shown on the bottom. The blue circles emphasise surface areas with artefacts due to dense tracking (in contrast to the compression artefacts).

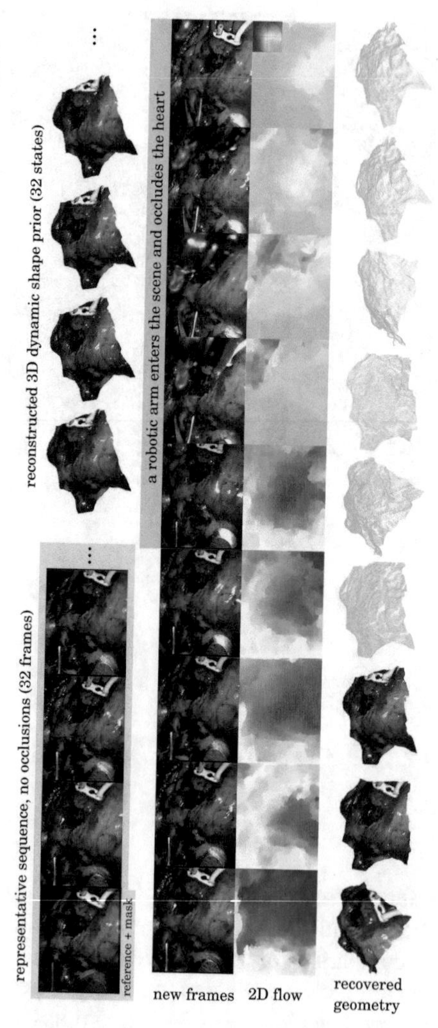

Figure 5.20: Application of DSPR in heart bypass surgery with pronounced reoccurring deformations. The representative sequence and exemplary DSP states are shown in the top row. The new incoming frames and calculated flow fields visualised with Middlebury colour scheme [29] (the colour wheel is attached to the rightmost flow) are given underneath. Our reconstructions from different perspectives are displayed in the bottom row.

Table 5.6: The summary of the experiment for the compression of dynamic reconstructions with the achieved compression ratios.

μ	# of MSGD seeds	*face* [104]		*back* [237]	
		|DSP|	C	|DSP|	C
1.5	20	73	1.64	67	2.23
2.5	20	59	2.03	57	2.63
5	20	42	2.85	40	3.75
10.0	12	25	4.8	25	6
20.0	8	15	8	16	9.375
30.0	5	10	12	11	13.63
40.0	4	8	15	9	16.6

Application of DSPR to Shape Compression

We compare DSPR and CDF [116] — which is explicitly designed for compressible representations — on the task of shape compression. Therefore, we use MFOF [105] point tracks of *face* and *back* sequences, and extra prepare perturbed measurements of the *face* sequence. For the latter, CDF achieves compression ratio $C = 7.0$ on clean tracks. On the noisy tracks with the perturbation magnitudes of 4.0 and 10.0 pixels, its C decays to 3.0 and 1.582, respectively (with $\varepsilon = 1.6E\text{-}3$). DSPR reaches $C = 8.0$ for $\mu = 20.0$ under 10.0 pixels of perturbations. If DSP is computed on clean reconstructions, the compression ratio is only weakly affected by the noise in point tracks, and only slight qualitative differences can be noticed. On the *back* sequence, CDF achieves $C = 4.0$ (with $\varepsilon = 8E\text{-}4$). At the same time, DSPR converges at $C = 9.375$ with $\mu = 20.0$. Recall that for DSPR, the longer an image sequence is, the higher are the compression ratios. The quality of compressions depends on how accurate the representative sequence for DSP generation reflects the state variety encountered in the interactive mode. Fig. 5.19 shows comparisons of reconstructions obtained by CDF and DSRP. The left column shows the resulting states on noisy tracks (10.0 pixels of perturbation magnitude). As also can be seen, especially with high compression ratios CDF causes noticeable compression artefacts (comparisons in the middle and on the right side on top). For DSPR, table 5.6 reports all combinations of the tested thresholds μ, corresponding DSP cardinalities, number of MSGD seeds and attained compression ratios for the *face* and *back* sequences.

DSPR in Heart Bypass Surgery

One scenario with temporally-disjoint state reoccurrence well-addressed by DSPR is a dynamic reconstruction of a beating human heart during an open heart bypass surgery. At the beginning of the sequence provided by Stoyanov [259], the heart in unoccluded for 97 frames, followed by a robotic arm entering the scene and partially occluding the heart for 43 frames. While contracting, the heart undergoes a series of repetitive deformations. This scene is cooperative and offers optimal conditions for DSPR. The course of the experiment is illustrated in Fig. 5.20. First, we compute MFOF on the first 32 frames — this duration corresponds to one complete cardiac cycle — and reconstruct a DSP with 32 states and $68k$ points per state. Note that the states during diastole (refilling) and systole (contraction) are different, and, hence, we *do not* perform state compression. Next, we compute TFOF [263] between the reference frame and every remaining frame, and execute DSPR achieving five frames per seconds on our system. We observe that the reconstruction follows the cardiac cycle *even if the robotic arm partially occludes the heart.*

5.3 Conclusion

In this chapter, we investigated static and dynamic shape priors obtained on the fly for NRSfM. In Sec. 5.1 we proposed the SPVA framework — a new approach for dense NRSfM which is able to handle severe occlusions. Thanks to the shape prior term, SPVA penalises deviations from a meaningful prior shape. The highest supported granularity is per frame per pixel. The shape prior is automatically obtained on the fly from several non-occluded frames under non-rigidity using the total intensity criterion. The new approach does not require any predefined template or a deformation model. Along with that, we analysed relation to the template-based monocular reconstruction and came to the conclusion that SPVA can be considered as a hybrid method. A new evaluation methodology was introduced allowing to jointly evaluate correspondence computation and non-rigid reconstruction. Experiments showed that the proposed framework can efficiently handle scenarios with large permanent occlusions. The SPVA pipeline outperformed the baseline occlusion-aware MFOF+VA in terms of accuracy and runtime.

In all experiments, the proposed method performed more accurate on uncorrected correspondences as the base scheme, with acceptable added runtime. In some cases, the proposed approach can even produce more realistic dynamic reconstructions, as in the case of the *heart surgery* and *ASL* sequences. An advantage compared to Taetz *et al.* [267] is that the correction of inaccuracies is not restricted to a predefined procedure based on Bayesian inference. Different methods can be used to generate occlusion tensor, also integrating prior knowledge about a scene. At the same time, the proposed pipeline is faster and more suitable for online operation in scenarios with severe occlusions. Moreover, the proposed scheme can enhance the accuracy of reconstructions when the occlusion-aware MFOF [267] fails to correct correspondences or correspondences cannot be computed anew.

A limitation of the proposed method lies in its pipeline nature — it can recover from the inaccuracies in the pre-processing steps only up to a certain degree. If the shape prior is corrupted, the overall accuracy can be deteriorated. Future work considers an extension to handle perspective distortions and a search for an optimal operation scheme for interactive processing.

In Sec. 5.2, we introduce a new method for hybrid NRSfM which can be classified as shape-from-DSP. For the first time, temporally-disjoint effects are investigated in *dense* monocular non-rigid reconstruction. In the first step, we reconstruct a representative set of views and generate DSP. Next, for new incoming dense point tracks, we solve a light-weight optimisation problem with a zero-norm which selects the closest shape from DSP while positioning it as observed in the measurements. Robustness to inaccurate correspondences (due to disturbing effects of different origins in the images such as occlusions or brightness inconsistency), the possibility to use faster and less accurate dense flow fields, the highest compression ratios as well as the suitability of the proposed technique for challenging medical applications with repetitive deformations significantly broaden the scope of modern NRSfM. We have shown experimentally that DSPR successfully bridges the gap between the accuracy of dense correspondences and the accuracy of reconstructions.

There are multiple future work directions. We are currently building a layer on top of DSRP for multimedia and medical applications. Further relevant medical scenarios are endoscopic therapies in the vicinity of the lungs (due to the periodic movements caused by breathing). DSPR can be used on a low-power consumption device such as augmented reality glasses for generation of an environmental map with deforming objects. DSP signatures are worth of trying for object class and activity recognition. Furthermore, DSP can be used in the blend shape model fashion if an observation is identified as being likely not well representable by DSP, or provide initialisation for a hierarchical shape refinement.

6 Coherent Depth Fields with High Dimensional Space Model

THIS chapter describes a lightweight NRSfM method with a new spatial regulariser and a new deformation model for NRSfM. The motivation for the new spatial regulariser and the new deformation model come, on the one hand, from handling occlusions, and, on the other hand, from a requirement to be able to compress the reconstructions which is espacially relevant in endoscopic applications as well as for saving the reconstructions.

As has been shown in the previous chapter, handling large occlusions in NRSfM currently requires either an expensive correspondence correction or estimation of a shape prior on several non-occluded views. To save computational cost and remove the dependency on additional pre-processing steps, we introduce the concept of *depth fields*. With the proposed depth fields, NRSfM is interpreted as an alternating estimation of vector fields with fixed origins on the one side, and estimation of displacements of the origins along the depth dimension on the other. The core of the new energy-based Coherent Depth Fields (CDF) approach is the spatial smoothness coherency term (CT) applied on the depth fields. Having its origins in the Motion Coherence Theory, CT interprets data as a displacement vector field and penalises irregularities in displacements. Not only for handling occlusions but also for unoccluded scenes CT has multiple advantages compared to previously proposed regularisers such as total variation.

While existing methods rely on low-rank models, we propose the concept of High Dimensional Space Model (HDSM). In HDSM, time-varying geometry is encoded by a high-dimensional static structure projected into different metric subspaces. To express non-rigid deformations, instead of directly modelling in the 3D space, we gradually increase space dimensionality as the complexity of the scene increases. HDSM allows for a compact representation with deformation localisation and can be interpreted as a generalisation of the previously proposed models for NRSfM. We combine CDF with HDSM and show experimentally that Lifted CDF (L-CDF) achieves state-of-the-art accuracy in dense NRSfM including scenarios with long and large occlusions, inaccurate correspondences as well as inaccurate initialisations, without requiring any additional pre-processing steps. Thanks to HDSM, CDF allows for the fine-grained control and different compression ratios.

V. Golyanik, *Robust Methods for Dense Monocular Non-Rigid 3D Reconstruction and Alignment of Point Clouds*,

In Sec. 6.1, we introduce depth fields and CDF. In Sec. 6.2, we extend CDF with HDSM to L-CDF.

6.1 Depth Fields in NRSfM

6.1.1 Motivation and Significance of Depth Fields

Among real-world challenges in NRSfM are self- and external occlusions naturally arising while observing a scene as well as effects such as specularities or weakly textured areas. There are several approaches to deal with large occlusions. If a template is available, *i.e.,* an accurate reconstruction for at least a single view, occlusions can be handled efficiently [319]. If no template is available, a shape prior can be estimated on-the-fly from dense point correspondences obtained on several unoccluded views. The shape prior can then be used as a constraint for reconstruction of the occluded areas, both with and without available correspondences for the rest of the sequence [116]. Finally, correspondence based NRSfM methods can employ correspondence correction in the pre-processing step [267], although this approach works well for rather short-time disturbances.

6.1.2 Contributions

This Sec. introduces two novel concepts which allow to design a computationally efficient, accurate and easy to implement (practical) approach as well as overcome the dependency on the pre-processing steps in NRSfM while handling large occlusions.

The first concept is the notion of a *depth vector field* or, concisely, a *depth field.* A depth field is a 2D parametrisation of a surface embedded into 3D space so that every tracked 2D point is associated with a displacement along the depth dimension. This definition implies that all displacements are parallel to each other, or, in other words, a *depth field is an irrotational vector field.* Next, we propose *coherency term* (CT) as a new soft spatial regulariser on the adjacent depth vectors. CT derives its origin from the motion coherence theory (MCT) [320, 321] which studies principles of coherent motion and perception. MCT, in accordance to the human visual system states that neighboring structures tend to move coherently, *i.e.,* with a common velocity and direction. We call the proposed approach *Coherent Depth Fields* (CDF) and formulate it as an energy-minimisation problem with CT. The main reason lies in the expressiveness of energy-based methods — an energy functional explicitly encodes assumptions on the underlying physical processes and relates input data with the sought solution. We elaborate *efficient optimisation techniques* involving direct and inverse fast Fourier transforms (FT) and show experimentally that the proposed form of regularisation has several advantages (*e.g.*, ability to filter depth values flexibly without edge oversmoothing) both when dealing with occluded and unoccluded scenes. CDF achieves state-of-the-art accuracy on the joint evaluation benchmark with large occlusions [116], an actor benchmark [36] and several established datasets for qualitative evaluation (with and without occlusions).

6.1.3 Related Work

Given point tracks throughout multiple views, the purpose of NRSfM is to recover the lost depth component of an observed non-rigidly deforming scene. This initial problem statement has motivated us to introduce the notion of depth fields. The proposed interpretation bears a resemblance to Helmholtz decomposition which decomposes an arbitrary vector field into curl-free, divergence-free and lateral components.

The phenomenon of coherent motion was initially studied in visual perception — the Gestalt theory [148, 164]. The seminal works of Yuille and Crzywacz on MCT [320,321] introduced the notions of *coherency* and a Gaussian *radial basis function* in computer vision. Since then, MCT was applied to many tasks such as camera pose and correspondence estimation [178], tracking [256], motion segmentation [307], visual search [150], and extensively influenced non-rigid point set registration [70,157,203,249]. Thus, Coherent Point Drift [203] and offspring methods [19,302] adopt CT for topology-preserving transformations.

Since sparse NRSfM was introduced by Bregler *et al.* [53], several dense approaches emerged [11, 238]. In energy-based formulations, total variation (TV) was shown as an efficient spatial regulariser [104]. TV allows for discontinuities at depth edges while being scale-unaware which is a favourable characteristic in the monocular setting. However, the resulting energy is non-convex and optimisation is performed with a computationally expensive iterative scheme. In contrast, the CDF energy is convex, the method requires fewer operations and is also well parallelisable. The accuracy of the methods [11, 104, 238] degrades considerably when correspondences are obtained on scenes with long and large occlusions.

To overcome this limitation, Golyanik *et al.* proposed a hybrid approach with a shape prior obtained on-the-fly on several non-occluded frames [116]. Guided by an occlusions tensor, the shape prior is used as a depth regulariser in the occluded areas. The main limitation of hybrid NRSfM is the dependency on the accurate occlusion tensor and the shape prior. Taetz *et al.* proposed Bayesian inference framework to stabilise occluded point trajectories [267]. The proposed multi-frame optical flow approach works in two passes and allows to compensate for short-time disturbances. In contrast, *CDF does not require any pre-processing steps and can handle large and long occlusions with weaker assumptions and in less time.*

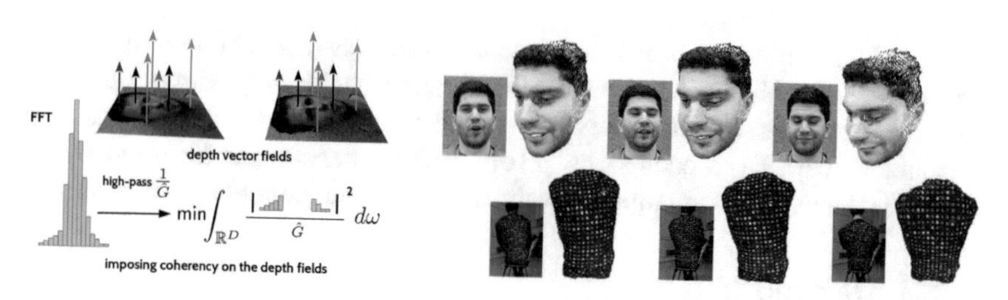

Figure 6.1: Coherency term penalises high-frequency component of a Fourier-transformed vector field, *i.e.*, it favors field homogeneities. CDF reconstructs depth vector fields and uses coherency term as a regulariser. On the right side, our reconstructions of the *face* [104] and *back* [237] sequences are shown.

Occlusions constitute a common reason for missing entries in the measurement matrix and several NRSfM methods can explicitly account for missing data [126, 216, 282]. Since we consider the dense case and track points with dense optical flow techniques [105, 267], measurement matrices are always complete in our case. Nonetheless, due to occlusions, the accuracy of point correspondences degrades. CDF assumes a complete measurement matrix and perhaps inaccurate correspondences.

6.1.4 Coherency Term

Suppose $v = v(x)$ is a displacement function, *i.e.*, for each element $x \in \Theta$ it outputs the corresponding displacement. MCT introduced a smoothing term for v which in a reproducing kernel Hilbert space can be written as a norm of a displacement field [112]:

$$\phi[v] = \int_{\mathbb{R}^D} \frac{|\hat{v}(\omega)|^2}{\hat{G}(\omega)} d\omega, \tag{6.1}$$

where $\hat{G}$ is a Fourier transformed reproducing kernel G, $\hat{v}$ is a Fourier transformed v and ω is a frequency variable. The right side of Eq. (6.1) — which we refer to as *coherency term* (CT) — applies two operators on $\hat{v}$. First, a high-pass filter $\frac{1}{\hat{G}}$ is applied. Second, L^2-norm[1] of the extracted component is taken. As a result, the norm $\phi[v]$ measures the total energy of the function at a high frequency. In other words, the less the function v oscillates, the smaller is the scalar $\phi[v]$ or, likewise,

[1]recall, $\|\psi\|^2_{L^2} = \langle \psi, \psi \rangle_{L^2} = \int |\psi(\omega)|^2 d\omega$.

the more *coherent* are the displacements. Note that CT regularises an extracted high-frequency component of the depth field and approaches low-pass filtering in functionality. Fig. 6.1 visualises CT as applied to a depth field, i.e., a vector field arising in the proposed CDF (see Sec. 6.1.5).

6.1.5 Coherent Depth Fields (CDF) Approach

Given coordinates of N points tracked throughout an image stream, CDF aims at recovery of 3D surface geometry of the observed scene $\mathbf{S}(t)$ and camera poses $\mathbf{R}(t)$. Suppose coordinates of the tracked points over F frames are stacked together row-wise in a measurement matrix

$$\mathbf{W}_{2F\times N} = \left[\begin{bmatrix} u_1^t & v_1^t & \dots \end{bmatrix}^\mathsf{T} \begin{bmatrix} u_2^t & v_2^t & \dots \end{bmatrix}^\mathsf{T} \dots \right], \tag{6.2}$$

with $t \in \{1,\dots,F\}$. Note that further on, the discrete and continuous notations $\mathbf{R} = \mathrm{D}\,\mathbf{R}(t)$, $\mathbf{S} = \mathrm{D}\,\mathbf{S}(t)$ and $\mathbf{W} = \mathrm{D}\,\mathbf{W}(t)$ are used interchangeably; D denotes a discretisation operator. Without loss of generality, we use an orthographic camera model and assume that the translation in the scene is resolved.

The observations $\mathbf{W}(t)$ are caused by imaging of the deformable geometry $\mathbf{S}(t) = [\mathbf{S}_1\mathbf{S}_2\dots\mathbf{S}_F]^\mathsf{T}$ by orthographic camera $\mathbf{R}(t)$:

$$\mathbf{W}(t) = \mathbf{R}(t)\,\mathbf{S}(t) = \begin{pmatrix} \mathbf{R}_1 & & \\ & \mathbf{R}_2 & \\ & & \ddots \end{pmatrix} \begin{pmatrix} \mathbf{S}_1 \\ \mathbf{S}_2 \\ \vdots \end{pmatrix}. \tag{6.3}$$

Note that we impose orthonormality on $\mathbf{R}_t$ matrices. NRSfM parameterised by a depth field is given by the energy functional

$$\mathbf{E}(\mathbf{R},\mathbf{S}) = \frac{1}{2}\|\mathbf{W}(t) - \mathbf{R}(t)\mathbf{S}(t)\|_{\mathscr{F}}^2 + \frac{\lambda}{2}\int_{\mathbb{R}^2} \frac{|\hat{\mathbf{S}}(s)|^2}{\hat{G}(s)} ds \tag{6.4}$$

$$\text{s.t. } \operatorname{rank}(\mathrm{P}(\mathbf{S})) = \tau, \tag{6.5}$$

where $\mathscr{F}$ stays for Frobenius or Schatten 2-norm, $\hat{\mathbf{S}}(s)$ denotes the Fourier transformed shape, $\hat{G}$ is a Fourier transformed reproducing kernel, s is the frequency domain variable, and operator $\mathrm{P}(\cdot)$ entangles rows of every submatrix $\mathbf{S}_i$ through reordering them in a single row. The right side of Eq. (6.28) contains the data term defined as elementwise 2-norm of the reprojections $\mathbf{R}(t)\,\mathbf{S}(t)$ and CT as a spatial regulariser. The rank-constraint expresses the assumption about the complexity of the deformations and steadily insures that at most τ shapes are linearly independent. This form of CT requires a detailed explanation.

Recall that 1) the number and ordering of the points in every frame are equal and 2) coordinates of all tracked points can be backprojected to the reference frame. In the ideal case, where there are no tracking inaccuracies in the pre-processing step, the objective of NRSfM — to reconstruct complete 3D coordinates of every point — can be simplified to the recovery of missing depths z only. Geometrically, this means that a static set of points induces a time-varying depth vector field $\mathfrak{X} = \mathfrak{X}(u, v, t)$ *with fixed origins*. We call such a depth vector field regularised by CT *coherent depth field*. Accordingly, we name the new algorithm CDF which emphasises the interpretation of data and the smoothness term. The concept of coherent depth field is visualised in Fig. 6.1. In other words, every $\mathbf{S}(t)$ can be comprehended as a vector field $\mathbf{S}\colon \mathbb{R}^2 \to \mathbb{R}$. Thus, the term $\frac{\lambda}{2}\int_{\mathbb{R}^2} \frac{|\hat{\mathbf{S}}(s)|^2}{\hat{G}(s)} ds$ imposes coherency on the neighboring elements of $\mathfrak{X}$ or *depths* (for the basic interpretation of CT cf. Sec. 6.1.4). It it worth nothing that in the real case, the origins of the depth fields can drift due to tracking inaccuracies — the same principles apply.

We minimise (6.28) by alternately fixing $\mathbf{S}$ and $\mathbf{R}$ and minimising for $\mathbf{R}$ and $\mathbf{S}$, respectively. While $\mathbf{S}$ is fixed, only the data term depends on $\mathbf{R}$. This subproblem can be efficiently solved in a closed form by projecting an affine update of the rotation matrix into the SO(3) group by normal equations, i.e, $\mathbf{R} = \mathbf{W}\mathbf{S}^{\mathsf{T}}(\mathbf{S}\mathbf{S}^{\mathsf{T}})^{-1}$ which is of comparable accuracy and faster than more computationally expensive non-linear optimisation schemes.

While $\mathbf{R}$ is fixed, the problem in Eq. (6.28) is convex in $\mathbf{S}$[2]. Nevertheless, as different norms are used in the data and smoothness terms, it can not be easily solved in a standard way (*e.g.*, by directly applying Euler-Lagrange differential equation). The problem is remedied by proximal splitting — through introduction of an auxiliary variable $\bar{\mathbf{S}}$ we split the problem in two subproblems. The original problem is thus equivalent to

$$\underset{\mathbf{S}}{\operatorname{argmin}} \frac{1}{2\theta}\|\mathbf{S} - \bar{\mathbf{S}}\|_{\mathscr{F}}^2 + \frac{\lambda}{2}\int_{\mathbb{R}^2} \frac{|\hat{\mathbf{S}}(s)|^2}{\hat{G}(s)} ds, \tag{6.6}$$

$$\underset{\bar{\mathbf{S}}}{\operatorname{argmin}} \frac{1}{2\theta}\|\mathbf{S} - \bar{\mathbf{S}}\|_{\mathscr{F}}^2 + \frac{1}{2}\|\mathbf{W} - \mathbf{R}\bar{\mathbf{S}}\|_{\mathscr{F}}^2 \tag{6.7}$$
$$\text{s.t. } \operatorname{rank}(P(\bar{\mathbf{S}})) = \tau$$

and solved through alternating optimisations of the functionals (6.30) and (6.31). (6.30) updates (filters) the depth field, and the x and y coordinates are fixed in this step. Given the updated depth field, (6.31) revises the complete shapes. We

[2] the low-rank constraint in Eq. (6.5) — which makes the whole minimisation objective non-convex when considered jointly — is imposed in a separate step after minimisation of Eq. (6.28).

reformulate the functional (6.30) as

$$\frac{1}{2\theta}\int_{\Omega}|\mathbf{S}(x)-\bar{\mathbf{S}}(x)|^2dx+\frac{\lambda}{2}\int_{\Omega}\frac{|\hat{\mathbf{S}}(s)|^2}{\hat{G}(s)}ds, \tag{6.8}$$

where Ω is the set of points considered for reconstruction (fixed depth field origins) and $x\in\Omega$. Next, FT of $\mathbf{S}$ is performed leading to

$$\frac{1}{2\theta}\int_{\Omega}\left|\int_{\mathbb{R}^2}\hat{\mathbf{S}}(s)e^{2\pi i\langle x,s\rangle}ds-\bar{\mathbf{S}}(x)\right|^2dx+\frac{\lambda}{2}\int_{\Omega}\frac{|\hat{\mathbf{S}}(s)|^2}{\hat{G}(s)}ds\,. \tag{6.9}$$

The energy in Eq. (6.9) is optimised w.r.t. $\hat{\mathbf{S}}$ whilst $\bar{\mathbf{S}}$ is fixed. To find the minimum, we take the partial derivative of $\mathbf{E}(\hat{\mathbf{S}})$ w.r.t. $\hat{\mathbf{S}}(t)$ and equate it to zero:

$$\frac{\partial E(\hat{\mathbf{S}})}{\partial\hat{\mathbf{S}}(t)}=\frac{1}{\theta}(\mathbf{S}(t)-\bar{\mathbf{S}}(t))\int_{\Omega}\frac{\partial\hat{\mathbf{S}}(s)}{\partial\hat{\mathbf{S}}(t)}e^{2\pi i\langle x,s\rangle}ds+\frac{\lambda}{2}\int_{\Omega}\frac{\partial\hat{\mathbf{S}}(s)^2}{\partial\hat{\mathbf{S}}(t)}\frac{1}{\hat{G}(s)}ds\overset{!}{=}0 \tag{6.10}$$

$$\Longrightarrow\ \frac{1}{\theta}(\mathbf{S}(t)-\bar{\mathbf{S}}(t))e^{2\pi i\langle x,t\rangle}+\lambda\frac{\hat{\mathbf{S}}(-t)}{\hat{G}(t)}\overset{!}{=}0. \tag{6.11}$$

Note that we introduce a new Fourier-space variable t to express a different integration area in contrast to those associated with the variable s. After inverse FT is applied, the multiplication alters to a convolution:

$$\mathbf{S}(t)=\frac{1}{\lambda\theta}G(t)*(\bar{\mathbf{S}}-\mathbf{S})(t). \tag{6.12}$$

This convolution equation is subsequently solved w.r.t $\mathbf{S}$, and the solution reads

$$\mathbf{S}=\mathscr{F}^{-1}\left(\mathscr{F}(\bar{\mathbf{S}})\circ\frac{\mathscr{F}(G)}{\lambda\theta 1_{m\times n}+\mathscr{F}(G)}\right), \tag{6.13}$$

where $1_{m\times n}$ is an all-ones matrix and $\circ$ is elementwise multiplication. The quotient on the right side of Eq. (6.32) is the resulting depth field filter (approaching the low-pass).

Next, we minimise (6.31) w.r.t $\bar{\mathbf{S}}$ whilst $\mathbf{S}$ is fixed. Therefore, we find the gradient w.r.t. $\bar{\mathbf{S}}$:

$$\nabla_{\bar{\mathbf{S}}}=(\frac{1}{\theta}+\mathbf{R}^{\mathsf{T}}\mathbf{R})\bar{\mathbf{S}}-(\mathbf{R}^{\mathsf{T}}\mathbf{W}+\frac{1}{\theta}\mathbf{S}). \tag{6.14}$$

The minimiser is obtained by demanding $\nabla_{\bar{\mathbf{S}}}\overset{!}{=}0$ as

$$\bar{\mathbf{S}}'=(\frac{1}{\theta}+\mathbf{R}^{\mathsf{T}}\mathbf{R})^{-1}(\mathbf{R}^{\mathsf{T}}\mathbf{W}+\frac{1}{\theta}\mathbf{S}). \tag{6.15}$$

To fulfil the rank constraint, the suboptimal $\bar{\mathbf{S}}'$ obtained by Eq. (6.15) is projected onto the subspace of τ-rank matrices using svd. Suppose

$$U\Sigma V^{\mathsf{T}} = \text{svd}\left(\mathrm{P}\left(\left(\frac{1}{\theta} + \mathbf{R}^{\mathsf{T}}\mathbf{R}\right)^{-1}\left(\mathbf{R}^{\mathsf{T}}\mathbf{W} + \frac{1}{\theta}\mathbf{S}\right)\right)\right). \tag{6.16}$$

The solution to the problem in Eq. (6.31) reads

$$\bar{\mathbf{S}} = \mathrm{P}^{-1}\left(U\Sigma_{\tau}V^{\mathsf{T}}\right), \tag{6.17}$$

where Σ_{τ} is the truncated diagonal matrix Σ with τ largest elements (singular values) preserved and zeroes otherwise. Once $\bar{\mathbf{S}}$ is recovered, $\mathbf{S}$ can be updated according to Eq. (6.32).

CDF expects $\mathbf{W}$ and four parameters (λ, θ, τ and σ — the variance of the Gaussian kernel) as an input. The entire algorithm is summarised in Alg. 3. An expensive part is $\mathbf{S}$ computation of $\mathcal{O}(FN \log N)$ complexity in Eq. (6.32) — it requires an FT, an inverse FT and an element-wise multiplication. Fortunately, it can be accomplished efficiently on a GPU. Otherwise, $\mathbf{R}^{\mathsf{T}}\mathbf{W}$ is fully parallelisable and svd is performed on 3×3 matrices twice per alternation. We initialise $\mathbf{S}$ under rigidity with the Tomasi-Kanade approach [281]. Convergence criteria for the inner and outer loops are defined as $\|\bar{\mathbf{S}} - \mathbf{S}\|_{\mathcal{F}} < \varepsilon$ and $\mathbf{E}(\mathbf{R},\mathbf{S})^{i} - \mathbf{E}(\mathbf{R},\mathbf{S})^{i+1} < \xi$, respectively; ε and ξ are scalars.

CDF is implemented in C++; external dependencies include `fttw3` library for fast FT and inverse fast FT [98] as well as `eigen3` for operations on matrices [100]. The test platform is composed of 32 GB RAM and Intel i7-6700K CPU running at 4 GHz. The next section describes the evaluation of the proposed approach on synthetic and real data.

Table 6.1: RMSE of VA [104], AMP [121] and the proposed CDF for the actor dataset [36].

method	conf. 1	conf. 2	optimal parameters
VA [104]	0.36762	0.33624	$\lambda = 5 \cdot 10^3, \theta = 10^{-4}$
AMP [121]	1.5058	1.509	$K = 7$
CDF (ours)	**0.20188**	**0.19638**	$\sigma = 4.4$, $\lambda = 0.4, \theta = 10^{-2}$, $\tau = 20$

6.1.6 Experiments

For datasets with ground truth (*flag mocap* [105], *actor mocap* [36], *synthetic face* [104]) we compare per sequence average normalised root-mean square error

Algorithm 3 Coherent Depth Fields

Input: measurements $\mathbf{W}$, parameters λ, θ, τ, σ
Output: time varying non-rigid shapes $\mathbf{S}$

1: **Initialisation:** $\mathbf{S} = \left[\mathbf{S}_r\mathbf{S}_r\ldots\mathbf{S}_r\right]^\mathsf{T}$, where $\mathbf{S}_r$ is factorisation under rigidity assumption [281]
2: **while** not converge **do**
3: **step 1: fix S, update R**
4: $\text{svd}(\mathbf{WS}(\mathbf{SS}^\mathsf{T})^{-1}) = U\Sigma V^\mathsf{T}$
5: $\mathbf{R} = UCV^\mathsf{T}$, where $C = \text{diag}(1,1,\ldots,1,\text{sign}(\det(UV^\mathsf{T})))$
6: **step 2: fix R, update S; initialise $\bar{\mathbf{S}} = \mathbf{S}$**
7: **while** not converge **do**
8: $U\Sigma V^\mathsf{T} = \text{svd}\left(\mathrm{P}\left(\left(\frac{1}{\theta}\mathbf{I}+\mathbf{R}^\mathsf{T}\mathbf{R}\right)^{-1}\left(\frac{1}{\theta}\mathbf{S}+\mathbf{R}^\mathsf{T}\mathbf{W}\right)\right)\right)$
9: $\bar{\mathbf{S}} = \mathrm{P}^{-1}\left(U\Sigma_{trunc}V^\mathsf{T}\right)$
10: $\mathbf{S} = \mathscr{F}^{-1}\left(\mathscr{F}(\bar{\mathbf{S}}) \circ \frac{\mathscr{F}(G)}{\lambda\theta 1_{m\times n}+\mathscr{F}(G)}\right)$ // $\circ$ denotes elementwise multiplication
11: **end while**
12: **end while**

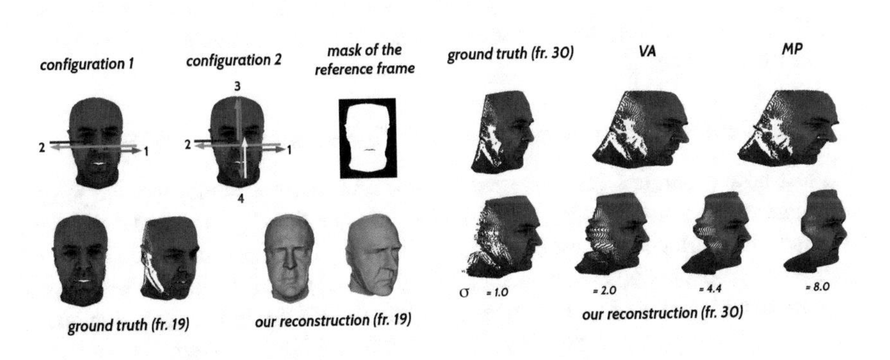

Figure 6.2: For the 3D actor mocap dataset [36], we created ground truth measurements through imaging by a virtual orthographic camera following two trajectories (upper left). Shown are exemplary reconstructions by VA [104], AMP [121] and our CDF on the new sequence. For CDF, we show surface evolution depending on σ (variance of the Gaussian).

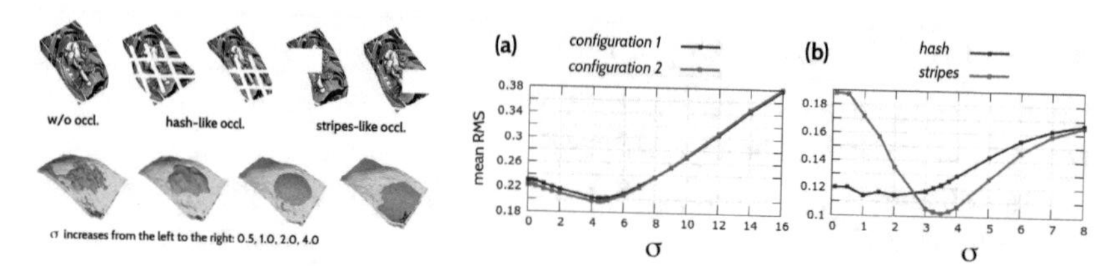

Figure 6.3: (Left): *Flag mocap* sequences with large external occlusions [116]; evolution of reconstructed occluded regions for different σ (reference is shown in green, reconstructions are shown in cyan). (Right): mean RMSE as a function of σ on the (a): *actor mocap*, for two camera trajectories and (b): *flag mocap* with large occlusions (we evaluate how close reconstructions obtained under occlusions approach reconstructions on the accurate tracks).

(RMSE) — defined as $e_{3D} = \frac{1}{F}\sum_{f=1}^{F} \frac{\left\|\mathbf{S}_f^{ref} - \mathbf{S}_f\right\|_{\mathscr{F}}}{\left\|\mathbf{S}_f^{ref}\right\|_{\mathscr{F}}}$ ($\mathbf{S}_f^{ref}$ are ground truth surfaces) — of several approaches supporting dense setting, *i.e.*, VA [104], Metric Projections (MP) [216] or Accelerated Metric Projections (AMP) [121] and the proposed CDF. In doing so, every reconstructed shape is registered to the ground truth with Procrustes analysis.

6.1.6.1 Synthetic Sequences and Joint Evaluation with MFOF

To test how accurate CDF performs on inaccurate correspondences, we jointly evaluate dense point tracking and NRSfM. We use the *flag mocap* dataset with ground truth surfaces, correspondences and rendered images with added *hash* and *stripes* large occlusions patterns (see Fig. 6.3-(left)) [105, 116]. For both cases, several combinations of multi-frame optical flow (MFOF) (either occlusion-aware MFOF or multi-frame subspace flow (MFSF)) and NRSfM methods are tested. The results are summarised in Table 6.2. Reconstructions by AMP and VA on noisy correspondences (columns two and three) exhibit strong depth variations. In contrast, CDF compensates for tracking inaccuracies while not jeopardising the unoccluded parts. Remarkably that RMSE of the combination MFSF [104] + CDF — without additional pre-processing steps — is comparable to RMSE of the computationally expensive MFOF with point trajectory correction [267] + VA [104]. MFOF requires twice to triple the runtime of MFSF, as it improves point trajectories in a separate pass. Next, the waving flag is reconstructed with CDF both on the ground truth and inaccurate point tracks. Using the reconstructions on the ground truth correspondences as a reference, we measured the relative

Table 6.2: Average RMSE on the occluded flag sequences [116].

	MFSF [105] + **AMP** [216]	**MFSF** [105] + **VA** [104]	**MFOF** [267] + **VA** [104]	**MFSF** [105] + **CDF**
hash	0.297 (0.381)	0.239 (0.252)	**0.181 (0.219)**	**0.188 (0.212)**
stripes	0.460 (0.523)	0.341 (0.355)	**0.195 (0.209)**	**0.211 (0.216)**

RMSE for multiple σ values. This test reveals how close reconstructions on inaccurate correspondences due to occlusions are approaching the structure obtained on unoccluded data. Results are plotted in Fig. 6.3-(b).

Likewise, we evaluate CDF in a scenario with inaccurate initialisations. Therefore, we take the 4D *actor* motion capture dataset of Beeler *et al.* [36] and generate measurements by projecting individual 3D shapes by a virtual orthographic camera. Two different camera trajectories are choosen, see Fig. 6.2-(upper left). In the first setting, the camera observes the face frontally and then rotates to the right and left eventually returning to the initial position; in the second setting, the camera follows the right-left-up-down pattern. The movements are more rapid and the amplitude is smaller compared to the first setting (max. 30°). Both sequences contain 51 frames with $3.7 \cdot 10^4$ points each. Facial expressions of the actor are realistic and moderate (there are no exaggerated expressions as a strong cue) and the dataset is particularly challenging for monocular reconstruction. Table. 6.1 summarises the obtained RMSE and Fig. 6.2 contains some exemplary reconstructions. AMP achieves the RMSE of 1.506. VA improves the error by the factor of four and reconstructs more fine details. The test shows that both methods can only lightly recover from the inaccuracy in the initial depth estimation. CDF, starting from the same initialisation through rigid factorisation [281] as VA, achieves the lowest RMSE of 0.202, as it is capable to regularise depth fields. In the second camera setting, all algorithms lessen RMSE consistently over all tested parameter configurations so that the overall placement remains the same. The reason is a richer rotation cue in the scene. Overall, we tested multiple σ values and identified that CDF exhibits well-posedness w.r.t. the parameter choice, unless set too high, see Fig. 6.3-(a). The RMSE percent variance for a suboptimal σ does not exceed 20% in the range $[10^{-3}; 8.0]$ (λ and τ were fixed to 0.4 and 20, respectively, in the course of all experiments).

Additionally, we evaluated CDF on four dense synthetic benchmark face sequences [104]. Several methods [21, 96, 104, 125, 216] were compared on this dataset before [96]. CDF achieves RMSE of 8.03% (an average RMSE for all four datasets). For the sequences 1 and 2 with ten frames each, CDF achieves 7.54% and 6.64%, respectively, and RMSE increases for sequences 3 and 4 (99 frames each) to 8.87% and 9.04%, respectively. Qualitatively, our reconstructions

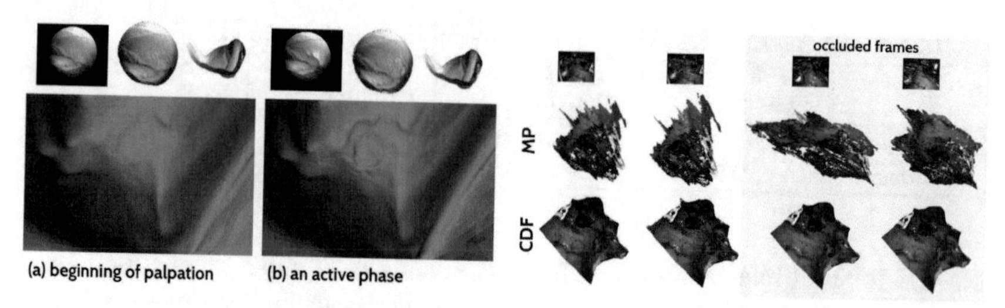

Figure 6.4: (Left): CDF reconstruction of the challenging laparoscopic sequence [17], (a): the palpation of the abdominal area begins, (b): an active phase of the palpation. In both sides, shown are the input image, a corresponding half-spherical reconstruction (frontal and side views) as well as inspection of shaded and zoomed in ROIs. (Right): results on the heart bypass surgery sequence with a robotic arm entering the scene. Our method can reconstruct the heart, AMP [121] is largely affected by the tracking inaccuracies, and VA [104] fails.

exhibit fewer surface fluctuations compared to [21] and [216]. Both VA and CDF rely on accurate initialisations. CDF's robustness against occlusions comes at cost of more flattened depth values in the case without occlusions, compared to VA. Table 6.5 provides a compact comparison of the four methods with average RMSE and standard deviation s for all synthetic face sequences jointly[3].

Table 6.3: Average RMSE and s on the synthetic faces [104] over four sequences jointly.

	TB [21]	**MP [216]**	**VA [104]**	**CDF (ours)**
RMSE / s	9.24 / 5.37	8.81 / 6.15	3.22 / 0.55	8.03 / 0.98

6.1.6.2 Real Sequences

Next, we show qualitative results on several challenging real-world image sequences —*face* (120 frames) [104], *back* (150 frames) [237], *heart* bypass surgery scene occluded by a robotic arm (40 frames) [259] and abdominal *laparoscopic* sequence (120 frames) [17]. Fig. 6.1-(right) shows selected reconstructions of the *face* and *back* sequences. The person's face is reconstructed up to the fine details such as

[3]the average RMSE for TB [21], MP [216] and VA [104] are taken from [104]

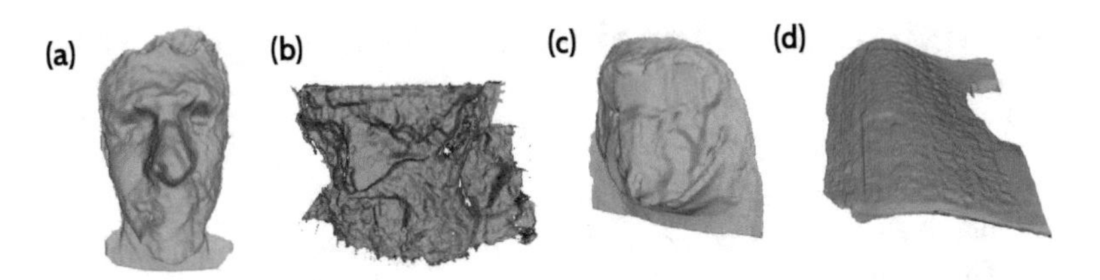

Figure 6.5: Examples of shaded surfaces reconstructed by CDF: (a) *new face*, (b) *shaman2*, (c) *owl* and (d) *notes* sequences.

closing and opening of the eyes. The back exhibits stronger large-scale non-rigid deformations and rotations which are likewise plausibly captured. All in all, both reconstructions appear highly realistic.

Laparoscopic sequence depicts palpation of the abdominal cavity of a rabbit. During palpation the soft tissues are deformed by human fingers and deformations are observed by a compact camera inside the body. Fig. 6.4-(left) shows two selected reconstructions. The moment when almost no pressure on the soft tissues is exerted is shown in Fig. 6.4-(a). In Fig. 6.4-(b), deformations of the tissues are the strongest. CDF successfully reconstructs the scene and provides the means for detailed deformation analysis. After meshing and shading of the resultant point clouds, deformations can be visually identified and analysed in a virtual fly-by along the captured dynamic surfaces.

Selected results on the *heart* sequence are given in Fig. 6.4-(right). In the course of the surgery, a robotic arm enters the scene and occludes up to 50% of the region of interest. Due to inaccurate initialisations, AMP outputs unlikely depth tensions of the structure whereas VA fails to reconstruct recognisable surfaces. Our approach obtains plausible structures, although the geometry differs from those obtained on the *heart* sequence without occlusions[4]. It is noteworthy that 1) unoccluded parts are reconstructed more accurately as one would intuitively expect from CDF and 2) strong occlusions do not ruin surface recovery. Finally, we reconstruct several other real image sequences (*new face* (120 frames) [116], *shaman2* (50 frames) [61], *owl* (202 frames) [82] and *notes* (139 frames) [119]). Exemplary shaded reconstructions are assembled into Fig. 6.5. The runtime of CDF depends on multiple factors (number of frames and number points in a sequence, σ, *etc.*). For the *back* sequence it amounts to 1322 seconds (8 alternations, 10 primal-dual iterations and $\sigma = 3.5$).

[4] the heart sequence *without occlusions* was reconstructed in multiple previous works [96, 104, 116, 267].

6.2 High Dimensional Space Model for NRSfM

6.2.1 Motivation

Templateless monocular surface recovery or NRSfM exploits motion and deformation cues for unsupervised learning of 3D shapes from 2D point correspondences. Three models for NRSfM have been proposed so far, *i.e.*, Low-Rank Space Model (LRSM) [53], Trajectory Space Model (TSM) [22] and Force Space Model (FSM) [13]. The common for all of them is the modelling of shapes or point trajectories as a linear combination of basis elements. Both the recovered shapes and the basis reside in 3D space. It has been shown that LRSM, TSM and FSM are dual to each other [13, 16].

In this section, we propose to model time-varying 3D surface geometry by projecting a multi-dimensional *static* structure into different 3D subspaces. The resulting lifted representation — High-Dimensional Space Model (HDSM) — generalises previously proposed models for NRSfM and allows designing an algorithm with unique properties.

Further on, we refer to this high-dimensional rigid structure $\mathbf{\Phi}$ as *lifted geometry*. The main motivation for HDSM is to stay as close as possible to the raw data representation.

HDSM is tightly related to compressed scene representations which follows from the main idea of HDSM, see Fig. 6.6. The compression in HDSM can be implemented on multiple levels (*e.g.*, frame-to-frame or point trajectory levels in the lifted space). Thus, HDSM is furnished with *lift-compress* and *decompress-expand* operators. Lift-compress allows to pass into the lifted space, detect frame-to-frame redundancies and suppress small local deformations and noise whereas decompress-expand generates human-interpretable 3D representations. The lifted geometry provides cues for deformation localisation (by analysing point trajectory patterns) and segmentation from deformation (segmenting the scene into rigid and non-rigid regions). A coarse-to-fine or wavelet-decomposition-like effect is also theoretically possible when applying multiple lift-compress operators.

To the best of our knowledge, the idea of compressible representations in the context of NRSfM remained unexplored in the literature so far. We believe, however, that especially in the dense case compression is important, as plain representations can occupy gigabytes of memory. In embedded systems, a several-times reduction of transmitted or saved data is particularly relevant. Due to the factorisation into camera pose and compact lifted geometry occurring in our approach, the reconstructed scene can be naturally compressed with the compression ratio of 10 and higher.

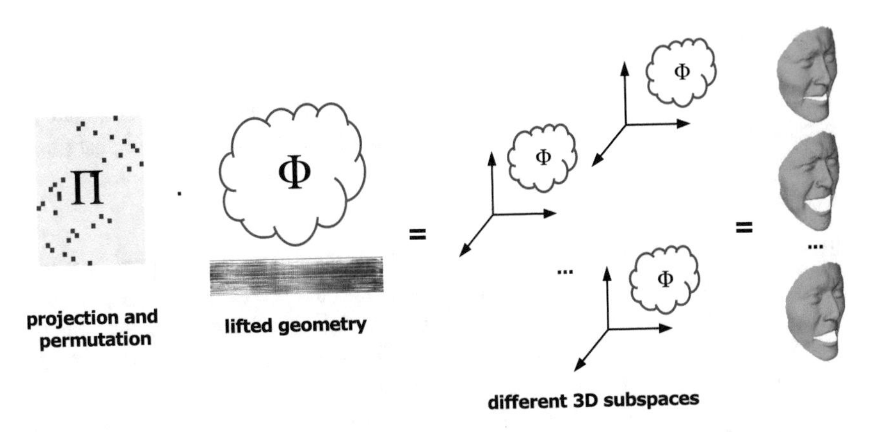

Figure 6.6: An overview of the main idea of the section. We propose to model different 3D states of a non-rigid shape by projecting a high-dimensional structure $\mathbf{\Phi}$ into different 3D subspaces. $\mathbf{\Pi}$ is a permutation matrix which relates projective 3D subspaces with the input frames.

6.2.2 Contributions

The introduction of HDSM and an initial investigation of its properties is the main contribution and focus of this section. We show that HDSM can be interpreted as a generalisation of the previously proposed models for NRSfM (Sec. 6.2.4). Based on HDSM, we design a new energy-based template-free method for monocular surface recovery — Lifted Coherent Depth Fields (L-CDF, Sec. 6.2.5). Our method includes two terms — the data term and the coherency term (CT) — a new spatial regulariser [117]. L-CDF requires a minimal number of priors. The low-rank constraint in 3D space is imposed implicitly, *i.e.,* we do not require any terms accounting for the low-rank nature of deformations in 3D space. We achieve this property by maintaining an optimal non-redundant lifted geometry in every optimisation step.

CT originates in the motion coherence theory [320] and imposes a soft constraint of coherent displacements on adjacent points in the reproducing kernel Hilbert space. Besides the properties inherited from HDSM, L-CDF is well parallelisable, robust and easy to implement due to CT.

6.2.3 Related Work

In this section, related works which concern localised modelling, compact and lifted representations as well as coarse-to-fine recovery in NRSfM are discussed.

6.2.3.1 Localised Modelling

To more accurately model non-linear deformations, Rabaud *et al.* proposed local LRSM where basis shapes are valid for a shape neighbourhood [226]. Related approaches are based on union of subspaces, *i.e.*, different 3D shapes lie in a disjunction of linear subspaces [166, 171, 332]. Localised modelling is another way to deal with complex deformations. A recent approach of Agudo and Moreno-Noguer [15] segments surfaces and models deformations of non-overlapping segments with different LRSM.

A particular type of nonlinearities arises in scenarios with large deformations. Piecewise methods [89,173,273] address this scenario well. Most of them follow the pipeline of splitting the scene into overlapping patches, reconstructing the patches and eventually imposing global policies to assemble individual patches into the meaningful reconstructions. The local methods, however, do not allow deformation localisation in space, as our approach does. Due to the structure compression, we automatically detect regions with the highest deformability and cluster the structure into the rigid (up to the noise level prior) and non-rigid segments. In computer graphics, a similar technique is used to model localised and human interpretable shape deformations. One example is the work of Neumann *et al.* [205].

6.2.3.2 Compressed and Compact Representations

Zhu *et al.* [331] combine LRSM and a compact shape representation in the sense of compressive sensing [95], where every shape is represented as a sparse linear combination of basis shapes (with few nonzero coefficients). Due to the sparsity constraints on the coefficients, the recovered basis shapes were shown to be closer to the elements of the subspace they span. A similar idea is pursued in a more recent work. Assuming 3D structure compressibility, [166] pose NRSfM as a block sparse dictionary learning problem involving ℓ_0-norm constraint on the coefficient vector.

Agudo *et al.* proposed a deformation model with a reduced basis and a direct physical interpretation [12]. Their approach involves spectral analysis on point distance matrices and computes eigenmodes of deformation which are used to model point displacements relative to the estimated shape at rest.

In this section, we mean structure compression in the sense of compact representation and the theory of data compression [244]. Especially in the case of dense reconstructions, the size of the reconstructions can be prohibitive for embedded and interactive applications. In contrast to the sparse settings, this aspect can not be ignored.

6.2.3.3 Coarse-to-Fine Recovery

[33] proposes to estimate basis shapes one at a time imposing the constraint on every new mode to express as much of remaining (more and more high frequent) deformations as possible. The method computes the mean shape and iteratively adds deformation modes. The search for a new mode leads to the energy decrease as much as possible, and this is equivalent to the coarse-to-fine effect. In the proposed L-CDF method, the coarse-to-fine effect is attained through the iterative thresholding. Also, the method can influence the desirable deformation scale. At the same time, we do not have explicit unknowns accounting for the deformation modes.

6.2.3.4 Geometry Lifting

Several approaches employ manifold learning [226, 227, 270] and nonlinear dimensionality reduction (NLDR) [126] techniques. [126] proposed NLDR with a kernel trick. The method defines a nonlinear mapping into a high-dimensional space and uses a set of radial basis functions to locally approximate high non-linearities in 3D. Compared to the manifold-learning and NLDR, a more precise designation for HDSM is *lifting*, and the main motivation is to stay as close as possible to the raw data. HDSM allows to express unions of subspaces and local LRSMs (though, modelling non-linear deformations is not in the scope of this section); frame-to-frame redundancies can also be modelled (and detected by our method automatically) — all by projecting into different 3D subspaces. As applied to NRSfM, the idea of lifted representations remained unexplored in the literature so far, to the best of our knowledge.

6.2.4 High Dimensional Space Model (HDSM)

In this section, we explain the concept of HDSM starting from the rigid orthographic case and subsequently interpret it for the non-rigid case. For the convenience, we provide in Table 6.4 a non-exhaustive list of symbols used in this section. We use the same symbols in Secs. 6.2.4 and 6.2.5 to denote the same structures, though they refer to different mathematical derivations.

Table 6.4: A non-exhaustive list of symbols used in the section.

Symbol/s	Signification
$\mathbf{W}$	dense correspondences, measurement matrix
F	number of frames
N, p	number of tracked points
$\mathbf{R}, \mathbf{S}$	relative camera pose and 3D shape
$\mathbf{\Pi}$, $\mathbf{P}$	permutation matrix
$\mathbf{\Phi}$	lifted geometry (l dimensional)
θ	a scalar threshold (used in iterative thresholding)
$\mathbf{U}$, $\mathbf{\Sigma}$, $\mathbf{V}$	$\text{svd}(\mathbf{X}) = \mathbf{U}\mathbf{\Sigma}\mathbf{V}^{\top}$, $\mathbf{X}$ is an arbitrary matrix
$\mathbf{Q}$	a full rank 4×4 matrix (corrective transformation)
$\mathrm{P}(\cdot)$	permutation or shape dimension entangling operator
$\hat{\mathbf{S}}$	shape in the Fourier space
$\hat{G}$	Fourier-transformed reproducing kernel
s	a Fourier space variable

6.2.4.1 Considerations in the Rigid Case

Rigid factorisation-based structure from motion [281] is the problem of recovering a rigid 3D scene from 2D observations (point tracks). Such reconstruction represents the same shape in a different pose for every frame. Suppose $\mathbf{W}_{2\times p}$ is the measurement matrix with p points per frame, $\mathbf{S}_{3\times p}$ is the observed 3D structure, and $\mathbf{R}_{2\times 3}$ is the orthographic camera matrix. Then, $\mathbf{W}_{2\times p} = \mathbf{R}_{2\times 3}\mathbf{S}_{3\times p}$. Applying the relative camera pose to $\mathbf{S}$ leads to the complete transfer of the pose to the object:

$$\mathbf{S}(t) = \mathbf{R}'_{3\times 3}\mathbf{S}_{3\times p}. \tag{6.18}$$

We can always use Eq. (6.18), because of the ambiguity between the camera and object poses. $\mathbf{R}'$ is obtained from $\mathbf{R}$ by adding a third orthonormal row (a cross product of the first two rows). *Our key observation is that the rotated rigid structure* $\mathbf{S}(t)$ *can be lifted into the 4D space and interpreted as multiple projections of a 4D rigid body* $\mathbf{\Phi}$ *into the 3D space under different angles of view (by different 4D to 3D orthographic projection matrices* $\mathbf{R}_{3\times 4}$ *in this case)*. Formally, we can write:

$$\mathbf{S}(t) = \mathbf{R}_{3\times 4}\mathbf{\Phi}_{4\times p}. \tag{6.19}$$

The rigid 4D shape $\mathbf{\Phi}$ can be found using the similar principles as in the rigid factorisation approach, going from multiple observations in 3D to 4D. A natural question arises, whether it is possible to go from multiple 2D observations directly to a 4D $\mathbf{\Phi}$. The further analysis suggests:

$$\mathbf{W}_{2\times p} = \mathbf{P}_{2\times 3}\mathbf{R}_{3\times 4}\mathbf{\Phi}_{4\times p}, \tag{6.20}$$

where $\mathbf{P}_{2\times 3} = \mathbf{I}_{2\times 3}$ is a projection matrix. Factorising $\mathbf{W}_{2\times p}$ with svd leads to

$$\text{svd}(\mathbf{W}_{2\times p}) = \mathbf{U}_{2\times p}\mathbf{\Sigma}^{\frac{1}{2}}_{p\times p}\mathbf{\Sigma}^{\frac{1}{2}}_{p\times p}\mathbf{V}^{\mathsf{T}}_{p\times p}, \tag{6.21}$$

with $\mathbf{\Phi} = \mathbf{\Sigma}^{\frac{1}{2}}_{4\times p}\mathbf{V}^{\mathsf{T}}_{p\times p}$ and $\mathbf{P}_{2\times 3}\mathbf{R}_{3\times 4} = \mathbf{U}_{2\times p}\mathbf{\Sigma}^{\frac{1}{2}}_{4\times 4}$. From Eq. (6.21) follows that $\mathbf{\Phi}$ is ambiguous. Similar to the 3D case, a corrective transformation $\mathbf{Q}$ is required, since an invertible matrix can be inserted between $\mathbf{M} = \mathbf{PR}$ and $\mathbf{\Phi}$:

$$\mathbf{W}_{2\times p} = \underbrace{\mathbf{P}_{2\times 3}\mathbf{R}_{3\times 4}}_{\mathbf{M}}\mathbf{Q}_{4\times 4}\mathbf{Q}^{-1}_{4\times 4}\mathbf{\Phi}_{4\times p}. \tag{6.22}$$

In the rigid case, recovering a 4D representation does not bring advantages. Eq. (6.22) provides the first evidence in favour of HDSM: *a 4D rigid body can exist so that its projections into the 3D space under different angles can encode rigid transformations of a 3D rigid body.*

6.2.4.2 Considerations in the Non-Rigid Case

In the non-rigid case, HDSM instantiates as follows:

$$\mathbf{W} = \underbrace{\begin{bmatrix} \mathbf{R}^1_{2\times 3} & 0 & \dots & 0 \\ 0 & \mathbf{R}^2_{2\times 3} & \dots & 0 \\ \vdots & \vdots & \ddots & \vdots \\ 0 & 0 & \dots & \mathbf{R}^F_{2\times 3} \end{bmatrix}}_{\mathbf{R}_{2F\times 3F}} \underbrace{\begin{bmatrix} \mathbf{P}^1_{3\times l} \\ \mathbf{P}^2_{3\times l} \\ \vdots \\ \mathbf{P}^F_{3\times l} \end{bmatrix}}_{\mathbf{\Pi}_{3F\times l}} \underbrace{\begin{bmatrix} \mathbf{\Phi} \end{bmatrix}}_{l\times p}. \tag{6.23}$$

In Eq. (6.23), $\mathbf{\Phi}$ denotes the high-dimensional structure with l rows and p points, $\mathbf{P}^i_{3\times l}$ are permutation matrices stacked into the block-matrix $\mathbf{\Pi}$, and $\mathbf{R}^i_{2\times 3}$ are orthographic projection matrices stacked into $\mathbf{R}$ (quasi-block-diagonal matrix).

$\mathbf{P}^i_{3\times l}$ always fetch three rows of $\mathbf{\Phi}$. As long as the structure remains unaltered, $\mathbf{\Phi}$ preserves its dimensionality and $\mathbf{P}^i_{3\times l}$ fetches the same 3D subspace over multiple frames. If deformations occur, $\mathbf{\Phi}$ is expanded, *i.e.*, three additional rows accounting for the new state need to be added to $\mathbf{\Phi}$ in the general case. This can lead to a redundancy as $\mathbf{\Phi}$ can already contain the newly observed structure. On the contrary, most of the newly observed deformations are localised, *i.e.*, they do not affect the whole 3D surface but rather its regions (*e.g.*, some areas can be rigid throughout the entire image sequence). The latter observation allows for further compression of $\mathbf{\Phi}$ (both lossless and lossy).

6.2.4.3 HDSM and Other Deformation Models for NRSfM

Consider the relation between LRSM — the most widely used deformation model for NRSfM — and HDSM. In LRSM, an observed $\mathbf{W}_i$ can be written as [52]

$$\begin{aligned}\mathbf{W}_i = [c_{i1}\mathbf{R}_i\; c_{i2}\mathbf{R}_i\; \ldots\; c_{ik}\mathbf{R}_i][\mathbf{B}_1\mathbf{B}_2\ldots\mathbf{B}_k]^\top = \\ \mathbf{R}_i(c_{i1}\mathbf{B}_1 + c_{i2}\mathbf{B}_2 + \ldots + c_{ik}\mathbf{B}_k) = \mathbf{R}_i\mathbf{S}_i,\end{aligned} \tag{6.24}$$

with k basis shapes $\mathbf{B}_i$ and scalar coefficients c_{ij}, $j \in \{1,\ldots,k\}$. Stacking all $\mathbf{W}_i$ together leads to

$$\mathbf{W}_i = \mathbf{I}_{2\times3}\mathbf{R}_{3\times3F}\mathbf{\Pi}_{3F\times3F}\mathbf{S}_{3F\times p} \;\Rightarrow \tag{6.25}$$

$$\begin{bmatrix}\mathbf{W}_1\\ \mathbf{W}_2\\ \vdots\\ \mathbf{W}_F\end{bmatrix} = \begin{bmatrix}\mathbf{I}_{2\times3} & 0 & \ldots & 0\\ 0 & \mathbf{I}_{2\times3} & \ldots & 0\\ \vdots & \vdots & \ddots & \vdots\\ 0 & 0 & \ldots & \mathbf{I}_{2\times3}\end{bmatrix}\begin{bmatrix}\mathbf{R}^1_{3\times3F} & 0 & \ldots & 0\\ 0 & \mathbf{R}^2_{3\times3F} & \ldots & 0\\ \vdots & \vdots & \ddots & \vdots\\ 0 & 0 & \ldots & \mathbf{R}^F_{3\times3F}\end{bmatrix}\begin{bmatrix}\mathbf{I}_{3F\times3F}\\ \mathbf{I}_{3F\times3F}\\ \vdots\\ \mathbf{I}_{3F\times3F}\end{bmatrix}\begin{bmatrix}\mathbf{S}_1\\ \mathbf{S}_2\\ \vdots\\ \mathbf{S}_F\end{bmatrix}, \tag{6.26}$$

where $l = 3F$ and $\mathbf{R}^1_{3\times3F} = [\mathbf{R}_1|0\ldots0]$, $\mathbf{R}^2_{3\times3F} = [0|\mathbf{R}_2|0\ldots0]$. In other words, *a set of 3D shapes stacked together forms a high dimensional rigid body; its projections into different 3D subspaces lead to the initial 3D shapes*. Thus, LRSM can be written and interpreted in terms of HDSM. The form in Eq. (6.26) represents a special case and will not hold when designing a method based on HDSM.

Akhter *et al.* revealed that TSM is a dual representation of LRSM [22]. In support of this relation, Agudo *et al.* showed that LRSM, TSM and the recently proposed FSM are all dual to each other [13, 16]. *Thus, the connection of HDSM to all three previous models is established.* Besides, it would be perhaps more correct to describe it not as equivalency, but rather a generalisation, since shapes and camera poses in LRSM can always be rearranged to agree with HDSM, but the opposite does not generally hold.

Similarly, the formulation of variational approach (VA) [104] agrees with HDSM, where individual shapes are stacked together into the combined shape matrix $\mathbf{S}$:

$$\mathbf{W} = \underbrace{\begin{bmatrix} \mathbf{R}^1_{2\times 3} & 0 & \dots & 0 \\ 0 & \mathbf{R}^2_{2\times 3} & \dots & 0 \\ \vdots & \vdots & \ddots & \vdots \\ 0 & 0 & \dots & \mathbf{R}^f_{2\times 3} \end{bmatrix}}_{(\mathbf{R\Pi})_{2F\times 3F}} \underbrace{\begin{bmatrix} \mathbf{S}_1 \\ \mathbf{S}_2 \\ \vdots \\ \mathbf{S}_F \end{bmatrix}}_{\mathbf{\Phi}_{3F\times N}} \tag{6.27}$$

In VA, the low-rank constraint is imposed in every optimisation step by minimising the nuclear norm of $\mathbf{S}$. In contrast to VA, we initialise $\mathbf{\Phi}$ without duplication and keep it compact and compressed in every optimisation step.

6.2.5 Lifted Coherent Depth Fields with HDSM

We assume that dense correspondences for an image batch are given, and the translation is resolved (an object is centred relative to the camera). To reconstruct a deformable structure from F uncalibrated views, we propose to minimise the energy functional of the form

$$\mathbf{E}(\mathbf{R},\mathbf{\Pi},\mathbf{\Phi},l) = \frac{1}{2}\|\mathbf{W} - \mathbf{R\Pi}\mathrm{C}^{-1}(\mathbf{\Phi})\|^2_{\mathscr{F}} + \frac{\lambda}{2}\int_{\mathbb{R}^2}\frac{|\hat{\mathbf{S}}(s)|^2}{\hat{G}(s)}ds, \tag{6.28}$$
$$\text{subject to } \mathrm{rank}(\mathrm{P}(\mathbf{\Pi}\mathrm{C}^{-1}(\mathbf{\Phi}))) = \tau,$$

with C^{-1} standing for an expansion operator, $\hat{\mathbf{S}}$ denoting a Fourier transformed decompressed and expanded surface geometry $\mathbf{\Pi}\mathrm{C}^{-1}(\mathbf{\Phi})$, and $\hat{G}$ being a Fourier transformed reproducing kernel; $\mathscr{F}$ denotes Frobenius norm, τ and s are a natural number (a parameter) and a Fourier space variable, respectively. Let $\mathbf{S} = \mathbf{\Pi}\mathrm{C}^{-1}(\mathbf{\Phi})$. The operator P entangles dimensions of the expanded shapes:

$$\mathrm{P}(\mathbf{S}_{3F\times N}) = \left[\mathbf{S}'_1\mathbf{S}'_2\dots\mathbf{S}'_F\right]^\top_{F\times 3N}, \tag{6.29}$$

with $\mathbf{S}'_i = (\mathrm{vec}(\mathbf{S}_i^\top))^\top$. Note that we combine discrete and continuous notations in our formulation.

The objective (6.28) with the rank constraint on $\mathrm{P}(\mathbf{\Pi}\mathrm{C}^{-1}(\mathbf{\Phi}))$ is a nonconvex multi-dimensional optimisation problem. As no closed-form solution to this type of problems exists, we minimise it alternatingly. At the beginning of every optimisation step, we perform decompression-expansion of $\mathbf{\Phi}$. Afterwards, we alternately optimise for $\mathbf{R}$ and $\mathbf{S}$ while fixing $\mathbf{S}$ and $\mathbf{R}$, respectively. Optimisation for $\mathbf{R}$ is performed in a closed form by projecting an unconstrained update to the SO(3) group by normal equations. Optimisation for $\mathbf{S}$ is carried out by an introduction of an auxiliary variable $\bar{\mathbf{S}}$ and splitting the objective into two subproblems:

$$\min_{\mathbf{S}} \mathbf{E}_1(\mathbf{S}) = \min_{\mathbf{S}} \frac{1}{2\theta}\|\mathbf{S}-\bar{\mathbf{S}}\|^2_{\mathscr{F}} + \frac{\lambda}{2}\int_{\mathbb{R}^2} \frac{|\hat{\mathbf{S}}(s)|^2}{\hat{G}(s)} ds, \tag{6.30}$$

$$\min_{\bar{\mathbf{S}}} \mathbf{E}_2(\bar{\mathbf{S}}) = \min_{\bar{\mathbf{S}}} \frac{1}{2\theta}\|\mathbf{S}-\bar{\mathbf{S}}\|^2_{\mathscr{F}} + \frac{1}{2}\|\mathbf{W}-\mathbf{R}\bar{\mathbf{S}}\|^2_{\mathscr{F}}, \quad \text{s.t. } \operatorname{rank}(\mathrm{P}(\bar{\mathbf{S}})) = \tau. \tag{6.31}$$

The subproblem (6.30) results in a convolution equation which is solved with respect to $\mathbf{S}$:

$$\mathbf{S} = \mathscr{F}^{-1}\left(\mathscr{F}(\bar{\mathbf{S}}) \circ \frac{\mathscr{F}(G)}{\lambda\theta\mathbf{J}_{m\times n} + \mathscr{F}(G)}\right), \tag{6.32}$$

where $\mathbf{J}_{m\times n}$ is an all-one matrix and $\circ$ is an elementwise multiplication.

The subproblem (6.31) is solved for $\bar{\mathbf{S}}$. First, an unconstrained $\bar{\mathbf{S}}'$ is obtained as

$$\bar{\mathbf{S}}' = (\frac{1}{\theta} + \mathbf{R}^\mathsf{T}\mathbf{R})^{-1}(\mathbf{R}^\mathsf{T}\mathbf{W} + \frac{1}{\theta}\mathbf{S}). \tag{6.33}$$

We subsequently impose the rank constraint on $\bar{\mathbf{S}}'$ by preserving τ largest singular values of $\mathrm{P}(\bar{\mathbf{S}}')$ and reassembling the matrix. The whole algorithm is summarised in Alg. 4. It includes lifting-compression as well as decompression-expansion operations which are explained further in this section. Please refer to Sec. 6.1.5 for the details on the derivation of Eq. (6.32).

6.2.5.1 Lifting-Compression of S

In every iteration, we perform geometry lifting and compression. Taking the expanded updates of $\mathbf{S}$, we automatically detect frame-to-frame redundancies, *i.e.*, generate an ordered map of pairs $x = (\|\mathbf{S}_i\|_{\mathscr{F}}, \mathbf{S}_i)$. Next, we sort x and compare Frobenius norms. If the difference in Frobenius norms for the consecutive shapes is below a scalar value μ, one of the shapes is considered as redundant and not added into $\mathbf{\Phi}'$. The correct assignment of projective 3D subspaces and the shapes $\mathbf{S}_i$ is ensured by constantly updating $\mathbf{\Pi}_i$. Note that the same principles can be applied to the separate dimensions (imagine the case when two shapes share the same x and y coordinates, and differ solely in the z coordinate). If only parts of the shape change, the structure can be split into several regions. To obtain $\mathbf{S}$ from $\mathbf{\Phi}'$, we need to compute $\mathbf{\Pi}\,\mathbf{\Phi}'$.

To compress along the temporal direction, we analyse point trajectories and apply state-to-state compression. Since the shapes are sorted (either in ascending or

Algorithm 4 Lifted Coherent Depth Fields (L-CDF)

Input: measurements $\mathbf{W}$, parameters λ, θ, τ, σ, μ, ε
Output: permutation matrix $\mathbf{\Pi}$ and lifted geometry $\mathbf{\Phi}$

1: **Initialisation:** $[\mathbf{\Pi}, \mathbf{\Phi}] = \mathrm{C}(\mathbf{S}_{\text{rigid}})$ (Alg. 5),
2: $l = 3$, $\mathbf{\Pi} = \mathbf{I}_{3\times 3} \otimes \left[1\,1\,1\,\ldots\right]^{\mathsf{T}} \in \mathbb{R}^{F\times 1}$,
3: $\mathbf{R}_{2F\times 3F} = \operatorname{diag}\left[\mathbf{R}_{2\times 3}^{1,\text{rigid}}\,\mathbf{R}_{2\times 3}^{2,\text{rigid}}\,\ldots\,\mathbf{R}_{2\times 3}^{\text{F, rigid}}\right]$
4: **while** not converge **do**
5: **decompression, Alg. 6:** $\mathbf{S} = \mathbf{\Pi}\,\mathrm{C}_{\mu}^{-1}(\mathbf{\Phi})$
6: **fix S, update R:**
7: $\operatorname{svd}(\mathbf{W}\mathbf{S}(\mathbf{S}\mathbf{S}^{\mathsf{T}})^{-1}) = U\Sigma V^{\mathsf{T}}$
8: $\mathbf{R} = UCV^{\mathsf{T}}$, where
$C = \operatorname{diag}(1, 1, \ldots, 1, \operatorname{sign}(\det(UV^{\mathsf{T}})))$
9: **fix R, update S:**
10: initialise $\bar{\mathbf{S}} = \mathbf{S}$
11: **while** not converge **do**
12: $U\Sigma V^{\mathsf{T}} = \operatorname{svd}\left(\mathrm{P}\left(\left(\frac{1}{\theta}\mathbf{I} + \mathbf{R}^{\mathsf{T}}\mathbf{R}\right)^{-1}\left(\frac{1}{\theta}\mathbf{S} + \mathbf{R}^{\mathsf{T}}\mathbf{W}\right)\right)\right)$
13: $\bar{\mathbf{S}} = \mathrm{P}^{-1}\left(U\Sigma_{trunc}V^{\mathsf{T}}\right)$
14: $\mathbf{S} = \mathscr{F}^{-1}\left(\mathscr{F}(\bar{\mathbf{S}}) \circ \frac{\mathscr{F}(G)}{\lambda\theta\mathbf{J}_{m\times n} + \mathscr{F}(G)}\right)$
15: **end while**
16: **compression, Alg. 5:** $[\mathbf{\Pi}, \mathbf{\Phi}] = \mathrm{C}_{\varepsilon}(\mathbf{S})$
17: **end while**

descending order according to the Frobenius norms), the point trajectories in $\mathbf{\Phi}'$ are temporally smooth with a high probability. Thus, we save only non-redundant values, up to an ε value. The resulting $\mathbf{\Phi}$ is a sparsified matrix. By saving only non-zero elements, high compression ratios are possible. We denote compression by the operator $\mathrm{C}(\cdot)$. An overview of the lift-compress operator is provided in Alg. 5.

Since in our approach, rigid transformations are decoupled from shape deformations, trajectory analysis provides a cue for clustering the points into rigid, nearly rigid and non-rigid subsets. By analysing groups of trajectories, we can perform a *segmentation-from-deformation* and detect localised and correlated deformations.

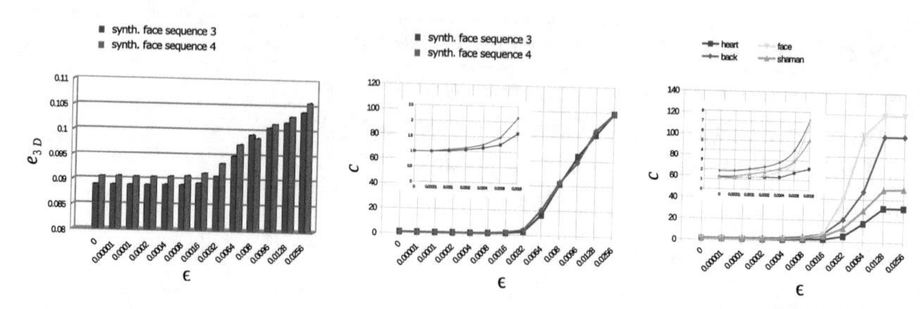

Figure 6.7: e_{3D} and $\mathfrak{c}$ values as functions of ε, in the experiment with the synthetic faces [104] (left, middle) and real image sequences (right).

Algorithm 5 Lifting-Compression of $\mathbf{S}$

Input: $\mathbf{S}$, frame-to-frame threshold μ, granularity parameter ε
Output: $[\mathbf{\Pi}, \mathbf{\Phi}] = \mathrm{C}_\mu(\mathbf{S})$

1: **Initialisation:** $\mathbf{\Pi} = \mathbf{0}$
2: **step 1, generate an uncompressed** $\mathbf{\Phi}'$**:**
3: $x = (\|\mathbf{S}_i\|_{\mathscr{F}}, \mathbf{S}_i),\ i \in \{1, \ldots, F\}$
4: sort x in ascending order based on the Frobenius norms
5: insert non-redandant frames into $\mathbf{\Phi}'$, generate $\mathbf{\Pi}$
6: **step 2, compress point trajectories:**
7: **for** points $p \in \{1, \ldots, N\}$ **do**
8: **initialise the seed:** $s = \mathbf{\Phi}'(p,1)$, $\mathbf{\Phi}(p,1) = \mathbf{\Phi}'(p,1)$
9: **for** point trajectory $\mathbf{\Phi}'(p,j), j \in \{1, \ldots, F\}$ **do**
10: **if** $|\,\|s\| - \|\mathbf{\Phi}'(p,j)\|\,| < \varepsilon$ **then**
11: $\mathbf{\Phi}(p,j) = \mathbf{0}_{3\times 1}$
12: **else**
13: $s = \mathbf{\Phi}'(p,j)$, $\mathbf{\Phi}(p,1) = \mathbf{\Phi}'(p,1)$
14: **end if**
15: **end for**
16: **end for**

6.2.5.2 Decompression-Expansion of the Lifted Geometry

The reverse operator to lift-compress is decompress-expand. It replicates the seed (non-zero) values along the point trajectories in the lifted space (decompression) and applies $\mathbf{\Pi}$ to $\mathbf{\Phi}'$ (expansion). The entire procedure is summarised in Alg. 6.

Algorithm 6 Decompression-Expansion of $\mathbf{\Phi}$

Input: compressed lifted geometry $\mathbf{\Phi}$, permutation matrix $\mathbf{\Pi}$
Output: expanded shape matrix $\mathbf{S} = \mathbf{\Pi}\,\mathrm{C}^{-1}(\mathbf{\Phi})$
1: **Initialisation:** $\mathbf{\Phi}' = \mathbf{0}$
2: **step 1, point trajectory completion,** $\mathbf{\Phi}' = \mathrm{C}^{-1}(\mathbf{\Phi})$:
3: **for** points $p \in \{1,\ldots,N\}$ **do**
4: **initialise the seed:** $s = \mathbf{\Phi}(p,1)$
5: **for** point trajectory $\mathbf{\Phi}(p,j), j \in \{1,\ldots,F\}$ **do**
6: **if** $\mathbf{\Phi}(p,j)$ is not a seed **then**
7: $\mathbf{\Phi}'(p,j) = s$
8: **else**
9: $s = \mathbf{\Phi}(p,j), \mathbf{\Phi}'(p,j) = \mathbf{\Phi}(p,j)$
10: **end if**
11: **end for**
12: **end for**
13: **step 2, expansion:** $\mathbf{S} = \mathbf{\Pi}\mathbf{\Phi}'$

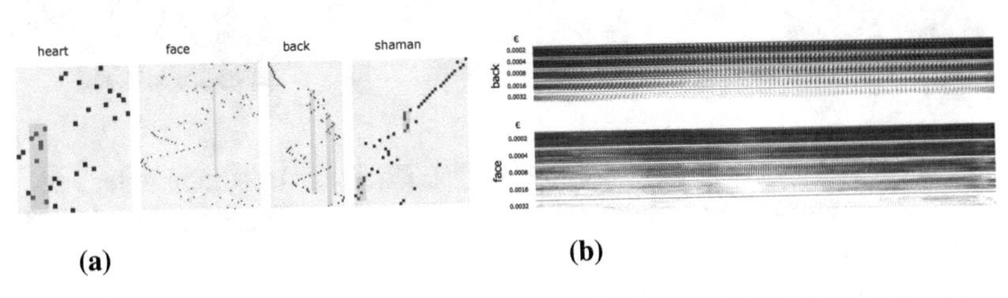

(a) (b)

Figure 6.8: Visualisations of (a): final permutation matrices $\mathbf{\Pi}$ (qualitative evaluation); (b): series of final $\mathbf{\Phi}s$ for the face [104] and back [237] sequences.

6.2.5.3 Solution Initialisation

We initialise $\mathbf{S}$ under rigidity assumption with [281]. $\mathbf{P}$ initially always fetches $\mathbf{S}_1$, *i.e.*, the only available element. The threshold ε is a scalar user-specified setting reflecting the noise level expectation, but also serving as a granularity level parameter.

6.2.6 Experimental Results

The primary purpose of this section consists in showing the validity of HDSM and evaluating L-CDF on synthetic and real world image sequences. All experiments

Table 6.5: Joint average e_{3D} and σ_e for the synthetic faces [104].

	TB [21]	MP [216]	VA [104]	L-CDF
e_{3D} / σ_e	9.24 / 5.37	8.81 / 6.15	3.22 / 0.55	8.03 / 0.98

are performed on a server with 32 GB RAM and Intel i7-6700K/4 GHz CPU. We do not require dedicated circuits, though multiple steps of the algorithm can be accelerated on parallel hardware (*e.g.*, matrix multiplications, Fourier transforms, lift-compress and decompress-expand operators).

6.2.6.1 Synthetic Face Sequences

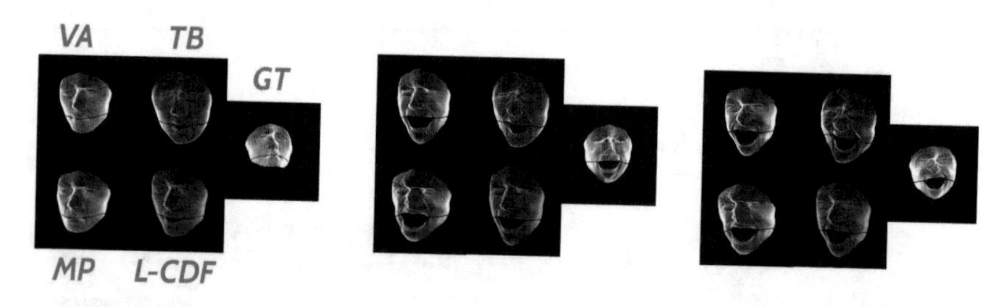

Figure 6.9: Reconstructions of frames 12, 22, 32 of the synthetic face sequence [104] by VA [104], MP [216], TB [22] and proposed L-CDF.

We conduct quantitative evaluation on four synthetic face sequences from [104]. Sequences 1 and 2 depict different facial expressions and are ten frames long each. Sequences 3 and 4 represent interpolated transitions between ten facial expressions and are both 99 frames long. Table 6.5 reports joint average root-mean-square errors (RMSE) e_{3D} and standard deviation σ_e for the synthetic face sequences and several approaches [22, 104, 216] supporting dense reconstruction, including the proposed L-CPD. RMSE is defined as $e_{3D} = \frac{1}{F}\sum_{f=1}^{F} \frac{\left\|\mathbf{S}_f^{ref} - \mathbf{S}_f\right\|_{\mathscr{F}}}{\left\|\mathbf{S}_f^{ref}\right\|_{\mathscr{F}}}$, with $\mathbf{S}_f^{ref}$ denoting the ground truth surfaces. Since all methods reconstruct the relative camera poses, we register the reconstructions and corresponding $\mathbf{S}_f^{ref}$ with Procrustes analysis. As the point ordering in reconstructions differs across the methods, we use non-rigid point set registration techniques [122] to establish correspondences between reconstructions and $\mathbf{S}_f^{ref}$. L-CDF achieves the second best result, after VA [104]. Every reconstruction contains $2.9 \cdot 10^4$ points per frame. Exemplary reconstructions are shown in Fig. 6.9. With 10 alternations (between estimation of $\mathbf{R}$ and $\mathbf{S}$) and

10 inner primal-dual iterations, the runtime of L-CDF amounts to 985 seconds, with a potential for improvement. In this experiment, we set $\varepsilon = 0$ and disable the compression of $\mathbf{\Phi}$.

Remark. CDF can explicitly regularise depth values which makes it especially robust against inaccurate correspondences. This comes at the cost of reduced e_{3D} in the cases with clean correspondences.

For the case with the activated compression, we set $\lambda = 10^{-1}$, $\theta = 10^{-2}$, $\sigma = 10^{-5}$, $\tau = 15$ and $\mu = 10^{-4}$ in further experiments with the sequences 3 and 4. The value of ε varies, and we measure its influence on e_{3D} and compression ratio $\mathfrak{c} = \frac{v_{\text{uncomp.}}}{v_{\text{comp.}}}$, with $v_{\text{uncomp.}}$ and $v_{\text{comp.}}$ denoting sizes of the uncompressed and compressed dynamic reconstructions, respectively. The results are summarised in Fig. 6.7-(left, middle). With the increase in ε from 0.0 to $1.6 \cdot 10^{-3}$, e_{3D} does not noticeably drop. At the same time, $\mathfrak{c}$ exceeds 2.1 for the sequence 4. Thus, a nearly lossless compression with $\mathfrak{c} = 2.1$ is observed. With further increase of ε, compression ratios of up to 100 are possible, accompanied by noticeable compression artefacts and, as a result, the drop of e_{3D}. In Fig. 6.6, several meshed instances of the synthetic face sequence 3 are shown. Compression artifacts in the form of curves on the reconstructed surfaces can be seen with zooming in.

6.2.6.2 Real and Naturalistic Image Sequences

Properties of L-CDF can be appreciated on real and naturalistic image sequences depicting real-world objects. Those are often prone to frame-to-frame redundancies (either consecutive or repetitive) and self-occlusions. Moreover, often only parts of a surface deform, whereas remaining areas are transformed rigidly. We evaluate L-CDF on three real image sequences, *i.e.*, *face* [104], *back* [237], *heart* [259] as well as *shaman* sequence from the SINTEL dataset [61]. In all cases, correspondences are computed with [267] in a single pass without an explicit occlusion handling. Fig. 6.7-(right) shows compression ratios as the functions of ε, for all four sequences. Because of the real setting, more redundancies — frame-to-frame as well as localised in space — are detected for the same ε, compared to the synthetic face sequences. As a result, higher compression ratios for lower ε values are observed. An exception is the *heart* sequence, since is deforms entirely up to a few frames in the diastolic phase. However, due to the periodic nature of the deformations and a fixed camera, frame-level compression is possible (recall that similar states do not necessarily need to be consecutive). Furthermore, *back* achieves the highest $\mathfrak{c}$, since rigid motion is dominant and non-rigid deformations are rather scarce in this sequence. Compared to the synthetic faces [104], the real *face* and the naturalistic *shaman* achieve somewhat higher $\mathfrak{c}$ values, as expected.

Reconstructions of all sequences are of the same order of magnitude in scale. This allows comparison of the compression ratios achieved for the same ε values.

Fig. 6.8a visualises the final $\mathbf{\Pi}$ matrices for all four image sequences and $\varepsilon = 8 \cdot 10^{-4}$. The cyan marks zero entries, and blue stays for $\mathbf{I}_{3\times3}$ matrices. Thus, every $\mathbf{\Pi}$ fetches a $3 \times N$ submatrix of $\mathbf{\Phi}$. Note the irregular structure of $\mathbf{\Pi}s$. Since $\mathbf{\Phi}$ is a compact lifted representation, $\mathbf{\Pi}$ serves as a key for the assignment of projective 3D subspaces to the frames. Besides, $\mathbf{\Pi}s$ carry information about the frame-level compression of the sequence. The grey bars mark areas with multiple reconstructions originating from the same 3D projections (and thus, detected as redundant by L-CDF, up to the μ parameter). Fig. 6.8b shows the series of visualised compressed $\mathbf{\Phi}s$ for different ε values, with the white and blue areas marking zero and non-zero values, respectively. Note, first, the repetitive structure of $\mathbf{\Phi}s$, and, second, how $\mathbf{\Phi}s$ are sparsified with increasing ε. Every point trajectory in $\mathbf{\Phi}$ reveals how often the respective point remains rigid in the sequence, up to the ε parameter. Moreover, by analysing the trajectory patterns, it is possible to detect the areas undergoing similar deformations.

Finally, Fig. 6.10 shows exemplary reconstructions from all four sequences. Reconstructions are of a high quality, with the advantage of the reduced storage requirement. For the *face* sequence, we visualise point-to-point distances between the uncompressed and compressed reconstructions. With moderately high compression ratios, compression artifacts become more noticeable. Those, however, do not disrupt the general perception of the scene if $\mathfrak{c} \in \{5,\ldots,30\}$. In the case of $\mathfrak{c} \approx 80$ and higher, the dynamic scene reduces to almost a single state, with the most of the deformations lost. It is intuitive that in a sequence with, suppose 100 frames, the compression ratio of 100 would imply a single most dominant or an average state preserved.

6.3 Conclusion

In the first section of this chapter, we introduced the CDF algorithm for dense NRSfM based on two central novel concepts, *i.e.*, a depth field and a new, in the context of NRSfM, spatial smoothness term — the coherency term. CDF proves itself as an accurate and robust to occlusions algorithm which can efficiently utilise available cues in a scene. On the challenging actor motion capture sequences with small deformations, we obtain the lowest RMSE among several approaches. We believe that CDF is the first method which can compensate for severe occlusions (dozens of frames with 50% or more of a scene eclipsed) without an explicit correspondence correction in the preprocessing step, a learned deformation model or a shape prior. Though CDF can oversmooth fine structures if variance of the

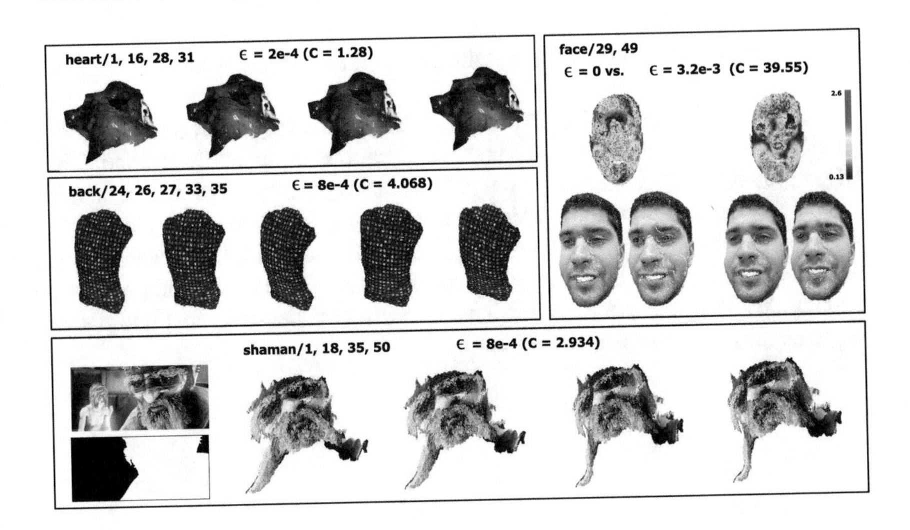

Figure 6.10: Exemplary reconstructions of real and synthetic image sequences. Frame numbers (comma separated), compression thresholds ε and the compression ratios $\mathfrak{c}$ are listed for every sequence. For the *shaman* sequence [61], the reference frame and the mask are shown on the left. For the *face* [104], point-to-point distances between the uncompressed and compressed reconstructions are visualised using Red>Yellow>Green>Blue colour code.

Gaussian is chosen suboptimally, we determined that the approach is well-posed w.r.t. parameter choice. We showed that CDF can be applied in a variety of real-world scenarios such as heart surgery, minimally invasive diagnostics as well as high-quality facial motion capture. Furthermore, CDF is easy to implement, has a high portion of parallelisable code and is suitable for implementation on embedded hardware. Regarding the coherency term, we believe that the set of tasks which could take advantage of it is not exhausted by NRSfM.

In the second section of this chapter, we propose a new expressive model for monocular surface recovery with geometry lifting — the HDSM. Besides generalising the previously proposed models for NRSfM — which was shown theoretically — it can serve as a foundation for a practical approach enabling compression of dense non-rigid reconstructions. The proposed variational energy-based L-CDF approach is an extension of CDF with HDSM. It achieves high reconstruction accuracy and enables nearly lossless compression with the compression ratios of $\approx 5-7$ on real image sequences.

A current limitation of the CDF and L-CDF is handling of large deformations. HDSM can support those, but this is currently not experimentally verified. Thus, HDSM can naturally model a union of subspaces. Another step towards the solution could be an initialisation policy which more accurately approximates the local states. An important direction of the future work is designing an algorithm which does not require expansion, *i.e.*, inferring $\mathbf{\Phi}$ directly from 2D observations and performing updates in the lifted space. A further interesting avenue is to test compressed representations for shape recognition.

7 Monocular Non-Rigid Surface Reconstruction with Learned Deformation Model

CURRENT techniques for monocular reconstruction either require dense correspondences and rely on motion and deformation cues, or assume a highly accurate reconstruction (referred to as a template) of at least a single frame given in advance and operate in the manner of non-rigid tracking. Accurate computation of dense point tracks often requires multiple frames and can be computationally expensive. Availability of a template is a very strong prior which restricts system operation to a pre-defined environment and scenarios (see chapter 3.1 for the detailed introduction to the field and the state of the art).

In this chapter, we propose deep neural network (DNN) based deformation model for monocular non-rigid 3D reconstruction (MNR). We train DNN with a new synthetically generated dataset covering the variety of smooth and isometric deformations occurring in the real world (*e.g.*clothes deformations, waving flags, bending paper and, to some extent, biological soft tissues). The proposed DNN architecture combines supervised learning with domain-specific loss functions. Our approach with a learned deformation model — Hybrid Deformation Model Network (HDM-Net) — surpasses performances of the evaluated state-of-the-art NRSfM and template-based methods by a considerable margin. We do not require dense point tracks or a well-positioned template. Our initialisation-free solution supports large deformations and copes well with several textures and illuminations. At the same time, it is robust to self-occlusions and noise. In contrast to existing DNN architectures for 3D, we directly regress 3D point clouds (surfaces) and depart from depth maps or volumetric representations.

In the context of MNR methods, our solution can be seen as a template-based reconstruction with considerably relaxed initial conditions and a broader applicability range per single learned deformation model. Thus, it constitutes a new class of methods — instead of a template, we rather work with a weak shape prior and a shape at rest for a known scenario class.

We generate a new dataset which fills a gap for training DNNs for non-rigid scenes and perform series of extensive tests and comparisons with state-of-the-art

V. Golyanik, *Robust Methods for Dense Monocular Non-Rigid 3D Reconstruction and Alignment of Point Clouds*,

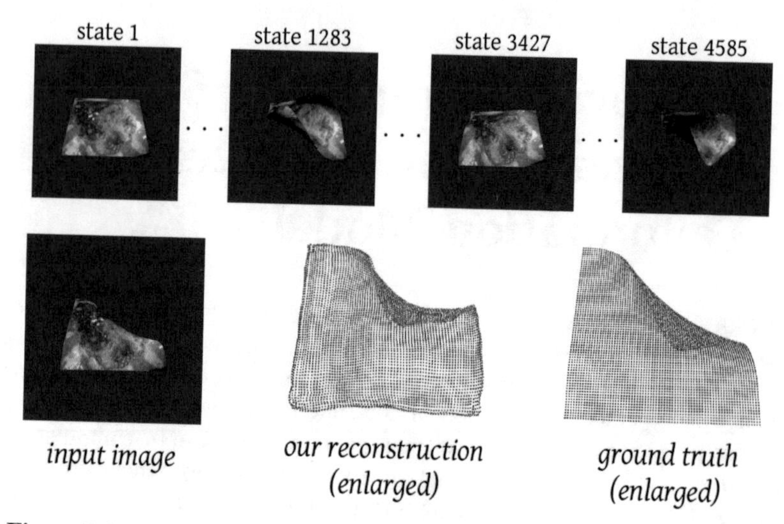

Figure 7.1: Reconstruction of an endoscopically textured surface with the proposed HDM-Net. The network is trained on a textured synthetic image sequence with ground truth geometry and accurately reconstructs unseen views in a small fraction of a second. Our architecture is potentially suitable for real-time augmented reality applications.

MNR methods. Fig. 7.1 provides an overview of the proposed approach — after training the network, we accurately infer 3D geometry of a deforming surface. Fig. 7.2 provides a high-level overview of the proposed architecture.

Related Work

Approaches for monocular 3D reconstruction based on supervised learning are usually specialised. Compared to template-based reconstruction, they are dedicated to the reconstruction of single object classes like human faces [43, 106, 252, 266] or human bodies [133, 298]. They do not use a single prior state (a template), but a whole space of states with feasible deformations and variations. The models are learned from extensive data collections showing a wide variety of forms, expressions (poses) and textures. In almost all cases, reconstruction with these methods means projection into the space of known shapes. To obtain accurate results, post-processing steps are required (*e.g.*for transferring subtle details to the initial coarse estimates). In many applications, solutions with predefined models could be a right choice, and their accuracy and speed can be sufficient.

HDM-Net bears a resemblance to DNN-based regressors which use encoder-decoder architecture [275]. In contrast to many DNN-based 3D regressors [68, 88, 275], our network does not include fully connected layers as they impede generalisability (lead to overfitting) as applied to MNR. As most 3D reconstruction approaches, it contains a 3D loss.

In many cases, isometry is an effective and realistic constraint for TBR, as shown in [67, 223]. In HDM-Net, isometry is imposed through training data. The network learns the notion of isometry from the opposite, *i.e.*by not observing other deformation modes. Another strong constraint in TBR is contour information which, however, has not found wide use in MSR, with only a few exceptions [134, 291]. In HDM-Net, we explicitly impose contour constraints by comparing projections of the learned and ground truth surfaces.

Under isometry, the solution space for a given contour is much better constrained compared to the extensible cases. The combined isometry and contour cues enable efficient occlusion handling in HDM-Net. Moreover, contours enable texture invariance up to a certain degree, as a contour remains unchanged irrespective of the texture. Next, through variation of light source positions, we train the network for the notion of shading. Since for every light source configuration, the underlying geometry is the same, HDM-Net acquires awareness of varying illumination. Besides, contours and shading in combination enable reconstruction of texture-less surfaces. To summarise, our framework has unique properties among MSR methods which are rarely found in other MNR techniques, especially when combined.

7.1 Architecture of the Hybrid Deformation Model Network (HDM-Net)

We propose a DNN architecture with encoder and decoder depicted in Fig. 7.2 (a general overview). The network takes as an input an image of dimensions 224×224 with three channels. Initially, the encoder extracts contour, shading and texture deformation cues and generates a compact latent space representation of dimensions $28 \times 28 \times 128$. Next, the decoder applies a series of deconvolutions and outputs a 3D surface of dimensions $73 \times 73 \times 3$ (a point cloud). It lifts the dimensionality of the latent space until the dimensionality of activation becomes identical to the dimensionality of ground truth. The transition from the implicit representation into 3D occurs on the later stage of decoder through a deconvolution. Fig. 7.3 provides a detailed clarification about the structures of encoder and decoder.

As can be seen in Fig. 7.2 and 7.3, we skip some connections in HDM-Net to avoid vanishing gradients, similar to *resnet* [139]. Due to the nature of convolutions, our deep network could potentially lose some important information in the forward path which could be advantageous in the deeper layers. Thus, connection skipping compensates for this side effect — for each convolution layer — which results in the increased performance. Moreover, in the backward path, shortcut connections help to overcome the vanishing gradient problem, *i.e.*a series of numerically unstable gradient multiplications leading to vanishing gradients. Thus, the gradients are successfully passed to the shallow layers.

Fully connected (FC) layers are often used in classification tasks [168]. They have more parameters than convolution layers and are known as a frequent cause of overfitting. We have tried FC layers in HDM-Net and observed overfitting on the training dataset. Thus, FC layers reduce generalisation ability of our network. Furthermore, spatial information is destroyed as the data in the decoder is concatenated before being passed to the FC layer. In our task, needless to say, spatial cues are essential for 3D regression. In the end, we omit FC layers and show generalisation ability of 3D reconstruction on the test data.

7.1.1 Loss Functions

Let $\mathbf{S} = \{\mathbf{S}_f\}$, $f \in \{1,\ldots,F\}$ denote predicted 3D states, and $\mathbf{S}^{GT} = \{\mathbf{S}_f^{GT}\}$ is the ground truth geometry; F is the total number of frames and N is the number of points in the 3D surface. In HDM-Net, contour similarity and the isometry constraint are the key innovations and we apply three types of loss functions summarised into the loss energy:

$$\mathbf{E}(\mathbf{S},\mathbf{S}^{GT}) = \mathbf{E}_{3D}(\mathbf{S},\mathbf{S}^{GT}) + \mathbf{E}_{iso}(\mathbf{S}) + \mathbf{E}_{cont.}(\mathbf{S},\mathbf{S}^{GT}). \tag{7.1}$$

3D error: The 3D loss is the main loss in 3D regression. It penalises the differences between predicted and ground truth 3D states and is common in training for 3D data:

$$\mathbf{E}_{3D}(\mathbf{S},\mathbf{S}^{GT}) = \frac{1}{F}\sum_{f=1}^{F} \|\mathbf{S}_f^{GT} - \mathbf{S}_f\|_{\mathscr{F}}^2, \tag{7.2}$$

where $\|\cdot\|_{\mathscr{F}}$ denotes the Frobenius norm. Note that we take an average of the squared Frobenius norms of the differences between the learned and ground truth geometries.

Isometry prior: To additionally constrain the regression space, we embed isometry loss which enforces the neighbouring vertices to be located close to each other. Several versions of inextensibility and isometry constraints can be found in MSR —

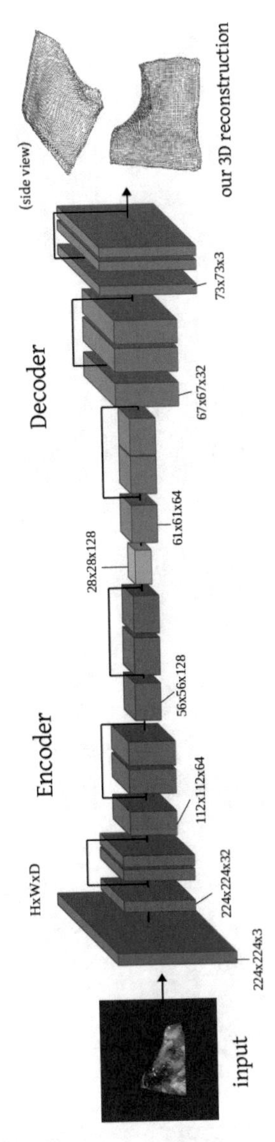

Figure 7.2: An overview of the architecture of the proposed HDM-Net with encoder and decoder. The input of HDM-Net is an image of dimensions 224×224 with three channels, and the output is a dense reconstructed 3D surface of dimensions $73 \times 73 \times 3$ (73^2 3D points).

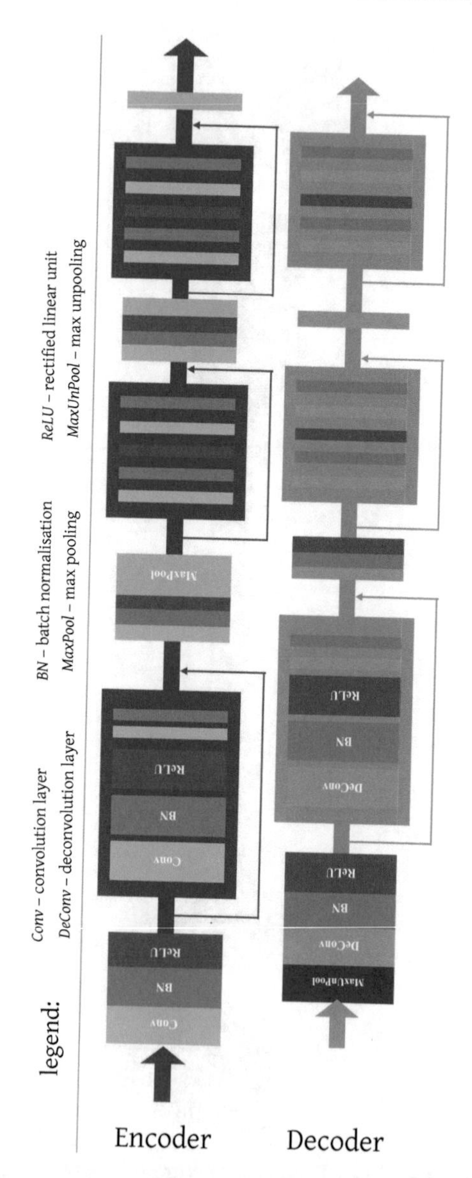

Figure 7.3: Architecture of the proposed HDM-Net: detailed clarification about the structures of the encoder and decoder.

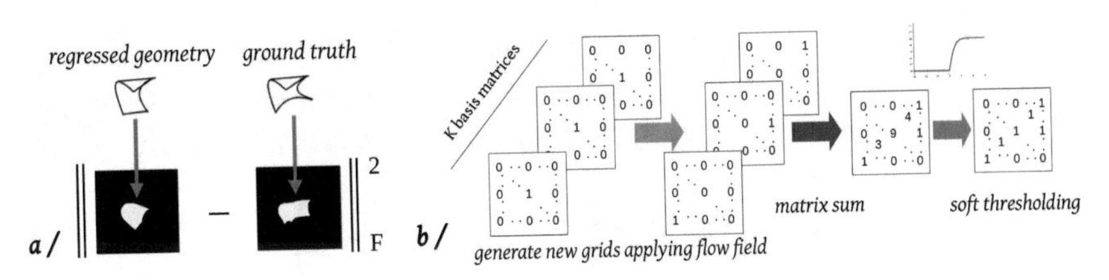

Figure 7.4: Our contour loss penalises deviations between reprojections of the regressed geometry and reprojections of the ground truth.

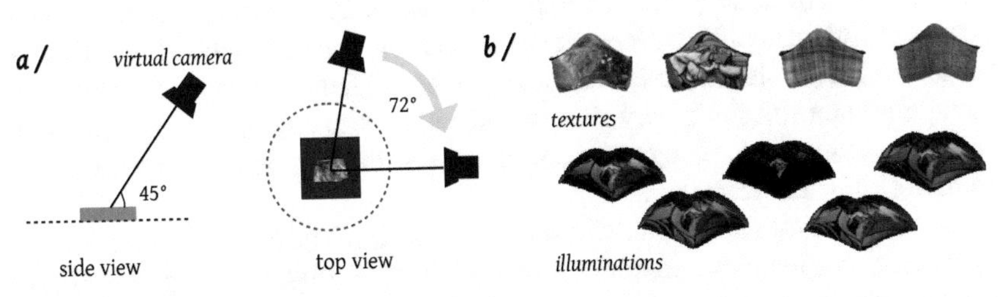

Figure 7.5: Camera poses used for the dataset generation (a); different textures applied to the dataset: *endoscopy*, *graffiti*, *clothes* and *carpet* (b-top) and different illuminations (b-bottom).

a common one is based on differences between Euclidean and geodesic distances. For our DNN architecture, we choose a differentiable loss which performs Gaussian smoothing of $\mathbf{S}_f$ and penalises the difference between the unembellished and smoothed version $\hat{\mathbf{S}}_i$:

$$\mathbf{E}_{iso}(\mathbf{S}) = \frac{1}{F}\sum_{f=1}^{F} \|\hat{\mathbf{S}}_f - \mathbf{S}_f\|_{\mathscr{F}}, \tag{7.3}$$

with

$$\hat{\mathbf{S}}_f = \frac{1}{2\pi\sigma^2}\exp\left(-\frac{x^2+y^2}{2\sigma^2}\right) * \mathbf{S}_f, \tag{7.4}$$

where $*$ denotes a convolution operator and σ^2 is the variance of Gaussian.

Contour loss: If the output of the network and the ground truth coordinates are similar, the contour shapes after projection onto a 2D plane have to be similar as well. The main idea of the reprojection loss is visualised in Fig. 7.4-(a). After the inference of the 3D coordinates by the network, we project them onto the 2D plane and compute the difference between the two projected contours. If focal lengths f_x,

f_y as well as the principal point (c_x, c_y) of the camera are known (the $\mathbf{K}$ used for the dataset generation is provided in Sec. 7.2), observed 3D points $\mathbf{p} = (p_x, p_y, p_z)$ are projected to the image plane by the projection operator $\pi : \mathbb{R}^3 \rightarrow \mathbb{R}^2$:

$$\mathbf{p}'(u, v) = \pi(\mathbf{p}) = \left(f_x \frac{p_x}{p_z} + c_x, f_y \frac{p_y}{p_z} + c_y \right)^{\mathsf{T}}, \tag{7.5}$$

where $\mathbf{p}'$ is the 2D projection of $\mathbf{p}$ with 2D coordinates u and v. Otherwise, we apply an orthographic camera model.

A naïve shadow casting of a 3D point cloud onto a 2D plane is not differentiable, *i.e.*the network cannot backpropagate gradients to update the network parameters. The reason is twofold. In particular, the cause for indifferentiability is the transition from point intensities to binary shadow indicators with an ordinary step function (the numerical reason) using point coordinates as indexes on the image grid (the framework-related reason).

Fig. 7.4-(b) shows how we circumvent this problem. The first step of the procedure is the projection of 3D coordinates onto a 2D plane using either a perspective or an orthographic projection. As a result of this step, we obtain a set of 2D points. We generate $K = 73^2$ translation matrices $\mathbf{T}_j = \begin{pmatrix} 1 & 0 & u \\ 0 & 1 & v \end{pmatrix}$ using 2D points and a flow field tensor of dimension $K \times 99 \times 99 \times 2$ (the size of each binary image is 99×99). Next, we apply bilinear interpolation [152] with generated flow fields on the replicated basis matrix $\mathbf{B}$ K times and obtain K translation indicators. $\mathbf{B}_{99 \times 99}$ is a sparse matrix with only a single central non-zero element which equals to 1. Finally, we sum up all translation indicators and softly threshold positive values in the sums to ≈ 1, *i.e.*our shadow indicator. Note that to avoid indifferentiability in the last step, the thresholding is performed by a combination of a rectified linear unit (ReLU) and tanh function (see Fig. 7.4-(b)):

$$\tau(\mathscr{I}(\mathbf{s}_f(n))) = \max(\tanh(2\mathbf{S}_f(n)), 0), \tag{7.6}$$

where $n \in \{1, \ldots, N\}$ denotes the point index, $\mathbf{s}_f(n)$ denotes a reprojected point $\mathbf{S}_f(n)$ in frame f, and $\mathscr{I}(\cdot)$ fetches intensity of a given point. We denote the differentiable projection operator and differentiable soft thresholding operator by the symbols $\pi^\dagger(\cdot)$ and $\tau(\cdot)$, respectively. Finally, the contour loss reads

$$\mathbf{E}_{cont.}(\mathbf{S}, \mathbf{S}^{GT}) = \frac{1}{F} \sum_{f=1}^{F} \| \tau(\pi^\dagger(\mathbf{S}_f)) - \tau(\pi^\dagger(\mathbf{S}_f^{GT})) \|_{\mathscr{F}}^2. \tag{7.7}$$

Note that object contours correspond to 0-1 transitions.

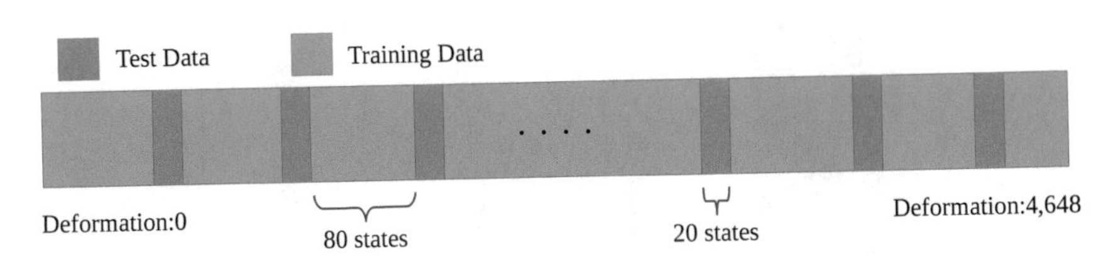

Figure 7.6: The pattern of the training and test datasets.

7.2 Dataset and Training

For our study, we generated a dataset with a non-rigidly deforming object using *Blender* [44]. In total, there are 4648 different temporally smooth 3D deformation states with structure bendings, smooth foldings and wavings, rendered under Cook-Torrance illumination model [72] (see Fig. 7.1 for the exemplary frames from our dataset). We have applied five different camera poses, five different light source positions and four different textures corresponding to the scenarios we are interested in — *endoscopy*, *graffiti* (it resembles a waving flag) *clothes* and *carpet* (an example of an arbitrary texture). The endoscopic texture is taken from [111]. Illuminations are generated based on the scheme in Fig. 7.5-(a), the textures and illuminations are shown in Fig. 7.5-(b). We project the generated 3D scene by a virtual camera onto a 2D plane upon Eq. (7.5), with $\mathbf{K} = \begin{pmatrix} 280 & 0 & 128 \\ 0 & 497.7 & 128 \\ 0 & 0 & 1 \end{pmatrix}$. The background in every image is of the same opaque colour. We split the data into training and test subsets in a repetitive manner, see Fig. 7.6 for the pattern. We train HDM-Net jointly on several textures and illuminations, with the purpose of illumination-invariant and texture-invariant regression. One illumination and one texture are reserved for the test dataset exclusively. Our images are of the dimensions 256×256. They reside in 15.2 Gb of memory, and the ground truth geometry requires 1.2 Gb (in total, 16.4 Gb). The hardware configuration consists of two six-core processors Intel(R) Xeon(R) CPU E5-1650 v4 running at 3.60GHz, 16 GB RAM and a GEFORCE GTX 1080Ti GPU with 11GB of global memory. In total, we train for 95 epochs, and the training takes two days in *pytorch* [220, 221]. The evolution of the loss energy is visualised in Fig. 7.11-(a). The inference of one state takes ca. 5 ms.

7.3 Geometry Regression and Comparisons

We compare our method with the template-based reconstruction of Yu *et al.* [319], variational NRSfM approach (VA) of Garg *et al.* [104] and NRSfM method of

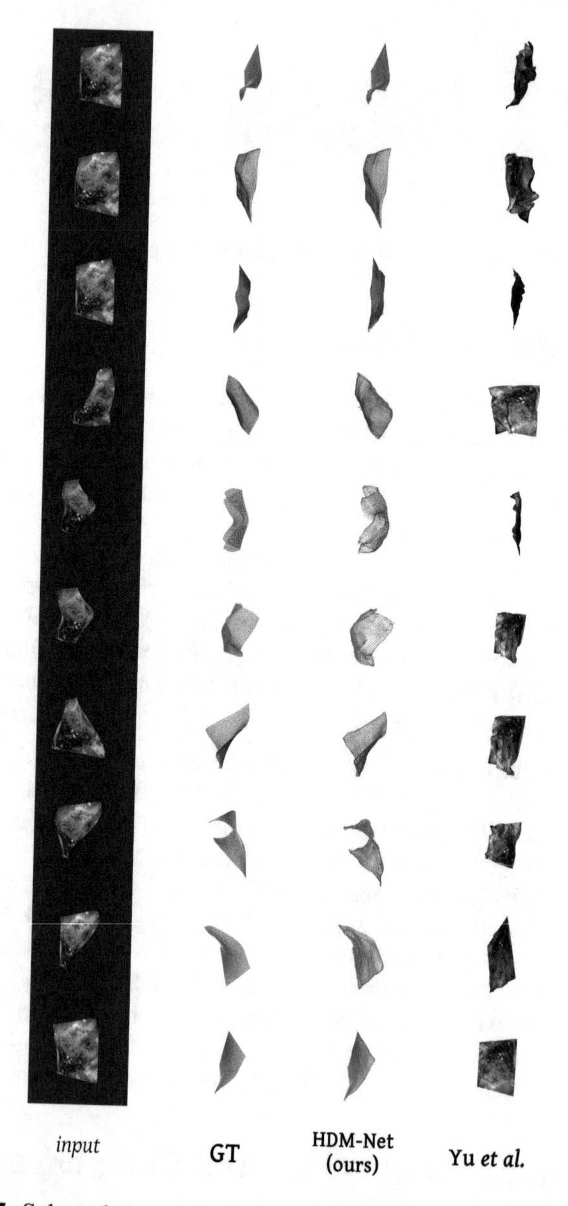

Figure 7.7: Selected reconstruction results on endoscopically textured surfaces for HDM-Net (our method) and Yu *et al.* [319].

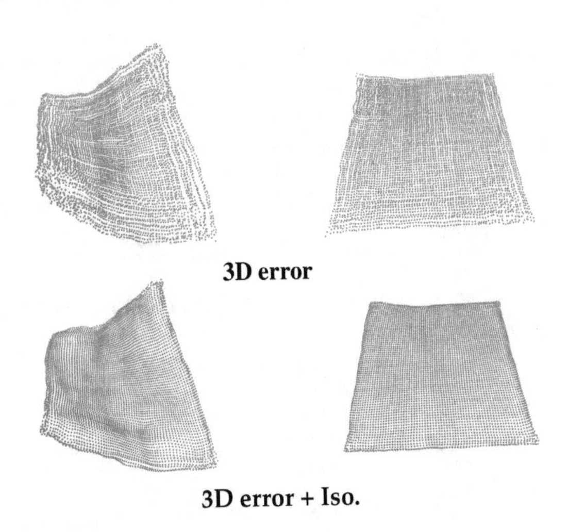

Figure 7.8: Comparison of 3D reconstruction with 3D error (top row) and 3D error + Isometry prior (bottom row)

Golyanik *et al.* [121] — Accelerated Metric Projections (AMP). We use an optimised heterogeneous CPU-GPU version of VA written in C++ and CUDA C [210]. AMP is a C++ CPU version which relies on an efficient solution of a semi-definite programming problem and is currently one of the fastest batch NRSfM methods. For VA and AMP, we compute required dense point tracks. Following the standard praxis in NRSfM, we project the ground truth shapes onto a virtual image plane by a slowly moving virtual camera. Camera rotations are parametrised by Euler angles around the *x*-, *y*- and *z*-axes. We rotate for up to 20 degrees around each axis, with five degrees per frame. This variety in motion yields minimal depth changes required for an accurate initialisation in NRSfM. We report runtimes, 3D error

$$e_{3D} = \frac{1}{F}\sum_{f=1}^{F}\frac{\|\mathbf{S}_f^{GT} - \mathbf{S}_f\|_{\mathscr{F}}}{\|\mathbf{S}_f^{GT}\|_{\mathscr{F}}} \tag{7.8}$$

and standard deviation σ of e_{3D}. Before computing e_{3D}, we align $\mathbf{S}_f$ and the corresponding $\mathbf{S}_f^{GT}$ with Procrustes analysis.

Runtimes, e_{3D} and σ for all three methods are summarised in Table 7.1. AMP achieves around 30 *fps* and can execute only for 100 frames per batch at a time. However, this estimate does not include often prohibitive computation time of dense correspondences with multi-frame optical flow methods such as [267]. Note that

Table 7.1: Per-frame runtime t in *seconds*, e_{3D} and σ comparisons of Yu *et al.* [319], AMP [121] and HDM-Net (proposed method).

	Yu *et al.* [319]	AMP [121]	VA [104]	HDM-Net
t, s	3.305	0.035	0.39	**0.005**
e_{3D}	1.3258	1.6189	0.46	**0.0251**
σ	**0.0077**	1.23	0.0334	0.03

Table 7.2: Comparison of 3D error for different textures and the same illumination (number 1).

	endoscopy	*graffiti*	*clothes*	*carpet*
e_{3D}	**0.0485**	0.0499	0.0489	0.1442
σ	**0.01356**	0.022	0.02648	0.02694

Table 7.3: Comparison of 3D error for different illuminations.

	illum. 1	*illum. 2*	*illum. 3*	*illum. 4*	*illum. 5*
e_{3D}	0.07952	0.0801	0.07942	**0.07845**	**0.07827**
σ	**0.0525**	0.0742	0.0888	0.1009	0.1123

runtime of batch NRSfM depends on the batch size, and the batch size influences the accuracy and ability to reconstruct. VA takes advantage of a GPU and executes with 2.5 *fps*. Yu *et al.* [319] achieves around 0.3 *fps*. In contrast, HDM-Net processes one frame in only 5 ms. This is by far faster than the compared methods. Thus, HDM-Net can compete in runtime with rigid structure from motion [281]. The runtime of the latter method is still considered as the lower runtime bound for NRSfM[1].

At the same time, the accuracy of HDM-Net is the highest among all tested methods. Selected results with complex deformations are shown in Fig. 7.7. We see that Yu *et al.* [319] copes well with rather small deformations, and our approach accurately resolves even challenging cases not exposed during the training. In the case of Yu *et al.* [319], the high e_{3D} is explained by a weak handling of self-occlusions and large deformations. In the case of NRSfM methods, the reason for the high e_{3D} is an inaccurate initialisation. Moreover, VA does not handle foldings and large deformations well.

[1] when executed in a batch of 100 frames with 73^2 points each, a C++ version of [281] takes 1.47 ms per frame on our hardware; for 400 frames long batch, it requires 5.27 ms per frame.

Table 7.4: Comparison of effects of loss functions.

	3D	3D + Con.	3D + Iso.	3D + Con. + Iso.
e_{3D}	0.0698	**0.0688**	0.0784	0.0773
σ	**0.0761**	0.0736	0.0784	0.0789

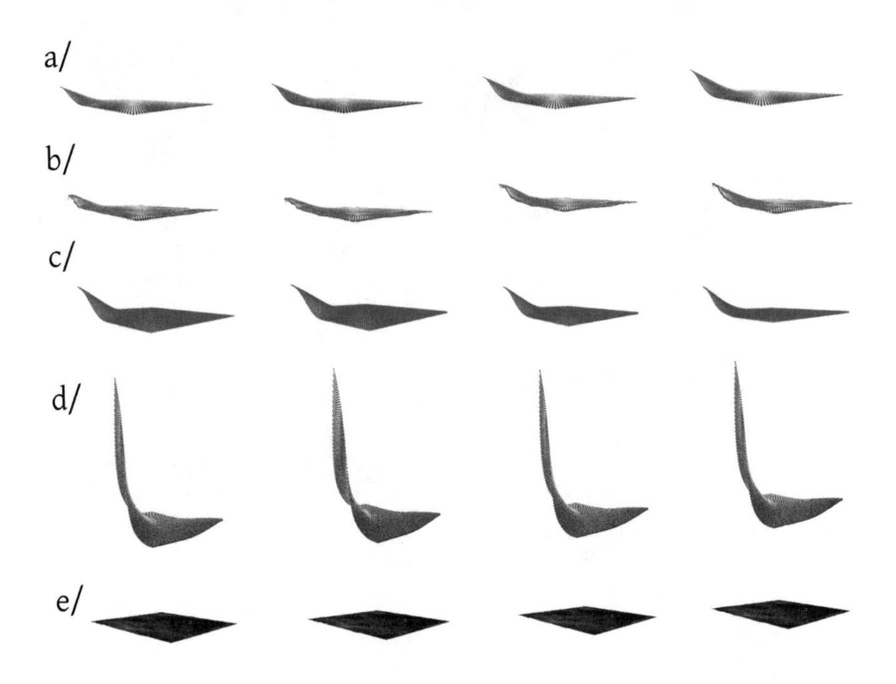

Figure 7.9: Qualitative comparisons of ground truth (a), HDM-Net (proposed method) (b), AMP [121] (c), VA [104] (d) and Yu *et al.* [319] (e) on several frames of our test sequence from the first 100 frames (each column corresponds to one frame).

Table 7.3 summarises e_{3D} for our method under different illumination conditions. We notice that our network copes well with all generated illuminations — the difference in e_{3D} is under 3%. Table 7.2 shows e_{3D} comparison for different textures. Here, the accuracy of HDM-Net drops on the previously unseen texture by the factor of three, which still corresponds to reasonable reconstructions with

Figure 7.10: Exemplary reconstructions from real images obtained by HDM-Net (music notes, a fabric, surgery and an air balloon)

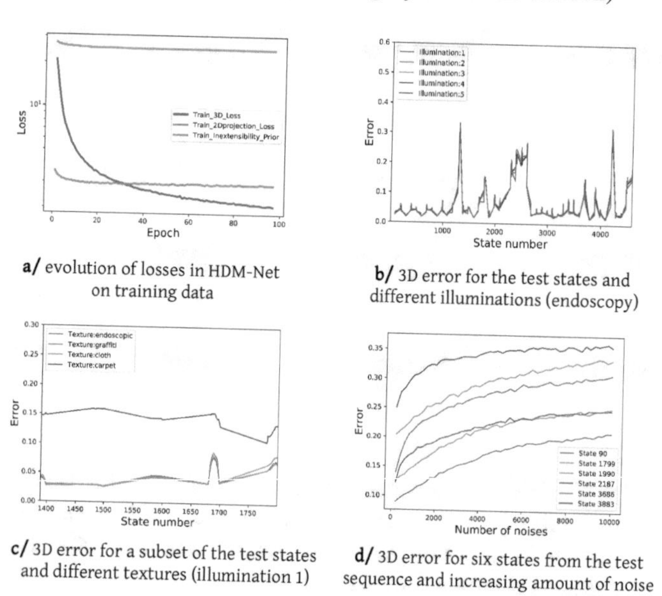

a/ evolution of losses in HDM-Net on training data

b/ 3D error for the test states and different illuminations (endoscopy)

c/ 3D error for a subset of the test states and different textures (illumination 1)

d/ 3D error for six states from the test sequence and increasing amount of noise

Figure 7.11: Graphs of e_{3D} for varying illuminations (for *endoscopy* texture), varying textures (for illumination 1) as well as six states under increasing amount of noise. Note that in b/ and c/, only the errors obtained on the test data are plotted. For c/, HDM-Net was trained on a subset of training states (three textures and one illumination).

the captured main deformation mode. Another quantitative comparison is shown in Fig. 7.9. In this example, all methods execute on the first 100 frames of the sequence. AMP [121] captures the main deformation mode with $e_{3D} = 0.1564$ but struggles to perform a fine-grained distinction (in Table 7.1, e_{3D} is reported over the sequence of 400 frames, hence the differing metrics). VA suffers under an inaccurate initialisation under rigidity assumption and Yu *et al.* [319], by contrast, does not recognise the variations in the structure. All in all, HDM-Net copes well with self-occlusions. Graphs of e_{3D} as functions of the state index under varying illuminations and textures can be found in Fig. 7.11-(b,c). Table 7.4 shows the comparison of e_{3D} using networks trained with various combinations of loss functions. *3D + Con.* shows the lowest e_{3D} and applying *isometry prior* increases e_{3D}. Since *isometry prior* is smoothing loss, the 3D grid becomes smaller in comparison to the outputs without *isometry prior* hence higher e_{3D}. However, as shown in Fig. 7.8, isometry prior allows the network to generate smoother 3D geometries.

Next, we evaluate the performance of HDM-Net on noisy input images. Therefore, we augment the dataset with increasing amounts of uniform salt-pepper noise. Fig. 7.11-(d) shows the evolution of the e_{3D} as a function of the amount of noise, for several exemplary frames corresponding to different input difficulties for the network. We observe that HDM-Net is well-posed w.r.t noise — starting from the respective values obtained for the noiseless images, the e_{3D} increases gradually.

We tested HDM-Net on several challenging real images. Fig. 7.10 shows the tested images and our reconstructions. We recorded a music note image for an evaluation of our network in real-world scenario. Despite different origin of the inputs (music notes, a fabric [276], an endoscopic view during a surgery [111] and an air balloon [258]), HDM-Net produces realistic and plausible results. Note how different are the regressed geometries which suggests the generalisation ability of the proposed solution.

In many real-world cases, HDM-Net produces acceptable results. However, if the observed states differ a lot from the states in the training data, HDM-Net could fail to recognise and regress the state. This can be addressed by an extension or tailoring of the dataset for specific cases. Adding training data originating from motion and geometry capture of real objects could also be an option.

7.4 Concluding Remarks

We have presented a new monocular surface recovery method with a deformation model replaced by a DNN — HDM-Net. The new method reconstructs time-varying geometry from a single image and is robust to self-occlusions, changing illumination and varying texture. Our DNN architecture consists of an encoder, a latent space and a decoder, and is furnished with three domain-specific losses. Apart from the conventional 3D data loss, we propose isometry and reprojection losses. We train HDM-Net with a newly generated dataset with ca. four an a half thousands states, four different illuminations, five different camera poses and three different textures. Experimental results show the validity of our approach and its suitability for reconstruction of small and moderate isometric deformations under self-occlusions. Comparisons with one template-based and two template-free methods have demonstrated a higher accuracy in favour of HDM-Net. Since HDM-Net is one of the first approaches of the new kind, there are multiple avenues for investigations and improvements. One apparent direction is the further augmentation of the test dataset with different backgrounds, textures and illuminations. Next, we are going to test more advanced architectures such as generative adversarial networks and recurrent connections for the enhanced temporal smoothness. Currently, we are also investigating the relevance of HDM-Net for medical applications with augmentation of soft biological tissues.

8 Probabilistic Point Set Registration with Prior Correspondences

EMBEDDING additional prior knowledge into point set registration is a way to constraint the solution space and disambiguate the problem. The amount of prior knowledge which can be embedded into point set registration should not exceed some "critical mass" in the sense that otherwise, the algorithmic class will change. For instance, registration of watertight meshes is known as mesh registration. One of the possibilities is embedding of prior correspondences, see Fig. 8.1. *Prior correspondences or matches* are affinity relations between points from different point sets. If available, every prior match is furnished with a reliability criterion. In contrast, *landmarks* refer to highly reliable prior matches.

In this chapter, we show how to integrate prior correspondences into probabilistic rigid and non-rigid point set registration. To the best of our knowledge, the proposed methods were the first at the time of publication which allowed considering prior matches in probabilistic point set registration. Our approaches build upon the Coherent Point Drift (CPD) [203] and broaden its scope. Thus, we combine a method of state-of-the-art accuracy with additional prior knowledge in the form of the prior correspondences. *In the first section*, we introduce Extended Coherent Point Drift (ECPD) in the general form (Sec. 8.1.1) and then elaborate it for the rigid and non-rigid cases.

In the second section (Sec. 8.1), we show how to embed prior correspondences into a rigid point set registration approach and present an elegant solution to joint pre-alignment and rigid point set registration with prior matches — rigid Extended Coherent Point Drift (R-ECPD). Instead of performing pre-alignment and the actual registration in the separate steps, prior matches explicitly influence the registration procedure in our approach. This results in several advantages. Firstly, R-ECPD solves the pre-alignment task — an approximate resolving of rotation and translation — with an insufficient number of prior correspondences, when other methods fail. Secondly, it produces more accurate rigid registrations of noisy point sets than the state-of-the-art Coherent Point Drift method. Combined with

V. Golyanik, *Robust Methods for Dense Monocular Non-Rigid 3D Reconstruction and Alignment of Point Clouds*,

application-specific methods for correspondence establishment, we demonstrate the superiority of R-ECPD in several synthetic and real-world scenarios.

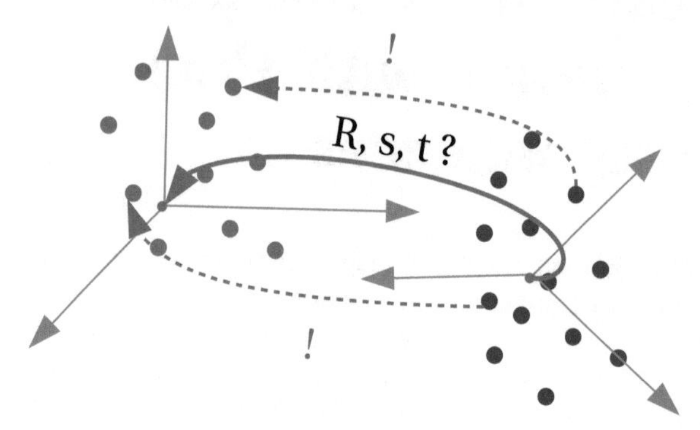

Figure 8.1: The objective of rigid point set registration (PSR) is the recovery of a rigid transformation along with the uniform scaling between point sets. In the most general case, the cardinalities of the input point sets are different, and the correspondences are assumed to be unknown. If a few correspondences are known in advance, they can be used to constrain the solution space and improve alignment accuracy. In this chapter, we show how to embed prior correspondences into probabilistic PSR, both for the rigid and non-rigid cases.

In the third section (Sec. 8.2), embedding of prior correspondences is performed for a more challenging non-rigid case. We call the resulting algorithm non-rigid Extended Coherent Point Drift (ECPD). Unlike in the rigid case, integrating prior knowledge into a registration algorithm is especially demanding in the non-rigid case due to the high variability of motion and deformation. ECPD enables, on the one hand, to couple prior correspondences into the dense registration procedure in a closed form and, on the other hand, to process large point sets in a reasonable time through adopting an optimal coarse-to-fine strategy. Combined with a suitable keypoint extractor in the preprocessing step, ECPD allows for non-rigid registrations of increased accuracy for point sets with structured outliers. We demonstrate the advantages of our approach in comparison to several other non-rigid point set registration methods in scenarios with synthetic and real data.

In the fourth section (Sec. 8.3), we propose a practical solution to the point cloud based registration of 3D human body scans and a 3D human template. Considering advances presented in sections two and three, we adopt rigid and non-rigid ECPD and design a fully automated registration framework. Our framework consists of several steps including establishment of prior matches, alignment of point clouds into a common reference frame, global non-rigid registration, partial non-rigid registration and post-processing. Thanks to the coarse-to-fine strategy, we can automatically handle large point clouds with significant variations in appearance and achieve high registration accuracy, which is shown experimentally. Finally, we demonstrate a pipeline for the treatment of social pathologies with animatable virtual avatars as an exemplary real-world application of the new framework.

8.1 Rigid Point Set Registration with Prior Correspondences

Generally, registration algorithms perform well on point sets representing noiseless objects. When this assumption is violated — (1) point sets represent partially overlapping parts of an object, (2) contain outliers or (3) differences in scene poses are significant — difficulties arise. All the violations often happen in practice. Although some approaches can handle different types of noise distributions [110, 203], it turns out that outliers often do not obey a particular probability distribution, but are rather *clustered*. Clustered outliers and regular points in a point set cannot be distinguished from each other during the registration. For instance, the main advantage of mixture model based methods [203, 204] over Iterative Closest Point (ICP) methods [40, 93, 236] lies in the soft assignment of correspondences via probabilities, proving to be more robust in the presence of outliers. However, if outliers are clustered, the methods can fail (see §10.1.4). Partial shapes can also be comprehended as point sets with large areas of clustered outliers. Consequently, violation (1) is a special case of violation (2).

To compensate for the aforementioned violations, preprocessing steps are required. To cope with clustered outliers, weighting the correspondences with respect to overlapping parts is possible [93]. However, the performance of the method [93] strongly depends on the initialisation. To cope with severe misalignment between point sets, registration is usually performed on a subset of reliable key points obtained by key point detectors [254, 328]. Through evaluation and comparison of the key points, correspondences are established [241]. Thus, the problem reduces to a transformation estimation problem, since *all* correspondences are known, and the number of points in both sets is equal. An overview and comparison of methods for solving transformation estimation problems can be found in [86].

As already mentioned, often one or several correspondences between the template and the reference — prior matches — are known in advance. They can constraint the

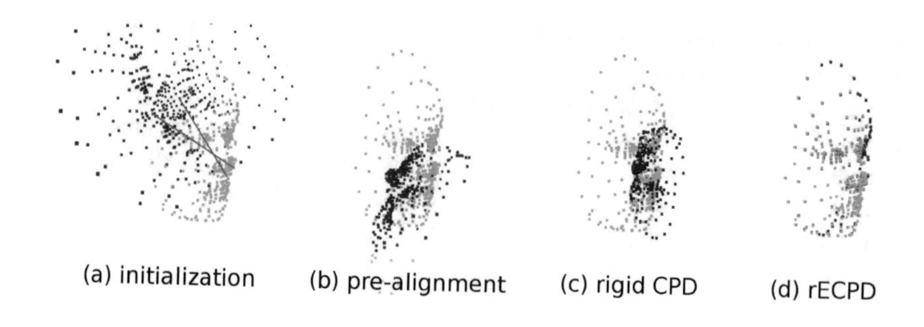

Figure 8.2: Rigid registration of two point clouds (representing human faces) with prior matches. (a): input with prior matches shown as red connecting lines; (b): pre-alignment with Kabsch algorithm; (c): registration results of the pre-aligned point sets by CPD; (d): registration results by our method, the R-ECPD. Given only two priors, our algorithm is capable of resolving rotation and rigid transformation correctly. In contrast, the Kabsch algorithm fails to pre-estimate rotation due to ambiguities in the solution space and CPD fails to register point sets due to severe initial misalignment.

solution space in a favourable way ensuring convergence to the desired range. At the same time, integration of prior matches is not straightforward: either prior matches are used in the pre-alignment step to assure an advantageous initial orientation, or a registration algorithm allows embedding of prior matches explicitly. In the first case, prior matches are decoupled from the registration procedure. To investigate embedding of prior matches into a rigid probabilistic point set registration algorithm is the main contribution of this section. More specifically: (1) we derive a solution for embedding prior correspondences into the CPD algorithm. This results in a robust and convenient approach, the *rigid Extended Coherent Point Drift* (R-ECPD) combining rough pre-alignment and robust dense rigid point set registration; (2) we reveal the synergy arising from the interdependency between the generic rigid registration and predominance of particular correspondences. If the number of prior matches $N_c < D$ (D is space dimensionality), instead of barely taking an insufficient amount of correspondences to pre-estimate rotation, our method additionally exploits the information contained in the point sets.

8.1.1 Extended Coherent Point Drift (ECPD) in General Form

Let $(\mathbf{y}_j, \mathbf{x}_k)$ be a set of correspondence priors with indices $(j,k) \in N_c \subset \mathbb{N}^2$. We model correspondence priors by a product of particular independent density functions

$$P_c(N_c) = \prod_{(j,k)\in N_c} p_c(\mathbf{x}_j, \mathbf{y}_k), \text{ with} \tag{8.1}$$

$$p_c(\mathbf{x}_j, \mathbf{y}_k) = \frac{1}{(2\pi\alpha)^{D/2}} \exp(-\frac{\|\mathbf{x}_j - T(\mathbf{y}_k, \theta)\|^2}{2\alpha^2}). \tag{8.2}$$

Due to the Gaussian form of the distribution, the parameter $\alpha > 0$ reflects the priors' degree of reliability. We incorporate correspondence priors into the CPD method by including the prior probability $P_c(N_c)$ into the GMM in Eq. (2.91) and obtain a modified GMM with the density function

$$\tilde{p}(\mathbf{x}) = P_c(N_c)\, p(x). \tag{8.3}$$

The modified energy function can be derived by considering the negative logarithm of the combined modified GMM:

$$\tilde{E}(\theta, \sigma^2) = -\log\left(P_c(N_c)\prod_{i=1}^{N} p(\mathbf{x}_n)\right) = E(\theta, \sigma^2) - \sum_{(j,k)\in N_c} \log(p_c(\mathbf{x}_j, \mathbf{y}_k)). \tag{8.4}$$

An objective function $\tilde{Q}$ can now be derived utilising the same derivation as in Eq. (2.94). Rewriting the last term of Eq. (8.4) and leaving out the constants the modified objective function reads

$$\tilde{Q} = Q + \frac{1}{2\alpha^2} \sum_{(j,k)\in N_c} \|\mathbf{x}_j - T(\mathbf{y}_k, \theta)\|^2 = Q + \frac{1}{2\alpha^2} \sum_{n=1}^{N} \sum_{m=1}^{M} \tilde{P}_{m,n} \|\mathbf{x}_n - T(\mathbf{y}_m, \theta)\|^2 \tag{8.5}$$

with the $M \times N$ matrix $\tilde{\mathbf{P}}$ of entries

$$\tilde{p}_{j,k} = \begin{cases} 1 & \text{for } (j,k) \in N_c \\ 0 & \text{else} \end{cases}. \tag{8.6}$$

This matrix can be precomputed once. The EM algorithm of ECPD with correspondence priors now reads: 1) in the E-step compute the probability matrix (Eq. (2.95));

2) in the M-step, the modified objective function (Eq. (8.5)) has to be minimised w.r.t. (θ, σ^2). In Secs. 8.1.2 and 8.2.2, T is specified for the rigid and non-rigid cases, respectively.

8.1.2 Rigid Extended Coherent Point Drift (R-ECPD)

To specify the parameter set θ we impose the rules of rigid body dynamics on the GMM centroids. Thus, the R-ECPD objective function reads

$$\tilde{Q}(\mathbf{R}, \mathbf{t}, s, \sigma^2) = \underbrace{\frac{1}{2\sigma^2} \sum_{n=1}^{N} \sum_{m=1}^{M} P^{old}(m|\mathbf{x}_n) \|\mathbf{x}_n - s\mathbf{R}y_m - \mathbf{t}\|^2 + \frac{N_p D}{2} \log \sigma^2 +}_{Q} \\ + \frac{1}{2\alpha^2} \sum_{n=1}^{N} \sum_{m=1}^{M} \tilde{P}_{m,n} \|\mathbf{x}_n - s\mathbf{R}y_m - \mathbf{t}\|^2, \tag{8.7}$$

so that $\mathbf{R}^\mathsf{T}\mathbf{R} = \mathbf{I}$, $\det(\mathbf{R}) = +1$ and $N_p = \sum_{n,m} P^{old}(m|\mathbf{x}_n)$. We find a minimiser of the R-ECPD objective function in Eq. (8.7) taking advantage of lemma 1 from [202]. First, we reformulate Eq. (8.7) to match the form $\mathrm{tr}(\mathbf{A}^\mathsf{T}\mathbf{R})$. To eliminate the translation term from $\tilde{Q}$, we compute its first derivative w.r.t. $\mathbf{t}$ and equate it to zero. This yields the modified term

$$\mathbf{t} = \frac{\left[\mathbf{X}^\mathsf{T}\mathbf{P}^\mathsf{T}\mathbf{1} + \frac{\sigma^2}{\alpha^2}\mathbf{X}^\mathsf{T}\tilde{\mathbf{P}}^\mathsf{T}\mathbf{1} - s\mathbf{R}(\mathbf{Y}^\mathsf{T}\mathbf{P}\mathbf{1} + \frac{\sigma^2}{\alpha^2}\mathbf{Y}^\mathsf{T}\tilde{\mathbf{P}}\mathbf{1})\right]}{N_P^c}, \tag{8.8}$$

where $\mathbf{1} = (1, 1, \ldots, 1)^\mathsf{T}_{M\times 1}$ and $\mathbf{P}$ is a matrix with elements $p_{m,n} = P^{old}(m|x_n)$ (P^{old} is computed as in the original CPD). From Eq. (8.8) follows

$$N_P^c = \mathbf{1}^\mathsf{T}\mathbf{P}\mathbf{1} + \frac{\sigma^2}{\alpha^2}\mathbf{1}^\mathsf{T}\tilde{\mathbf{P}}\mathbf{1} = N_P + \frac{\sigma^2}{\alpha^2}N_{P_c} \tag{8.9}$$

with $N_{P_c} = \sum_{(j,k)\in N_c} p_c(\mathbf{x}_j, \mathbf{y}_k)$. Considering the mean vectors

$$\mu_x = \mathbf{E}(X) = \frac{\mathbf{X}^\mathsf{T}\mathbf{P}^\mathsf{T}\mathbf{1}}{N_P}, \text{ and} \tag{8.10}$$

$$\mu_y = \mathbf{E}(Y) = \frac{\mathbf{Y}^\mathsf{T}\mathbf{P}\mathbf{1}}{N_P} \tag{8.11}$$

we define the modified mean vectors

$$\mu_x^c = \frac{1}{N_P^c}\left[N_P\mu_x + \frac{\sigma^2}{\alpha^2}\mathbf{X}^\top\tilde{\mathbf{P}}^\top\mathbf{1}\right] \text{ and} \tag{8.12}$$

$$\mu_y^c = \frac{1}{N_P^c}\left[N_P\mu_y + \frac{\sigma^2}{\alpha^2}\mathbf{Y}^\top\tilde{\mathbf{P}}\mathbf{1}\right]. \tag{8.13}$$

By substituting $\mathbf{t}$ from Eq. (8.8) back into Eq. (8.7) and further defining centred point set matrices $\hat{\mathbf{X}}_c = \mathbf{X} - \mathbf{1}(\mu_x^c)^\top$ and $\hat{\mathbf{Y}}_c = \mathbf{Y} - \mathbf{1}(\mu_y^c)^\top$, we rewrite the objective function as

$$\tilde{Q} = -\underbrace{\frac{s}{\sigma^2}}_{>0}\underbrace{\operatorname{tr}\left[\left((\hat{\mathbf{X}}_c^\top\mathbf{P}^\top\hat{\mathbf{Y}}_c)^\top + \frac{\sigma^2}{\alpha^2}(\hat{\mathbf{X}}_c^\top\tilde{\mathbf{P}}^\top\hat{\mathbf{Y}}_c)^\top\right)\mathbf{R}\right]}_{=:\,[\mathbf{A}^\top\mathbf{R}]} \tag{8.14}$$

(to achieve this form, we also utilise the orthogonality of the rotation matrix and the fact that the trace is linear and invariant under cyclic matrix permutations). Minimisation of $\tilde{Q}$ is equal to maximisation of $\operatorname{tr}(\mathbf{A}^\top\mathbf{R})$ as defined in Eq. (8.14), so that $\mathbf{R}^\top\mathbf{R} = \mathbf{I}$ and $\det(\mathbf{R}) = +1$. We apply Lemma 1 from [202] and obtain the rotation matrix

$$\mathbf{R} = \mathbf{U}\mathbf{C}\mathbf{V}^\top \text{ with} \tag{8.15}$$

$$\mathbf{C} = \operatorname{diag}(1, 1, \ldots, \det(\mathbf{U}\mathbf{V}^\top)) \text{ and} \tag{8.16}$$

$$\mathbf{U}\mathbf{S}\mathbf{V}^\top = \operatorname{svd}\left(\hat{\mathbf{X}}_c^\top\mathbf{P}^\top\hat{\mathbf{Y}}_c + \frac{\sigma^2}{\alpha^2}(\hat{\mathbf{X}}_c^\top\tilde{\mathbf{P}}^\top\hat{\mathbf{Y}}_c)\right). \tag{8.17}$$

Analogously, in order to obtain the optimal s and σ^2, respective derivatives of $\tilde{Q}$ in Eq. (8.14) have to be computed and equated to zero. The whole method is summarised in Algorithm 7.

Now we are ready to investigate how R-ECPD contrasts from most closely related approaches. From CPD with a pre-alignment step, our approach differs in the way that prior matches influence the registration procedure explicitly in every EM iteration. If $\alpha = 1$, the prior matches are not valid, and our approach reduces to CPD. In opposite, if α positive and infinitely close to zero, R-ECPD operates similarly to the Kabsch algorithm [158]. Indeed, the term $(\hat{\mathbf{X}}_c^\top\tilde{\mathbf{P}}^\top\hat{\mathbf{Y}}_c)^\top$ in Eq. (8.14) can be rewritten as $(\hat{\mathbf{Y}}_c^\top\tilde{\mathbf{P}}\hat{\mathbf{X}}_c)$. At the same time, our approach differs from the Kabsch algorithm in several ways. Firstly, all points of both point sets are involved in the optimisation, whereas prior matches are usually several orders of magnitude stronger weighted. Secondly, different weighting and thus the uncertainty level adjustment for every distinct prior is possible. This allows combining priors from

Algorithm 7 Rigid Extended Coherent Point Drift

Input: Reference and template point sets $\mathbf{X}, \mathbf{Y}$, prior matches N_c

1: **Initialisation:** $\mathbf{R} = \mathbf{I}$, $\mathbf{t} = \mathbf{0}$, $s = 1$, $0 \leq w \leq 1$, $\sigma^2 = \frac{1}{DNM} \sum_{n=1}^{N} \sum_{m=1}^{M} ||x_n - y_m||^2$,

2: $\tilde{\mathbf{P}}$ precomputed as in Eq. (8.5).

3: **EM-optimisation**

4: **while** not converge **do**

5: E-step: compute $\mathbf{P}$ with $p_{m,n} = p^{old}(m|\mathbf{x}_n)$ as in the original CPD.

6: M-step: solve for $\mathbf{R}, s, \mathbf{t}, \sigma^2$

7: N_P^c as in Eq. (8.9) and μ_x^c, μ_y^c as in Eq. (8.12).

8: $\hat{\mathbf{X}}_c = \mathbf{X} - \mathbf{1}(\mu_x^c)^\mathsf{T}, \hat{\mathbf{Y}}_c = \mathbf{Y} - \mathbf{1}(\mu_y^c)^\mathsf{T}$

9: Compute $\mathbf{R}$ as in Eq. (8.15).

10: $s = \frac{\text{tr}[\mathbf{A}^\mathsf{T}\mathbf{R}]}{\text{tr}[\hat{\mathbf{Y}}_c^\mathsf{T} \text{diag}(\mathbf{P1})\hat{\mathbf{Y}}_c + \frac{\sigma^2}{\alpha^2}\hat{\mathbf{Y}}_c^\mathsf{T} \text{diag}(\tilde{\mathbf{P}}\mathbf{1})\hat{\mathbf{Y}}_c]}$ with $\mathbf{A}$ as in Eq. (8.15).

11: Compute $\mathbf{t}$ as in Eq. (8.8).

12: $\sigma^2 = \text{tr}[\hat{X}_c^\mathsf{T} \text{diag}(\mathbf{P1})\hat{X}_c] - s\, \text{tr}[\mathbf{A}^\mathsf{T}\mathbf{R}] + s^2 \frac{\sigma^2}{\alpha^2} \text{tr}[\hat{Y}_c^\mathsf{T} \text{diag}(\mathbf{P1})\hat{Y}_c]$

13: **end while**

Output: Aligned point set $T(\mathbf{Y}) = s\mathbf{Y}\mathbf{R}^\mathsf{T} + \mathbf{1}\mathbf{t}^\mathsf{T}$. Correspondence probabilities are given by $\mathbf{P}$.

various sources in a flexible manner. Depending on settings and number of prior matches, different effects are possible such as pre-alignment with an insufficient number of prior matches or robust registration in the presence of clustered outliers.

8.1.3 Evaluation

In this section, we evaluate the performance of R-ECPD in synthetic and real-world scenarios. We use publicly available implementation of CPD [203] as well as Matlab's implementation of the Kabsch algorithm [277]. We implemented R-ECPD in C++ and ran experiments on the system with 32 GB RAM and Intel Xeon E3-1245 processor.

8.1.3.1 Experiments with Synthetic Data

In the experiment on synthetic data, we take the sparse 3D face point set from [203], duplicate it and systematically change the orientation of the copy jointly around the x, y and z axes with the angle-step size of 36°. This results in 1000 different initial orientations of the point sets. The scaling factor and the translation vector are chosen randomly. In the preprocessing step, we establish prior matches between the point sets. Those are obtained through comparison of the Persistent Feature Histograms (PFHs) [241] at the 3D key points. We find the 3D key points with the Harris3D (H3D) [254] and the Intrinsic Shape Signatures (ISS) [328] 3D key point detectors. Only correspondences with the highest match scores are taken into account. Thus, we obtain two reliable matches in total. Note that in both cases the prior matches relate not exactly the same points.

For every initial orientation, we perform a preprocessing step with the Kabsch algorithm and eventually register the point sets with CPD and R-ECPD.In the latter case, prior matches are input directly (no preprocessing step is undertaken) as one of the algorithm's parameters. In Fig. 8.2 selected results are shown. CPD is able to restore correct rigid transformations in 242 cases out of 1000 (24% success rate), mostly when absolute values of the individual inclination angles do not exceed 72 degrees. CPD performs as if no rotation pre-estimate would be accomplished since two prior matches do not suffice to unambiguously pre-estimate rotation in 3D space ($N_c \leq D$). In opposite, R-ECPD is able to recover correct rigid transformations in all cases, achieving a 100% success rate in this experiment. This experiment shows that the synergetic effect of joint point set registration and explicit incorporation of prior matches results in the correct transformation in this underconstrained case. In other words, taking into account all other points allows compensating for missing information. After that, the above-stated experiment is repeated in different configurations, swapping reference and template point set and using only one prior match at a time. In 25% of the cases, CPD was able to resolve rotation correctly. Interestingly, rECPC achieves a 98% success rate with only one prior match. For the sake of completeness, we repeat the experiment also with three prior matches, obtaining a 100% success rate for both methods, as expected. In this experiment R-ECPD — both with one and two matches — clearly outperforms the state-of-the-art rigid registration algorithm CPD with the preprocessing step. The results also reveal that the performance of R-ECPD is equivalent to the performance of CPD with pre-alignment step if $N_c = D$. In this experiment point sets do not contain clustered outliers. Under real-world conditions, this prerequisite is not always fulfilled, and further advantages of R-ECPD are shown below.

8.1.3.2 Experiments with Real Data

In the experiment on real data, we evaluate our method on the scans form the Lion dataset (Fig. 8.3, (a); reference is located on the right). Both point clouds represent the same object, namely the statue of a lion. The scans were reconstructed with the PMVS-2 algorithm [99] with different parameters. Therefore, they are noisy and contain areas with clustered outliers constituting 1.5% of points in the reference

Figure 8.3: Rigid registration of lion dataset. (a): inputs with prior matches shown on red; (b): registration results by CPD (left) and cloud-to-cloud distance in Blue < Green < Yellow < Red scale (right), saturation point = 7.0 distance units, (mean error; std. deviation) = (0.940; 0.902); (c): registration results by R-ECPD and cloud-to-cloud distance in Blue < Green < Yellow < Red scale (right), saturation point = 7.0 distance units, (mean error; std. deviation) = (0.483; 0.855).

and 3% of points in the template point clouds. The reference point cloud contains $5 \cdot 10^5$ points and the template point cloud $1.5 \cdot 10^5$ points.

We perform the preprocessing step to obtain prior matches that are used to pre-align the scans (in the case of CPD) or as one of the input parameters (in the case of R-ECPD). Correspondence establishment follows similar steps as in the experiment on synthetic data. The ISS key point detector is used, and two correspondences with the highest scores are taken. Eventually, we register scans with CPD and R-ECPD. The running times amounted to 80 seconds and 72 seconds, respectively. Results are shown in Fig. 8.3. The influence of clustered outliers is substantial, and CPD fails to register the scans in the ROI correctly. In contrast, R-ECPD performs more accurately. Results are also reflected in the corresponding cloud-to-cloud comparisons (Fig. 8.3, (b), (c), right parts contain plots with root-mean-square errors). The measure reveals that R-ECPD, provided reliable prior matches, is more robust compared to CPD in the presence of clustered outliers in this experiment.

8.1.3.3 The Stadium Dataset

In this experiment, we evaluate our approach on partial scans. Therefore, we consider the Stadium dataset (Fig. 8.4). The scans were computed by the PMVS-2 algorithm [99]. The reference point set contains $7.78 \cdot 10^6$ points and the template point set $1.4 \cdot 10^6$ points. The scans are placed in the different reference frames and

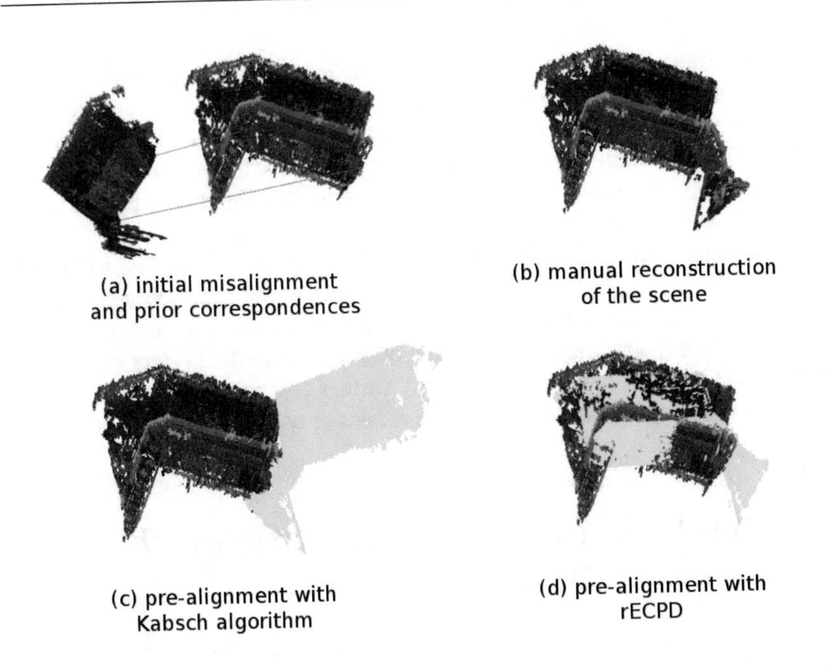

Figure 8.4: Registration of the point clouds representing partially overlapping shapes of a stadium. (a): initial misalignment and prior matches are shown by the red connecting lines; (b): manual reconstruction of the scene; (c): the result of the pre-alignment with the Kabsch algorithm; (d): the result of the pre-alignment with our method, the R-ECPD. In (c) and (d), the template is shown in cyan for better distinction.

are strongly misaligned (flipped by almost 180° and shifted). The reference point set (Fig. 8.4, (a), to the right) contains $\approx$35% of clustered outliers, and the template point set contains $\approx$19% of clustered outliers.

We detect ISS key points [328] in both point sets. We then compare Persistent Feature Histograms [241] at those key points and consider two correspondences with the highest scores as prior matches. Firstly, we estimate the transformation between the scans with the Kabsch algorithm (Fig. 8.4, (c)) and our approach (Fig. 8.4, (d)). Kabsch algorithm finishes in less than a second (Matlab version [277]), whereas R-ECPD needs 181 seconds (C++ version). The runtime discrepancy arises due to the different amount of points used by the algorithms, *i.e.,* in R-ECPD, all the points are involved in the computation.

In the former case (Kabsch), the transformation is not unique, and the result of the pre-alignment is erroneous — point clouds overlap minimally, and rotation could not be approximated (with the angular error of $\approx 110°$). In the latter case (R-ECPD), the rotation is pre-estimated correctly, though the entire transformation is still far from the correct. On Fig. 8.4, (b) the reconstruction of the scene through a manual interaction is shown.

In this experiment, R-ECPD was able to solve the transformation estimation problem more robustly compared to CPD. All points of the point sets compensated for missing priors. Nevertheless, clustered outliers are still present among these points, negatively affecting the overall result. To obtain the correct registration result (close to the shown in Fig. 8.4, (b)), further processing is needed, such as the establishment of additional correspondences or segmentation of overlapping regions.

8.2 Non-Rigid Point Set Registration with Prior Correspondences

In this section, we propose the Extended Coherent Point Drift (ECPD) algorithm allowing to include prior information in the form of point correspondences into the non-rigid registration process to influence it in a favourable way (see Fig. 8.12). It allows combining application-specific correspondence search algorithms operating on different information sources with the robust non-rigid point set registration algorithm. Furthermore, we adopt a coarse-to-fine registration strategy which maintains the impact of the correspondence priors and accelerates the registration process linearly.

Our contributions are stated as follows. We derive ECPD method by embedding the correspondence priors into CPD in a closed form and thus extend its scope. To the best of our knowledge, utilising the correspondence priors for point set registration in a closed-form was not shown in the literature so far, at least for the probabilistic case. We suggest how subsampling can be adopted in the context of non-rigid registration resulting in linear speedup as a function of the subsampling factor and point sets sizes. We also show an efficient implementation of ECPD in a heterogeneous environment with a GPU. Finally, we demonstrate several application-specific approaches for finding correspondence priors and show that their appropriate utilisation in combination with the proposed ECPD can address diverse issues, especially 1) resolving complex non-rigid deformations, not addressed by the original coherency constraint; 2) obtaining accurate registrations

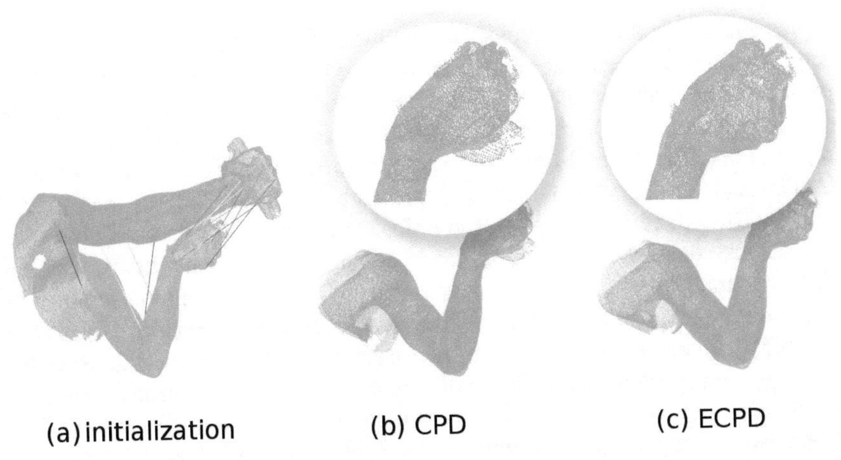

Figure 8.5: Non-rigid registration of two 3D point sets from [23] representing arms in different poses. (a): initial alignment with correspondences obtained through a comparison of the Persistent Feature Histograms at ISS 3D keypoints; (b): result of the non-rigid registration with CPD; (c): result of the non-rigid registration with ECPD using the correspondence priors. Complex non-rigid deformations (combination of supination, flection and abduction of the arm) in the area of the hand were resolved with higher precision, compared to (b). See Fig. 8.9 for a detailed comparison.

in the regions of interest (ROI) for point sets with structured outliers; 3) obtaining registration results linearly faster with respect to the baseline method CPD.

8.2.1 Related Work

In this section, we focus on general-purpose point set alignment and review recent works closely related to ECPD. Note that most of the discussed works appeared after the first publication of ECPD [122].

Motivated by the affordability and pervasiveness of RGB-D sensors, Saval-Calvo *et al.* [250] have recently proposed an extension of CPD, which takes into account colour information - Colour CPD (CCPD). If colours accurately match between two point clouds, the method can register with higher accuracy than CPD. On the contrary, if colours originate from an RGB-D sensor, brightness inconsistency will affect the registration accuracy of CCPD. Wang *et al.* [301] extract salient structures from the input point sets and use them to weight Gaussian fields. Point sets are represented by undirected graphs, and the method is computationally expensive (partially, due to combinatorial optimisation) so that it is suitable for registration of sparse and small point sets. Dependent Landmark Drift (DLD) [142] uses statistical shape models obtained by principal component analysis to guide the registration. The method operates for the learned object classes with available training data, and a template always represents a mean shape from the training set. In DLD, landmarks refer to the points in the training set. Availability of a statistical shape prior is a very strong assumption. In many cases, training data with inter-instance correspondences can not always be available in practice, and the template can substantially differ from scenario to scenario. Kolesov *et al.* [165] propose a stochastic point set registration (SPSR) approach with simulated annealing which allows for the integration of different types of constraints (smooth deformation without self-intersections, rigidity and stationarity constraints). The approach to integration of the prior correspondences in SPSR is different from our technique. In SPSR, prior correspondences are used to infer a transformation field for the sparse shape representation in the first step. For inferring deformation field from a sparse set of prior correspondences, a method [46] is used. Next, the transformation field is applied to the whole deformable point set, and the point sets are registered anew with the stationarity constraints on the prior correspondences. Thus, integration of prior correspondences in SPSR results in a multistep scheme and the soft point stationarity constraint in the core SPSR method. In Ma *et al.* [186], point-GMM membership probabilities are assigned based on matching of local feature descriptors [37, 240] calculated on both point sets, and radial basis functions are used for the parametrisation. Point set registration is then posed as an expectation conditional maximisation problem. Compared to CPD, [186] imposes a constraint on the preservation of local structures through the matching of local rotation-invariant features. Wang and Chen [299] address non-rigid point set registration in a similar manner, but propose fuzzy correspondences (allowing one-to-many mappings) instead of one-to-one assignments as in [186].

The difference of our technique compared to the methods discussed above is that *ECPD accepts correspondences in a unified form, abstracting from the origin of the correspondences*. Prior correspondences can originate from keypoint matches, intensity matches, manually selected or automatically detected landmarks, and, more importantly, the establishment of prior correspondences is decoupled from registration and is not a part of ECPD itself. We are not restricted by individual channels or techniques which can be efficient for one problem and less efficient for another problem. Depending on the problem, *ECPD can be used in combination with the respectively most efficient correspondence search approach*. In contrast to [165], we solve registration with prior correspondences in a single optimisation procedure, without requiring a two-step scheme with pre-processing.

8.2.2 Non-Rigid Extended Coherent Point Drift (ECPD)

In the proposed approach the case of general non-rigid alignment can be addressed by defining the displacement field for $\mathbf{Y}$ as

$$T(\mathbf{Y},v) = \mathbf{Y} + v(\mathbf{Y}). \tag{8.18}$$

We build upon the regularising prior on the displacement field of CPD [203]. In the Bayesian framework it can be formulated as $p(v) = \exp(-\frac{\lambda}{2}\phi(v))$ with a weighting parameter $\lambda \in \mathbb{R}$ and a regularisation function $\phi(v)$. By multiplying the density function (2.91) with $p(v)$ or equivalently adding the exponent to the negative likelihood function (2.93) we integrate the regularising prior into the GMM and obtain

$$f(v,\sigma^2) = E(v,\sigma^2) + \frac{\lambda}{2}\phi(v). \tag{8.19}$$

In this work, we extend this framework by including the additional correspondence priors (8.1) into Eq. (8.4) and obtain the following energy function

$$\tilde{f}(v,\sigma^2) = \tilde{E}(v,\sigma^2) + \frac{\lambda}{2}\phi(v). \tag{8.20}$$

To solve the M-step, the energy function (8.20) needs to be minimised w.r.t. v and σ^2. Following the same principles as in Eq. (2.94) we derive an upper bound for the energy function (8.20) that reads

$$\begin{aligned}\tilde{Q}(v,\sigma^2) = \frac{1}{2\sigma^2}\sum_{n=1}^{N}\sum_{m=1}^{M} P^{old}(m|\mathbf{x}_n)\|\mathbf{x}_n - T(\mathbf{y}_m,v,\sigma^2)\|^2 + \\ \frac{1}{2\alpha^2}\sum_{n=1}^{N}\sum_{m=1}^{M} \tilde{P}_{m,n}\|\mathbf{x}_n - T(\mathbf{y}_m,v,\sigma^2)\|^2 + \frac{N_P D}{2}\log\sigma^2 + \frac{\lambda}{2}\phi(v).\end{aligned} \tag{8.21}$$

The regularisation function of CPD can be written in the reproducing kernel Hilbert space as

$$\phi(v) = \int_{\mathbb{R}^D} \frac{|\tilde{v}(\mathbf{s})|^2}{\tilde{G}(\mathbf{s})}\, d\mathbf{s}. \tag{8.22}$$

Here, $\tilde{G}$ is the Fourier transform of a kernel function G which in turn describes a positive function that approaches zero as $\|\mathbf{s}\| \to \infty$. As in [203] we use a Gaussian kernel function $G(\mathbf{s}) = \exp(-\|\frac{\mathbf{s}}{\beta}\|^2)$. Furthermore, $\tilde{v}$ is the Fourier transform of the displacement field v in Eq. (8.18) and $\mathbf{s}$ is a frequency domain variable. This regularisation can be interpreted as low-pass filtering with respect to the displacement field v. The filtered frequencies can be adjusted via the parameter β. To obtain v, we minimise the objective function (8.21) with respect to v by treating

σ^2 as a constant. Note that due to our choice to integrate the correspondence priors (Eq. (8.1)) the minimising function for the corresponding energy (8.20) still has the form of a radial basis function as in [203, 204], *i.e.*,

$$v(\mathbf{z}) = \sum_{m=1}^{M} \mathbf{w}_m G(\mathbf{z} - \mathbf{y}_m). \tag{8.23}$$

See Sec. 8.2.5 for the proof of the proposition (8.23). To compute the displacement field $v(\mathbf{Y}) = \mathbf{GW}$ minimising the energy (8.20) (where in our case $\mathbf{G}$ is a symmetric Gram matrix with entries $G_{i,j} = \exp(-\|\frac{\mathbf{y}_i - \mathbf{y}_j}{\beta}\|^2)$), we substitute v into Eq. (8.21) and leave out all terms independent of v. This yields

$$\begin{aligned} \tilde{Q}(\mathbf{W}) = \frac{1}{2\sigma^2} \sum_{n=1}^{N} \sum_{m=1}^{M} P^{old}(m|\mathbf{x}_n) \|\mathbf{x}_n - T(\mathbf{y}_m, \mathbf{W})\|^2 + \\ + \frac{1}{2\alpha^2} \sum_{n=1}^{N} \sum_{m=1}^{M} \tilde{P}_{m,n} \|\mathbf{x}_n - T(\mathbf{y}_m, \mathbf{W})\|^2 + \frac{\lambda}{2} \mathbf{W}^\mathsf{T} \mathbf{G} \mathbf{W}. \end{aligned} \tag{8.24}$$

Minimising $\tilde{Q}$ with respect to $\mathbf{W}$ will minimise the energy function (8.20) with respect to v. Setting the derivative of $\tilde{Q}$ to zero in matrix form yields

$$\begin{aligned} \frac{\partial \tilde{Q}}{\partial \mathbf{W}} = \mathbf{G} \bigg(\frac{1}{\sigma^2} [\mathrm{diag}(\mathbf{P1})(\mathbf{Y} + \mathbf{GW}) - \mathbf{PX}] \\ + \frac{1}{\alpha^2} [\mathrm{diag}(\tilde{\mathbf{P}}\mathbf{1})(\mathbf{Y} + \mathbf{GW}) - \tilde{\mathbf{P}}\mathbf{X}] \bigg) + \lambda \mathbf{GW} = 0. \end{aligned} \tag{8.25}$$

Here, $\mathrm{diag}(\cdot)$ is a diagonal matrix. Multiplying the whole equation with $\mathbf{G}^{-1}\sigma^2$ (which always exists for a Gaussian kernel [204]) and rearranging it, we obtain

$$\begin{aligned} \left(\mathrm{diag}(\mathbf{P1})\mathbf{G} + \frac{\sigma^2}{\alpha^2} \mathrm{diag}(\tilde{\mathbf{P}}\mathbf{1})\mathbf{G} + \lambda \sigma^2 \mathbf{I} \right) \mathbf{W} = \\ \mathbf{PX} - \mathrm{diag}(\mathbf{P1})\mathbf{Y} + \frac{\sigma^2}{\alpha^2} (\tilde{\mathbf{P}}\mathbf{X} - \mathrm{diag}(\tilde{\mathbf{P}}\mathbf{1})\mathbf{Y}). \end{aligned} \tag{8.26}$$

The transformed positions of $\mathbf{y}_m$ are found according to Eq. (8.18) as $T(\mathbf{Y}, \mathbf{W}) = \mathbf{Y} + \mathbf{GW}$. Thereafter, we obtain σ^2 by setting the corresponding derivative of Eq. (8.21) to zero. This yields the same result as in [203], *i.e.*,

$$\begin{aligned} \sigma^2 &= \frac{1}{N_P D} \sum_{n=1}^{N} \sum_{m=1}^{M} \|\mathbf{x}_n - T(\mathbf{y}_m, \mathbf{W})\|^2 = \\ &= \frac{1}{N_P D} (\mathrm{tr}(\mathbf{X}^\mathsf{T} \mathrm{diag}(\mathbf{P}^\mathsf{T}\mathbf{1})\mathbf{X}) - 2\,\mathrm{tr}((\mathbf{PX})^\mathsf{T}\mathbf{T}) + \mathrm{tr}(\mathbf{T}^\mathsf{T} \mathrm{diag}(\mathbf{P1})\mathbf{T})). \end{aligned} \tag{8.27}$$

Algorithm 8 ECPD with Correspondence Priors

Input: Point sets $\mathbf{X}, \mathbf{Y}$ and index set for correspondence priors N_c
Output: Aligned point set $T(\mathbf{Y})$

- **Initialisation:** $\mathbf{W} = 0$, $\sigma^2 = \frac{1}{DMN} \sum_{m,n=1}^{M,N} \|\mathbf{x}_n - \mathbf{y}_m\|^2$,
- $0 \leq w \leq 1, \beta > 0, \lambda > 0$.
- Construct $\mathbf{G}$: $G_{i,j} = \exp^{(-\frac{1}{2\beta^2}\|\mathbf{y}_i - \mathbf{y}_j\|^2)}$,
- Precompute $\tilde{\mathbf{P}}$, as in Eq. (8.6).
- **EM optimisation**, repeat until convergence:
  - E-Step: compute $\mathbf{P}$, with $p_{m,n} = P^{old}(m|\mathbf{x}_n)$ as in Eq. (2.95).
  - M-Step: Solve for T, σ^2.
  - Solve Eq. (8.26) with respect to $\mathbf{W}$.
  - $N_P = \mathbf{1}^\top \mathbf{P} \mathbf{1}, \mathbf{T} = \mathbf{Y} + \mathbf{G}\mathbf{W}$.
  - $\sigma^2 = \frac{1}{N_P D}[\mathrm{tr}(\mathbf{X}^\top \mathrm{diag}(\mathbf{P}^\top \mathbf{1})\mathbf{X})$
  - $-2\,\mathrm{tr}((\mathbf{P}\mathbf{X})^\top \mathbf{T}) + \mathrm{tr}(\mathbf{T}^\top \mathrm{diag}(\mathbf{P}\mathbf{1})\mathbf{T})]$.
- The aligned point set is $T(\mathbf{Y}, \mathbf{W}) = \mathbf{Y} + \mathbf{G}\mathbf{W}$.
- The probability for correspondences is given by $\mathbf{P}$.

ECPD is summarised in Alg. 8. Convergence criteria are defined in terms of the maximum number of iterations and the smallest feasible step size of the EM algorithm.

8.2.3 Implementation

For fast registration of large point sets, several approaches were introduced in [203] for CPD. First, the *Fast Gauss Transform* (FGT) allows to efficiently compute sums of exponentials in $\mathcal{O}(M+N)$ time. It can be applied in exactly the same way for ECPD. The second method is the *low-rank matrix approximation* of the Gram matrix $\hat{\mathbf{G}} = \mathbf{Q}\Lambda\mathbf{Q}^\top$, where $\Lambda_{K\times K}$ denotes the diagonal matrix of the largest eigenvalues and $\mathbf{Q}_{M\times K}$ the corresponding eigenvectors in matrix form. Also, the Woodbury matrix identity can be applied to efficiently solve the linear system for the transformation $\mathbf{W}$. In our case we need to solve the modified system (8.26) and rewrite it as follows

$$\underbrace{([\text{diag}(\mathbf{P1}) + \frac{\sigma^2}{\alpha^2}\text{diag}(\tilde{\mathbf{P}}\mathbf{1})]}_{\text{diag}(\hat{\mathbf{P}}\mathbf{1})}\mathbf{G} + \lambda\sigma^2\mathbf{I})\mathbf{W} = \underbrace{\mathbf{PX} - \text{diag}(\mathbf{P1})\mathbf{Y} + \frac{\sigma^2}{\alpha^2}(\tilde{\mathbf{P}}\mathbf{X} - \text{diag}(\tilde{\mathbf{P}}\mathbf{1})\mathbf{Y})}_{F}$$

$$\Longleftrightarrow (\mathbf{G} + \lambda\sigma^2\text{diag}(\hat{\mathbf{P}}\mathbf{1})^{-1})\mathbf{W} = \text{diag}(\hat{\mathbf{P}}\mathbf{1})^{-1}F. \quad (8.28)$$

Therefore, the Woodbury matrix identity for solving the modified system (8.26) with the low-rank matrix approximation reads

$$(\mathbf{Q}\Lambda\mathbf{Q}^\top + \lambda\sigma^2\text{diag}(\hat{\mathbf{P}}\mathbf{1})^{-1})^{-1} = \frac{1}{\lambda\sigma^2}\left[F - \text{diag}(\hat{\mathbf{P}}\mathbf{1})\mathbf{Q}\left(\lambda\sigma^2\Lambda^{-1} + \mathbf{Q}^\top\text{diag}(\hat{\mathbf{P}}\mathbf{1})\mathbf{Q}\right)^{-1}\mathbf{Q}^\top F\right). \quad (8.29)$$

We aim at the non-rigid registration of large point sets ($> 10^5$ points) and thus opt for an optimised implementation of ECPD on a system with a multicore CPU and a GPU. The FGT was implemented as a heterogeneous function applying acceleration techniques similar to those described in [74]. Here, the tangible impact on the performance yields both the enabling of high occupancy and extensive use of faster shared and register memory for reusable data. For the computation of the eigenvalue decomposition, which the low-rank approximation of the Gram matrix G is based on, we rely on Implicitly Restarted Lanczos Method [62] (provided by the library ARPACK [1]) in combination with the FGT. In the final iterations of the EM optimisation — when the width of Gaussians becomes small — the algorithm switches to the truncated Gaussian approximation mode (truncated mode) and the complexity increases to $\mathcal{O}(MN)$. The acceleration of this mode is achieved by taking advantage of the parallel portable shared memory programming model offered by OpenMP [212]. Even though the implementation of the truncated mode is scalable with the number of multiprocessors, the improvement on a multicore CPU is still not sufficient to compensate for the increase in the computation time caused by the $\mathcal{O}(MN)$ complexity.

8.2.3.1 Coarse-To-Fine Strategy with Correspondence Preserving Subsampling

To reduce the registration time further, we introduce the procedure of *correspondence preserving subsampling* (CPS) and adopt a coarse-to-fine strategy. CPS is a joint subsampling procedure of two (generally k) point sets which reduces the number of points in one or several point sets according to some subsampling rule (*e.g.*, uniformly) and guarantees to preserve established correspondence priors between those sets. Suppose n_c to be the number of correspondence priors between the point

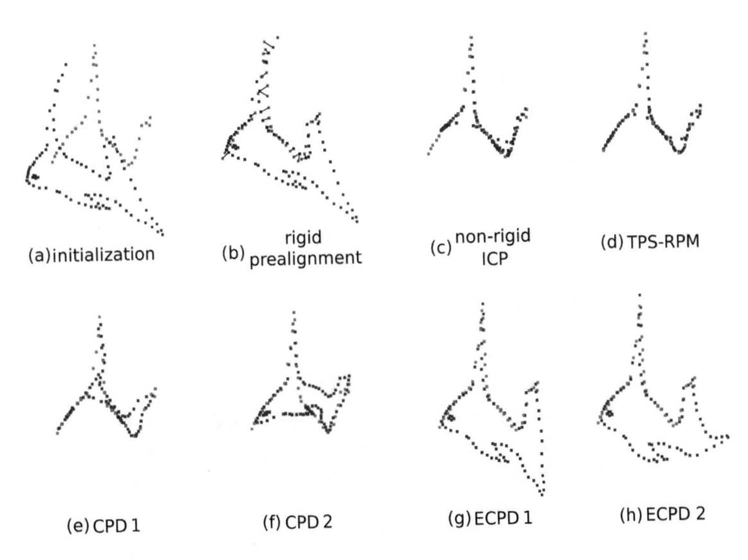

Figure 8.6: Non-rigid registration of a 2D "Fish" dataset. The reference is shown in red, the template is shown in blue. (a): initialisation of the point sets; (b): rigid pre-alignment with established correspondence priors; note that rigid CPD recovers prior correspondences by far not correct; results of the registrations with non-rigid ICP [93] (c); TPS-RPM [70] (d); CPD [203] with two different parameter sets: $\beta = 1.0$, $\lambda = 1.0$ (e) and $\beta = 2.0$, $\lambda = 2.0$ (f); proposed approach ECPD with two different parameter sets: $\beta = 14.0$, $\lambda = 14.0$, $\alpha = 10^{-8}$ (g) and $\beta = 8.0$, $\lambda = 0.8$, $\alpha = 10^{-8}$ (h).

sets and t the subsampling factor. Then, the subsampled template contains at most $\lfloor \frac{N}{t} \rfloor + n_c$ points.

ECPD in subsampled mode operates as follows: 1) perform the CPS on the template point set; 2) register the subsampled template point set with the reference point set; 3) register the registered subsampled template (the result of the second step) as a reference with the initial dense template. Thereby, the amount of points in the result remains unchanged, independently of the subsampling factor. We observe an essential property of the CPS. Since correspondences between the subsampled template and the initial template are known, they can be taken as strong correspondence priors on the final registration step. In the following, we give a rough comparison of the number of operations required to register a given dataset with and without CPS. For T_{plain} operations required by the plain non-rigid

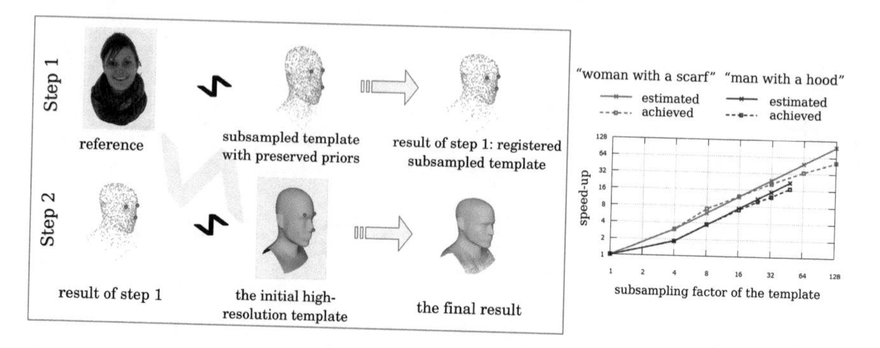

Figure 8.7: Acceleration scheme of ECPD. In the first step, the subsampled template is registered with the reference, while all original correspondences are preserved. In the second step, the registered subsampled template (the result of step one) as an intermediate reference is registered with the original unaltered template. Using the proposed coarse-to-fine acceleration scheme, twentyfold and higher speedups are possible without a perceptible decay in the accuracy. On the right side, a plot in a log-scale is given comparing theoretically estimated and practically achieved speedups for two use cases. Up to $s = 32$, our model very accurately reflects the achieved speedups.

registration it holds

$$T_{plain} = c_1 MN, \tag{8.30}$$

and for $T_{subs.}$ operations required by ECPD with subsampling it holds

$$T_{subs.} = c_2\left(M \cdot \frac{N}{t} + \frac{N}{t} \cdot N\right) = c_2 \frac{N(M+N)}{t}, \tag{8.31}$$

where c_1 and c_2 are constants. Thus, the speedup of the algorithm with subsampling can be estimated as

$$s(m,n,t) = \frac{T_{plain}}{T_{subs.}} = \frac{c_1 MN}{\frac{c_2 N(M+N)}{t}} = c_s \frac{Mt}{M+N} \tag{8.32}$$

To achieve a speedup s, according to Eq. (8.32), the subsampling factor t should be chosen as $\lceil \frac{s(M+N)}{M} \rceil$. Note that the given estimate does not consider the constant time spent for the template subsampling and approximation of the Gram matrix G for the template. A summary of the coarse-to-fine acceleration scheme together with the comparison of the theoretically estimated and practically achieved speedups for two use cases (see Sec. 10.1.4) is given in Fig. 8.7.

8.2.4 Evaluation

In this section, we detail the experimental evaluation of ECPD with synthetic and real data. The experiments are designed to verify the main advantages of ECPD — an improvement of registration quality — both in the presence of structured outliers and without those when correspondence priors are available. If appropriate, we compare our approach against CPD (which we consider the baseline) and other non-rigid registration methods. We run ECPD on a system with 32 GB RAM, Intel Xeon processor and NVIDIA GTX 660 Ti graphics card.

8.2.4.1 Experiments with Synthetic Data

In our first experiment, we compare the performances of ECPD, CPD as well as two widely used non-rigid registration methods (non-rigid ICP [93] and TPS-RPM [70]) on a synthetic 2D "Fish" dataset (Fig. 8.6, (a)). Implementations of the latter ones are taken from [70], and parameters of the algorithms are not altered. The reference point set represents a non-rigidly transformed section of the template point set. Accordingly, not all points of the template have valid correspondences in the reference point set. Points without valid correspondences are outliers which are not uniformly distributed but are rather structured. We are especially interested in preserving the topology of these points. In the preprocessing step, the point sets were pre-aligned using rigid CPD [203]. The rigid registration produces a rough estimation of correspondences (Fig. 8.6, (b)) and its outcome serves as an input for the non-rigid registrations. Fig. 8.6, (c), (d) shows registration results of non-rigid ICP and TPS-RPM algorithms, respectively. Both algorithms minimise energy functions without further constraints on the non-rigid transformations. This circumstance leads to flattening and breaking of the topology of the template point set. The experiments show that these algorithms perform poor in the presence of structured outliers. Segmentation of the point sets could improve performance in this case but would assume outlier detection beforehand. In contrast, the coherency constraint valid both in CPD and ECPD preserves the outliers from breaking up the topology. Though in the case of CPD, either the correspondences between inliers are assigned precisely while the topology of the outliers is broken (Fig. 8.6, (e)); or the topology of the outliers is preserved while a significant part of correspondences between the inliers still is not assigned precisely (Fig. 8.6, (f)). Through the synergetic effect of its constraints on the displacement field, ECPD is able to recover the non-rigid transformation between the inliers reliably (Fig. 8.6, (g), (h)). The synergetic effect originates in simultaneous the impact of appropriately weighted coherency constraint and prior correspondences. Crucial is that structured outliers are modelled as a part of the point sets. Prior matches enforce the registration to restrict itself to predefined areas, decreasing the influence of structured outliers. Concurrently, the coherency constraint ensures point topology to be preserved in all regions.

The above experiment was designed to illustrate the advantage of ECPD against other non-rigid registration methods in the presence of structured outliers. The parameters were chosen for emphasising particular effects peculiar to the algorithms in general. The correspondences obtained through rigid pre-alignment were assumed to be reliable. This assumption, of course, may not always be valid in practice. Prior knowledge about the underlying point sets (*e.g.*, shape prior) often allows extracting a sparse set of correspondences which is eventually used to pre-align point sets rigidly. However, robust correspondence priors establishment is application-specific. Most registration algorithms handle correspondences established through rigid pre-alignment as correspondence priors implicitly. For instance, non-rigid ICP assigns higher weights to those points from the beginning. In contrast, ECPD can additionally incorporate an arbitrary set of correspondences explicitly.

Experiment with SINTEL Dataset. Next, we perform an additional experiment with the synthetic SINTEL dataset [61]. We show how ECPD can be used to perform image registration with constraints or to compute optical flow. *Even though ECPD is not an optimal method for image registration, the goal of the experiment is to showcase some further properties of the method.* We take two RGB images of resolution 1024×436 pixels from the SINTEL *alley* 1 image sequence for optical flow estimation and convert them into the lifted point set representation. In the lifted representation, x and y coordinates originate from the regular sampling of the image, and the z coordinate is a function of the pixel intensities. Thus, we obtain two regular point clouds with structural variation along the z-axis. The function for the conversion of the intensities to the depth is

$$P_z(x,y) = 255 - I(x,y), \tag{8.33}$$

with $P_z(x,y)$ denoting the z coordinate of a point $\mathbf{P} = (P_x, P_y, P_z)$ and $I(x,y)$ denoting an intensity of the pixel (x,y). The values of the x and y coordinates are taken unaltered from the image. For convenience, we use $P_x = x$ and $P_y = y$ notation for the pixel and point cloud coordinates interchangeably. The resulting point clouds contain 446464 points each.

Next, we compute sparse optical flow between the input images with the Lukas-Kanade method [183] and convert the result into a set of prior correspondences (see Fig. 8.8-(a)). In total, 500 prior correspondences are identified. Next, we register the converted point clouds with ECPD with prior correspondences and project the

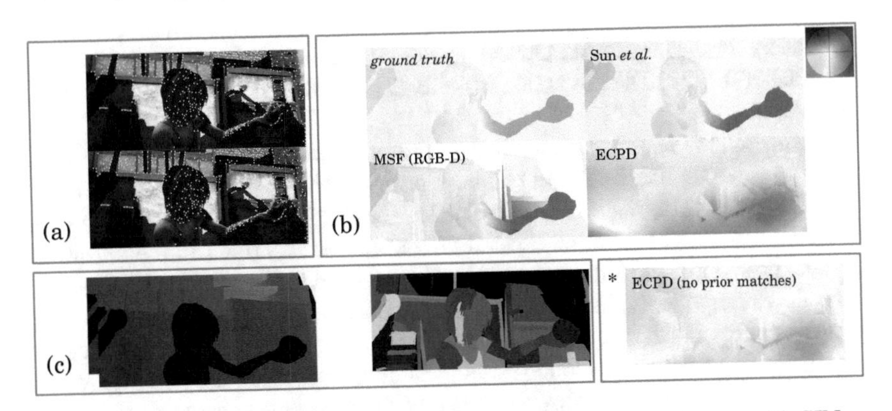

Figure 8.8: An overview and the results of the ECPD experiment with SINTEL dataset [61]. We cast an input pair of images from the *alley* 1 sequence to point clouds with a lifting function which relates the brightness to the depth. (a): Detected and tracked sparse correspondences in the pair of images by the sparse optical flow Lucas-Kanade method [183]; (b), row-wise from left to right: Ground truth optical flow, optical flow obtained by the method of Sun *et al.* [263], the projection of the RGB-D scene flow calculated by MSF [118] into the image plane (see Sec. 10.2), and the results of ECPD with prior correspondences on the images converted to the point clouds. Results of ECPD without prior correspondences are additionally shown underneath and marked with "$*$"; (c): Depth maps which MSF additionally requires as an input (on the left) and the segmentation of the reference frame performed on the depth channel in MSF (on the right).

recovered point displacements into the image plane, in a similar way as described in Sec. 10.2. We set $\beta = \lambda = 1.2$ since the deformations and displacements are local, and $\alpha = 10^{-5}$ since the correspondences are reliable. The resulting 2D vector field is the result of the image registration or optical flow. We compare the obtained optical flow with the ground truth optical flow and visualise it using the Middlebury colour scheme [29], see Fig. 8.8-(b).

We show the results of the optical flow of Sun *et al.* [263] and the projection of the RGB-D scene flow obtained by Multiframe Scene Flow (MSF) [118] introduced in this thesis in Sec. 10.2. Optical flow [263] operates on RGB images. MSF [118] additionally requires corresponding depth maps for the segmentation and initialisation (see Fig. 8.8-(c)). Unsurprisingly, dedicated methods perform more accurately for the target task. Optical flow [263] achieves an average endpoint error (AEPE) of 0.1437. The projection of RGB-D scene flow [118] achieves AEPE of 0.7295.

Compared to the registration without constraints, constrained registration leads to more accurate results. ECPD without prior matches which operates similarly to CPD [203] achieves AEPE of 2.0421. The AEPE of ECPD with 500 prior matches amounts to 1.3534. Several effects can be identified in the visualised optical flow. Most predominant are the effects related to the coherency regulariser. The flow component of the hand movement spreads into the surrounding areas. Consequently, the flow boundaries are blurred. Moreover, the hand movement creates drag of the point cloud resulting in displacement in the direction of the principal hand movement. Most prominently, the hand drag causes a displacement of the background in the opposite direction, *i.e.*, there are green spots which mark the opposite direction compared to the purple hue. Some of the observed effects are due to the form of the depth lifting function. The alignment of some depth variations which should result in a homogeneous flow field with only the lateral component present in 2D (*e.g.*, the visible flow of the background in light blue, see Fig. 8.8-(b)) causes locally inhomogeneous point set registrations so that the spurious boundaries appear in the background.

8.2.4.2 Experiments with Real Data

In the second experiment, we show how embedding the correspondence priors improves the accuracy of non-rigid registrations in real-world applications. We register 3D scans of arms in different poses. The scans were taken from [23]. First, initial correspondences between the scans are precomputed (Fig. 8.12, (a)). For this purpose, we use an approach similar to those described in [241]. We extract 3D keypoints on the point clouds with an Intrinsic Shape Signature descriptor [328], determine Persistent Feature Histograms (PFH) [241] at those keypoints and establish correspondences by comparing the PFH's. In the next step, we register the scans with CPD and ECPD. In both cases, the established correspondences are used to rigidly pre-align the point clouds, whereas in the case of ECPD, the correspondences are additionally applied as correspondence priors. In the case of CPD, we show the result with $\beta = 8.0$, $\lambda = 8.0$ and in case of ECPD with $\beta = 1.0$, $\lambda = 1.0$, $\alpha = 10^{-2}$ and $t = 20$ (Fig. 8.12, (b) and (c), respectively). Note that for CPD this is the best-achieved result (with the smallest cloud-to-cloud mean distance) and parameters β and λ differ from those of ECPD. Again, in the case of CPD, only the coherency constraint is applied, and complex non-rigid deformations (a combination of supination, flection and abduction of the arm) in the area of the hand cannot be resolved. Employing the correspondence priors

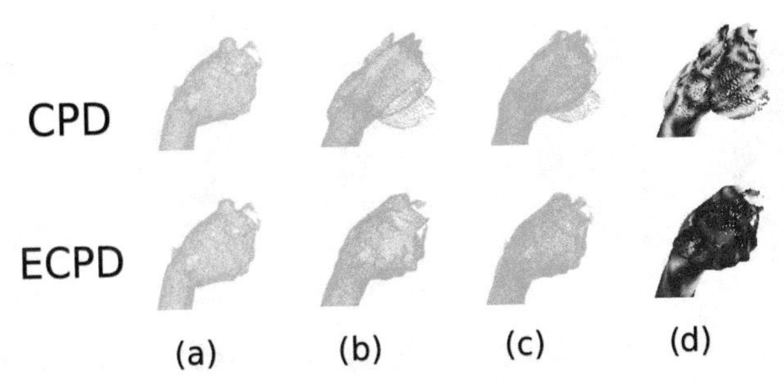

Figure 8.9: Comparison of the registration results of the hands with CPD (top row) and ECPD (bottom row) methods. (a): reference; (b): registered template; (b): overlapped view ((a) + (b)); (d): cloud-to-cloud distance in Blue < Green < Yellow < Red scale. Red corresponds to 0.0171 distance units. The mean distance and standard deviation amounts to $(0.0029; 0.003)$ distance units for CPD and $(0.0024; 0.0015)$ distance units for ECPD.

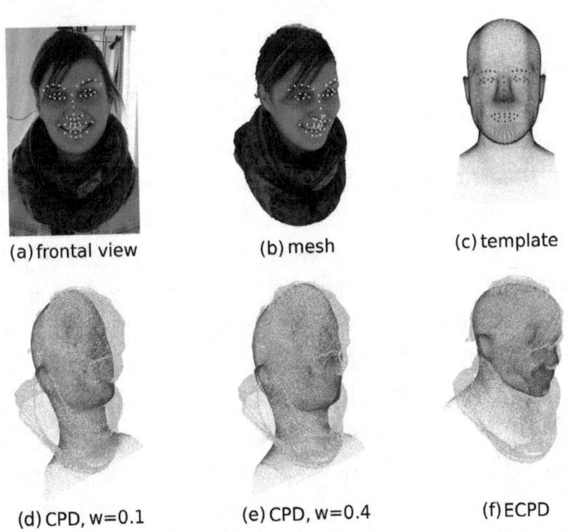

Figure 8.10: Results on the "woman with a scarf" dataset. (a): Keypoints extracted on the frontal view; (b): keypoints, transferred to the mesh as 3D keypoints; (c): template with predefined 3D keypoints; (d), (e): results of CPD, $w = 0.1$ and $w = 0.4$, respectively; (f): result of ECPD with prior matches, $\alpha = 10^{-6}$. For both methods, $\beta = 20.5$ and $\lambda = 20.5$. The mean distance and standard deviation amount to $(0.05; 0.043)$, $(0.048; 0.046)$ and $(0.023; 0.024)$ distance units for (d), (e) and (f), respectively.

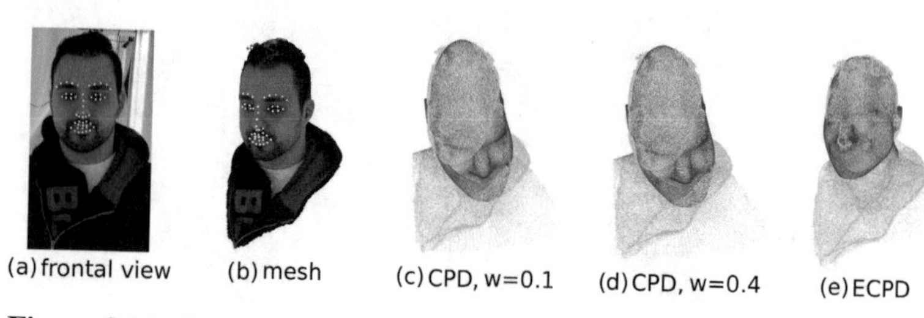

Figure 8.11: Scan-template non-rigid registration of the "man with a hood". (a): Keypoints, extracted on the frontal view; (b): keypoints, transferred to the mesh as 3D keypoints; (c), (d): results of the registration with CPD, $w = 0.1$ and 0.4, respectively; (e): result of ECPD with correspondence priors, $\alpha = 10^{-5}$. In this expermient $\beta = 8.0$ and $\lambda = 16.0$ are the same for CPD and ECPD. The mean distance and standard deviation amounts to $(0.049; 0.047)$, $(0.048; 0.047)$ and $(0.019; 0.018)$ distance units for (c), (d) and (e), respectively.

improves registration accuracy significantly. Moreover, the performance of ECPD is less sensitive to the selection of the algorithm's parameters in the latter case, as was ascertained in the course of the experiment. Fig. 8.9 displays the comparisons of the arm registrations in the area of the hands in more detail.

In another experiment, we show an application of ECPD in the scenario of scan-template registration of human heads recovered with [99] from multiple views under real-world conditions. Point sets representing scans of human heads exhibit a wide variety compared to other body parts. Often they contain areas with structured outliers (parts of the clothes, hair). We are especially interested in high registration quality in the facial area. The first dataset we analyse is the "woman with a scarf" shown in Fig. 8.10. The scan and the template contain $2.9 \cdot 10^5$ and $9 \cdot 10^4$ points, respectively. After the pre-alignment with rigid CPD, we run CPD and ECPD with correspondence priors. We obtain correspondence priors between the scan and the template in the following way. Positions of the 3D keypoints on the template are known, and the template is kept the same for all registrations (shown in Fig. 8.10, (c)). For detecting 3D keypoints on the scan, we start with extracting 2D facial keypoints with the Chehra face tracker [28] on one of the frontal views (Fig. 8.10, (a)). We obtain normalised texture coordinates by dividing 2D keypoint coordinates by the extent of the image in respective dimensions and project the frontal view onto the mesh (we know the projection matrices from the multi-view setting). Next, we determine a vertex with texture coordinate, closest to the keypoint's one, and retrieve the corresponding 3D keypoint on the scan (Fig. 8.10, (b)). Finally, the

Table 8.1: Speedup of ECPD as a function of the subsampling factor for the "woman with a scarf" dataset. Bold font emphasises the best results, whereas orange font (light bold if seen in the monochrome mode) emphasises threshold values when further raise of subsampling factor causes rapid degradation of the metrics.

subsampling factor	runtime, sec	speedup		mean distance	std. deviation
		estimated	achieved		
1	4124.0	1.0	1.0	**0.0228**	**0.0240**
4	1360.2	2.99	3.03	**0.0226**	**0.0236**
8	576.4	5.98	7.15	0.0235	0.0243
16	338.78	11.97	12.17	**0.0225**	**0.0240**
32	196.37	23.97	21.00	0.0236	0.0241
64	122.13	47.88	33.77	0.0236	0.0248
128	81.08	95.75	50.86	0.0293	0.0339

correspondences are established, since the order of the extracted keypoints both on the scan and on the template is known.

The result of the registration with CPD can be observed in Figs. 8.10, (d) and (e). As expected, the method is not able to register the facial area accurately since the outliers are not uniformly distributed. Eventually, we registered the point clouds with ECPD with facial correspondence priors. The results can be observed in Fig. 8.10, (f). We notice significantly enhanced alignment precision in the ROI.

We also evaluate the performance of ECPD on varying values of the subsampling factor t. We measure runtimes of the algorithm and compute the speedup and cloud-to-cloud metrics (mean distance error and standard deviation) of the results depending on different t. The speedup is measured as a ratio between the runtimes of ECPD without subsampling and with a particular value of t.

As expected, a discrepancy between estimated and achieved speedup is observed (see Table 8.1). For low values of t (4-16) the achieved speedup is larger than estimated, as (8.32) does not consider convergence criteria hidden in the multiplicative factor. For larger values of t (32-128) the speedup is smaller than estimated. With increasing values of t the time spent for subsampling and computation of the Gram matrix (which the speedup estimation (8.32) does not consider as well) is also increasing, relative to the algorithm's core computation time. For t varying from 4 to 64 results are qualitatively similar. Indeed, the smallest mean distance and sigma values are obtained with $t = 16$ and not without subsampling ($t = 1$) as one might expect. As the value of t exceeds 128, the amount of points in the subsampled template is not sufficient to capture the variation of the reference point cloud and the mean error/sigma increases. An optimal value of t in this experiment amounted to 64, leading to more than a thirtyfold speedup.

Table 8.2: Speedup of ECPD as a function of the subsampling factor for the "man with a hood" dataset. See caption of Table 8.1 for the colour scheme.

subsampling factor	runtime, sec	speedup		mean distance	std. deviation
		estimated	achieved		
1	1570.3	1.0	1.0	**0.0140**	**0.0148**
4	847.0	1.84	1.85	0.0169	0.0166
8	420.9	3.69	3.73	**0.0144**	**0.0145**
16	223.9	7.38	7.01	0.0152	0.0158
24	158.7	11.08	9.89	0.0170	0.0185
32	128.3	14.77	12.24	0.0166	0.0171
48	91.5	22.16	17.15	0.0177	0.0174

Another challenging dataset in this experiment was the "man with a hood" (the scan contained $7.65 \cdot 10^4$ points). It was processed analogously, and we show results in Fig. 8.11.

8.2.5 Proof of the Proposition

In this section, we provide the proof of the proposition essential for the derivation of the minimiser of the ECPD energy function.

Proposition. *Minimiser of the energy function of the form*

$$\tilde{f}(v,\sigma^2) = \tilde{E}(v,\sigma^2) + \frac{\lambda}{2}\phi(v), \tag{8.34}$$

with Eq. (2.92), (2.93), (8.1), (8.2), (8.4), (8.22) *in force has the form of a radial basis function*

$$v(\mathbf{z}) = \sum_{m=1}^{M} \mathbf{w}_m G(\mathbf{z} - \mathbf{y}_m). \tag{8.35}$$

Proof. This sketch of the proof follows in some parts the structure shown in [112, 204]. The whole energy function to be considered (Eq. (8.34)) reads

$$\tilde{f} = -\sum_{n=1}^{N} \log\left(w\frac{1}{N} + (1-w)\sum_{m=1}^{M} c_1 e^{-\frac{1}{2\sigma^2}\|\mathbf{x}_n-\mathbf{y}_m\|^2}\right) - \sum_{(i,j)\in N_c} \log(c_2 e^{-\frac{1}{2\alpha^2}\|\mathbf{x}_i-\mathbf{y}_j\|^2}) + \int_{\mathbb{R}^D} \frac{|\tilde{v}(\mathbf{s})|^2}{\tilde{G}(\mathbf{s})} ds, \tag{8.36}$$

with $c_1 = \frac{1}{2\pi\sigma^2 M}$ and $c_2 = \frac{1}{2\pi\alpha^2}$ (σ and α are constants). Consider $\mathbf{y}_m = \mathbf{y}_{0m} + v(\mathbf{y}_{m0})$ with $\mathbf{y}_{0m}$ being the initial position of $\mathbf{y}_m$ and $v(\mathbf{y}_{m0}) = \int_{\mathbb{R}^D} \tilde{v}(\mathbf{s})\exp(2\pi i < \mathbf{y}_{m0}, \mathbf{s} >)\, ds$ is a continuous function in terms of its Fourier transform $\tilde{v}$. By substituting v into Eq. (8.36) we obtain

$$\begin{aligned}\tilde{f}(\tilde{v}) =- &\sum_{n=1}^{N} \log(w\frac{1}{N} + (1-w)\sum_{m} c_1 e^{(-\frac{1}{2\sigma^2}\|\mathbf{x}_n-\mathbf{y}_{0m}-\int_{\mathbb{R}^D}\tilde{v}(\mathbf{s})\exp(2\pi i<\mathbf{y}_{m0},\mathbf{s}>)\,ds\|^2)})\\ - &\sum_{(i,j)\in N_c} \log(c_2 e^{(-\frac{1}{2\alpha^2}\|\mathbf{x}_i-\mathbf{y}_{0m}-\int_{\mathbb{R}^D}\tilde{v}(\mathbf{s})\exp(2\pi i<\mathbf{y}_{m0},\mathbf{s}>)\,ds\|^2)}) + \int_{\mathbb{R}^D} \frac{|\tilde{v}(\mathbf{s}|^2}{\tilde{G}(\mathbf{s})} d\mathbf{s}.\end{aligned} \tag{8.37}$$

To find the minimum of this functional, we take its functional derivative with respect to $\tilde{v}$, so that $\frac{\delta \tilde{f}(\tilde{v})}{\delta \tilde{v}(\mathbf{t})} = 0, \forall \mathbf{t} \in \mathbb{R}^D$:

$$\begin{aligned}\frac{\delta \tilde{f}(\tilde{v})}{\delta \tilde{v}(\mathbf{t})} =& -\sum_{n=1}^{N} \frac{\sum_{m=1}^{M} c_1 e^{-\frac{1}{2\sigma^2}\|\mathbf{x}_n-\mathbf{y}_m\|^2} \frac{1}{\sigma^2}(\mathbf{x}_n - \mathbf{y}_m)\int_{\mathbb{R}^D} \frac{\delta\tilde{v}(\mathbf{s})}{\delta\tilde{v}(\mathbf{t})} e^{2\pi i<\mathbf{y}_{0m},\mathbf{s}>} d\mathbf{s}}{w\frac{1}{N} + (1-w)\sum_{m=1}^{M} c_1 e^{(-\frac{1}{2\sigma^2}\|\mathbf{x}_n-\mathbf{y}_m\|^2)}} -\\ &- \sum_{(i,j)\in N_c} \frac{c_2 e^{-\frac{1}{2\alpha^2}\|\mathbf{x}_i-\mathbf{y}_j\|^2} \frac{1}{\alpha^2}(\mathbf{x}_i - \mathbf{y}_j)\int_{\mathbb{R}^D} \frac{\delta\tilde{v}(\mathbf{s})}{\delta\tilde{v}(\mathbf{t})} e^{2\pi i<\mathbf{y}_{0j},\mathbf{s}>} d\mathbf{s}}{\sum_{(i,j)\in N_c} c_2 e^{(-\frac{1}{2\alpha^2}\|\mathbf{x}_i-\mathbf{y}_j\|^2)}} +\\ &+\frac{\lambda}{2}\int_{\mathbb{R}^D} \frac{\delta}{\delta\tilde{v}(\mathbf{t})}\frac{|\tilde{v}(\mathbf{s})|^2}{\tilde{G}(\mathbf{s})} d\mathbf{s} = -\sum_{n=1}^{N} \frac{\sum_{m=1}^{M} c_1 e^{-\frac{1}{2\sigma^2}\|\mathbf{x}_n-\mathbf{y}_m\|^2} \frac{1}{\sigma^2}(\mathbf{x}_n - \mathbf{y}_m)e^{2\pi i<\mathbf{y}_{0m},\mathbf{t}>}}{w\frac{1}{N} + (1-w)\sum_{m=1}^{M} c_1 e^{(-\frac{1}{2\sigma^2}\|\mathbf{x}_n-\mathbf{y}_m\|^2)}} -\\ &- \sum_{(i,j)\in N_c} \frac{c_2 e^{-\frac{1}{2\alpha^2}\|\mathbf{x}_i-\mathbf{y}_j\|^2} \frac{1}{\alpha^2}(\mathbf{x}_i - \mathbf{y}_j)e^{2\pi i<\mathbf{y}_{0j},\mathbf{t}>}}{\sum_{(i,j)\in N_c} c_2 e^{(-\frac{1}{2\alpha^2}\|\mathbf{x}_i-\mathbf{y}_j\|^2)}} + \lambda\frac{\tilde{v}(-\mathbf{t})}{\tilde{G}(\mathbf{t})} = 0.\end{aligned} \tag{8.38}$$

We now define the coefficients a_{mn} as

$$a_{mn} := \sum_{n=1}^{N} \frac{\sum_{m=1}^{M} c_1 e^{-\frac{1}{2\sigma^2}\|\mathbf{x}_n - \mathbf{y}_m\|^2} \frac{1}{\sigma^2}(\mathbf{x}_n - \mathbf{y}_m)}{w\frac{1}{N} + (1-w)\sum_{m=1}^{M} c_1 e^{(-\frac{1}{2\sigma^2}\|\mathbf{x}_n - \mathbf{y}_m\|^2)}} + \sum_{(i,j)\in N_c} \frac{c_2 e^{-\frac{1}{2\alpha^2}\|\mathbf{x}_i - \mathbf{y}_j\|^2} \frac{1}{\alpha^2}(\mathbf{x}_i - \mathbf{y}_j)}{\sum_{(i,j)\in N_c} c_2 e^{(-\frac{1}{2\alpha^2}\|\mathbf{x}_i - \mathbf{y}_j\|^2)}} =$$

$$\sum_{n=1}^{N} \frac{\sum_{m=1}^{M} c_1 e^{-\frac{1}{2\sigma^2}\|\mathbf{x}_n - \mathbf{y}_m\|^2} \frac{1}{\sigma^2}(\mathbf{x}_n - \mathbf{y}_m)}{w\frac{1}{N} + (1-w)\sum_{m=1}^{M} c_1 e^{(-\frac{1}{2\sigma^2}\|\mathbf{x}_n - \mathbf{y}_m\|^2)}} + \sum_{n=1}^{N} \frac{\sum_{m=1}^{M} c_2 e^{-\frac{1}{2\alpha^2}\tilde{p}_{mn}\|\mathbf{x}_i - \mathbf{y}_j\|^2} \frac{1}{\alpha^2}\tilde{p}_{mn}(\mathbf{x}_i - \mathbf{y}_j)}{\sum_{m=1}^{M} c_2 e^{(-\frac{1}{2\alpha^2}\tilde{p}_{mn}\|\mathbf{x}_i - \mathbf{y}_j\|^2)}}. \tag{8.39}$$

Here we integrate the matrix $\tilde{P}$ with entries $\tilde{p}_{mn}$ as defined in Eq. (8.6). The functional derivative can now be rewritten as

$$-\sum_{m=1}^{M}(\sum_{n=1}^{N} a_{mn})e^{2\pi i<\mathbf{y}_{0m},\mathbf{t}>} + \lambda\frac{\tilde{v}(-\mathbf{t})}{\tilde{G}(\mathbf{t})} = 0 \tag{8.40}$$

Denoting $\mathbf{w}_m = \frac{1}{\lambda}\sum_{n=1}^{N} a_{mn}$ and changing $\mathbf{t}$ to $-\mathbf{t}$ we multiply the Eq. (8.40) by $\tilde{G}(\mathbf{t})$ and obtain

$$\tilde{v}(\mathbf{t}) = \tilde{G}(-\mathbf{t})\sum_{m=1}^{M} \mathbf{w}_m e^{-2\pi i<\mathbf{y}_{0m},\mathbf{t}>}. \tag{8.41}$$

We adopt the Gaussian kernel for the reasons argued in [203] so that $\tilde{G}$ is symmetric and its Fourier transform is real. Taking the inverse Fourier transform of the Eq. (8.41) we obtain the following result

$$v(\mathbf{z}) = G(\mathbf{z}) * \sum_{m=1}^{M} \mathbf{w}_m \delta(\mathbf{z} - \mathbf{y}_{0m}) = \sum_{m=1}^{M} \mathbf{w}_m G(\mathbf{z} - \mathbf{y}_{0m}). \tag{8.42}$$

Since $\mathbf{w}_m$ depends on v via the coefficients a_{mn} and $\mathbf{y}_m$, they have to be determined through the self-consistency equation equivalent to Eq. (8.26). □

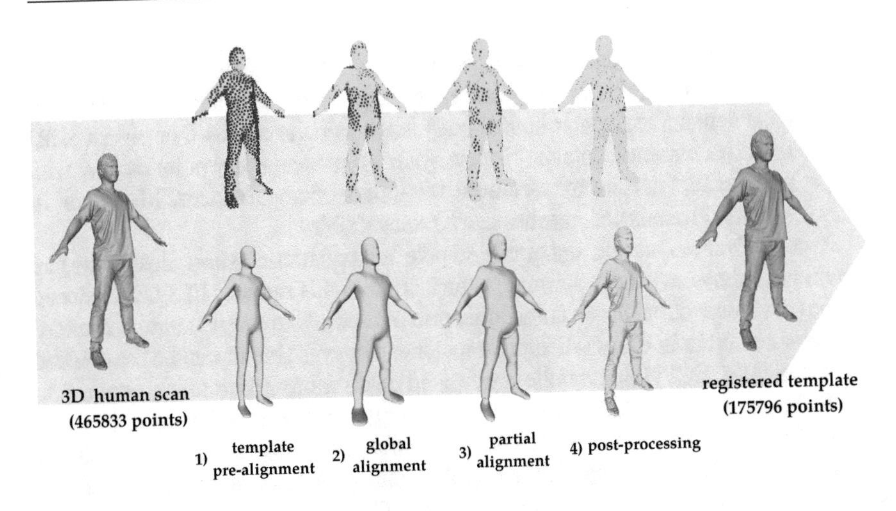

Figure 8.12: An overview of the proposed framework for a point cloud based 3D human scan-template registration with prior matches. On the left: the input 3D scan. The goal is to align it with a predefined full-body template. On the right: registration result. Middle: the scan overlayed with the results of the individual steps (top row) and the evolution of the deforming template progressing through the pipeline (bottom row).

8.3 An Application in a Pipeline for Human Appearance Transfer

Alignment of real-world 3D human scans is a challenging problem in computer vision. In this section, we demonstrate a pipeline for human appearance transfer and treatment of social pathologies as an application of rigid and non-rigid probabilistic point set registration with prior correspondences. Our goal is to align or register a given raw 3D scan with a 3D body template, *i.e.*, to recover correspondences and a displacement field of a 3D human template to a reference 3D human body scan. Accurate human body registration has various applications in statistical shape analysis, anthropometric measurement extraction and rehabilitation, to name a few.

If a human body scan and a template are represented by meshes, *i.e.*, polygonal networks with normals and defined point topology, the problem is specified as mesh registration. This is a well-studied area with many works covering articulated human body registrations [65]. If both a human scan and a template are represented by point clouds (bare 3D point coordinates), the problem remains generally unsolved, and there are only a few attempts to tackle this problem [107]. There is, although, a demand to register human body scans as point clouds. For instance, 3D reconstructions obtained on a multi-view system represent noisy point clouds, with structured outliers caused by variations in cloths and appearances. Moreover, in many real-world scenarios, meshes can be unavailable.

Recent advances in rigid and non-rigid point set registration allow closing the gap mentioned above, which is our main contribution in this section. ECPD introduced in the previous section allows embedding prior matches into a registration procedure. It was shown that in cases when prior matches are available or can be established automatically, ECPD can handle articulated cases more accurately compared to CPD. Thus, we adopt ECPD for point cloud based registration of human body scans and design a multiple-stage human body registration pipeline schematically shown in Fig. 8.12. To align a scan and a template into a common reference frame and to estimate scale, our pipeline contains a pre-alignment step. The pre-alignment is followed by a global non-rigid registration providing initialisation for partial non-rigid registration. On several stages of the registration pipeline, automatically established prior matches guide the registration procedure. At the same time, our method is semi-automatic, and neither depends on large datasets (in contrast to [182]) nor requires point topology.

8.3.1 Related work

Among previous works in the field, there are several approaches related to ours. Alignment of point sets involving complex displacement fields (with rigid and non-rigid components) constitutes the class of *articulated* point set registration methods. One of the early attempts was a generalisation of ICP [40] to articulated motion [222]. Especially for the case of articulated registration of 3D human body scans represented by point clouds, several approaches were recently proposed [107, 108]. These works offer efficient algorithms for registrations of the same bodies (or bodies very similar in appearance) in different poses. In contrast, we solve the problem of registering *dissimilar* human scans and a predefined template, with the goal of high accuracy, especially in the head and facial areas. In our case, 3D human scans can evince significant variation in clothes, hairstyle, body shape, proportions of individual body parts; we assume that poses of a scan and a template do not differ largely.

Dey *et al.* [79] proposed a markerless technique to align a human body mesh template to a new pose specified by a noisy point cloud of a human body. Similar to our approach, the method relies on body landmarks (head, hands and feet). However, we work with point clouds and do not solve the problem of pose alignment. The

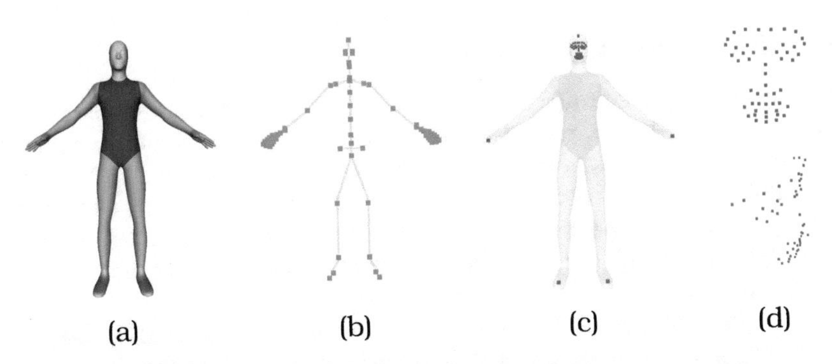

Figure 8.13: A full-body human template: (a) template with the pre-segmented body parts shown in different colours; (b) the skeleton of the template (can be optionally provided for the registration); (c) positions of the predefined facial and body landmarks on the template; (d) zoomed-in facial landmarks in a frontal and a side view.

idea of the alignment of a human body scan and a predefined template is used for anthropometric measurement extraction. Thus, in [287], a human scan and a template meshes are registered so that measurements can be consistently extracted on the template. A similar idea leading to high accuracy is adopted in [304], wherein a human body scan and a template are represented by point clouds. The method adopts global full body alignment, which is less accurate than our approach (though the accuracy suffices for the anthropometric measurement extraction). In contrast, we use a segmented template, apply partial non-rigid registration for higher accuracy in regions with high structural variation and optional post-processing.

8.3.2 The Proposed Framework

In this section, we describe in detail the proposed framework for registration of full 3D human body scans.

8.3.2.1 A 3D Human Body Template

The proposed approach relies on a human body template, *i.e.*, a segmented point cloud in the shape of a human body, see Fig. 8.13. The template was created by a designer and contains $1.75 \cdot 10^5$ points. Optionally, point topology and joints are available for it (Fig. 8.13-(b)). For an enhanced precision, our pipeline accepts prior matches — facial and body landmarks, if available (Fig. 8.13-(c), -(d)).

8.3.2.2 Overview of the Framework

Assume we are given two 3D point clouds: the reconstruction (reference) and the template. In ECPD, prior matches are provided by a set of indices with the uncertainty parameter α (see Sec. 8.3.2.3 for details on establishment of prior matches). In the rigid case, ECPD outputs parameters of rigid body motion and scaling to pre-align the template with a scan. In the non-rigid case, ECPD outputs a displacement field aligning template and a reference as well as the probability of correspondences P.

Thus, the first step of the framework — *template pre-alignment* — consists in the alignment using rigid ECPD. Note that the joints of the template rigging are included in the registration as additional points. The second step — *global alignment* — consists in a rough non-rigid registration using non-rigid ECPD on the whole template (without segmentation). As can be observed in Fig. 8.12, this step reasonably accounts for overall body shape variations to obtain preliminary correspondences between the body parts of the template (see Fig. 8.13) and the corresponding body parts of the reconstruction using the correspondences of the highest probabilities, *i.e.*, the nearest neighbours. Furthermore, due to the coherency constraint of the ECPD, the joints are automatically placed to the corresponding position w.r.t. the registered template. The correspondences on the surface of the template are subsequently used in the third step for the *partial alignment* of the corresponding parts of the scan. The partial alignment is much more accurate, in particular for all extremal parts of the body and the hand regions (see Sec. 8.3.2.5 for resolution of challenging cases). Due to the coherency constraint, the registered surfaces are still rather smooth. In step four — the *post-processing* — the proximity of the noise-free registered template and the initial reconstruction is used to recover fine details via the projection techniques explained in Sec. 8.3.2.4.

8.3.2.3 Landmark Extraction

Suppose a reconstructed scan in an arbitrary pose and original images of the reconstruction are given. Since reconstructions can be noisy, it is hard to extract semantics from the scan (determine positions of the body parts). Thus, we detect faces in the input images using the Chehra [28] facial feature detector, in a similar way as described in Sec. 8.2.4.2. The facial features robustly define a semantic coordinate system. The difference vector of the eye positions is sufficient to decide between left and right w.r.t the reconstruction, and the difference vector between an eye and the mouth is sufficient to decide between up and down direction.

Next, the visual hull of the scan is voxelised. The scan is rendered using orthogonal projection from the six principal directions (imposed by the semantic

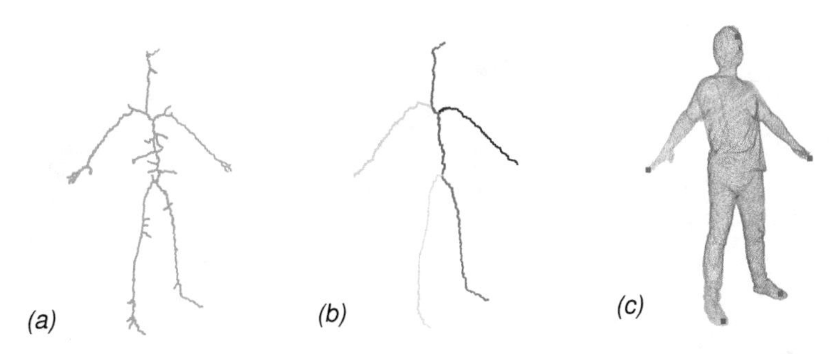

Figure 8.14: Extraction of the body landmarks. From left to right: (a) minimum spanning tree extracted from the scan using the Kruskal's algorithm [169]; (b): a clean skeleton obtained by recursively removing all branches that do not start with one of the five feature points; (c): the resulting skeleton with an overlayed scan.

vectors[1]). The voxelised version of the reconstruction is then thinned to the medial lines using a highly optimised implementation of [217]. This algorithm always reduces the volume to one-dimensional medial structures in contrast to other methods (*e.g.*, [194]) which also produce two-dimensional medial structures (medial surfaces). Furthermore, the algorithm is pleasingly robust to noise and does not generate overly many branches. Once the 8-connected thinning is computed, the resulting voxelised representation of the medial structure is transformed into a graph which is in turn transformed into a minimum spanning tree using Kruskal's algorithm [169], see Fig. 8.14-(a). Since the face position is roughly known from the landmark computation, we find a leaf node in the tree suited to represent the head position. Starting from this leaf, the hands and feet can be defined to be represented by those four leaves that are maximally far apart from the head and each other. Hence, once the head is determined, it is sufficient to identify four of those points to obtain candidates for hands and feet. Since the search of the feature points is based on the tree, the actual body pose during scanning does not change the result as long as no topological changes of the scan occur (*e.g.*, a hand touches the body). Next, four additional feature points are equipped with semantic information (*i.e.*, hand, feet, left and right). We separate the feet from the hands by choosing the two landmarks with the greatest tree-distance to the head as feet and the remaining two as hands. Left and right is decided by considering semantic vectors computed from the facial landmarks. Positional ambiguities introduced

[1] "semantic vector" is a vector giving meaning to a specific direction. The unit vector x does not have semantics but the vector "right" has a semantic, namely that it points to the right w.r.t a reference.

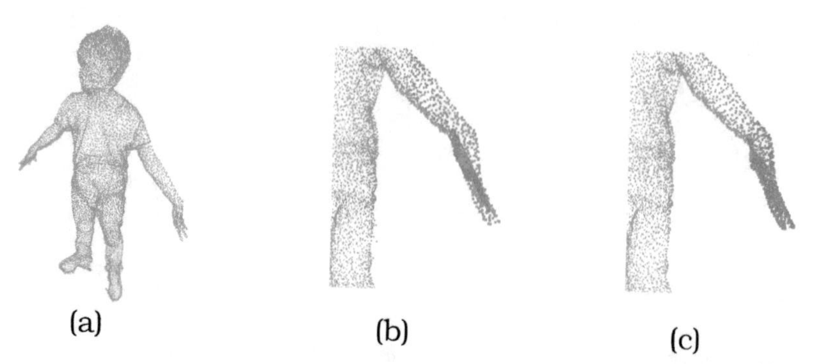

Figure 8.15: The figure shows a 3D scan (a), a result up to the step 3 (in orange) overlayed with the reconstruction (in cyan) (b) and the registration up to the step 3 (in orange) with registration correction (in cyan) (c).

by crossing the arms are solved by tracing back the tree towards the shoulders. Similarly, a decision between left and right leg is made. In the last step, the skeleton tree is pruned by recursively removing all branches that do not start with one of the five feature points resulting in a clean skeleton (Fig. 8.14-(b), -(c)).

8.3.2.4 Post-Processing

In the post-processing step, the projection of a partially refined template onto the scan is performed. Two variants of the projection are possible: nearest-neighbour and reverse nearest-neighbour. During the nearest-neighbour projection, every point of the template is projected onto the nearest point of the scan. This method often leads to point clustering. During the reverse nearest-neighbour projection, closest points on the template, for every point of the scan, are projected to the scan. This variant allows reducing point clustering effects and positively affects the overall registration quality. Fig. 8.16 illustrates differences in results depending on the respective projection algorithm.

8.3.2.5 Handling Variety in Hand Poses

Accurate registration is particularly challenging in the hand region, due to the low point density compared to the amount of detail, and possible pose differences with the template. Moreover, missing data and different hand/finger configurations frequently occur. One example of a challenging case due to a bent hand can be

Table 8.3: The parameters for the core steps of the proposed pipeline summarised.

Step	α	templ. s	scan s	theor. speedup	runtime, sec
1) pre-alignment	10^{-5}	1 (no subs.)	1 (no subs.)	1	61.8
2) global non-rigid registration	10^{-3}	5	10	13.7	1744.9
3) partial non-rigid registration	10^{-5}	5	1 (no subs.)	$\{1.2; 3\}$	289.0

observed in Fig. 8.15. In this case, non-rigid ECPD flattens the hand (see Fig. 8.15-(b)) and, consequently, the registration accuracy decreases.

A possible remedy is to use the hand of the rigidly registered template and perform non-rigid registration w.r.t. corresponding point cloud of the hand that was previously segmented via the framework. The result after applying the described technique can be observed in Fig. 8.15-(c).

It is noteworthy that the underlying ECPD algorithm is topology preserving. We exploit this property in multiple ways, *i.e.*, in partial registration, in handling various topology, applying a topology transfer of the template, a straightforward transfer of vertex qualities like texture coordinates or skinning weights to the registered point cloud as well as the co-registration of auxiliary points (*e.g.*, joint positions) allowing the transfer of rigging and skinning data between models. Furthermore, registration against the same 3D template guarantees a 1:1 mapping between all registration results, which is advantageous for keyframe based animations.

8.3.3 Experimental Results

We run the proposed pipeline on a server under operating system Debian GNU 8.0 (codename Jessie), Intel Xeon E3-1245 V2 (3.4 GHz) processor and NVIDIA GeForce 660Ti graphics card (GK104-300-KD-A2 GPU). In the non-rigid case, we use the correspondence preserving subsampling strategy proposed in Sec. 8.2.3.1 — instead of registering a reference with a scan directly, the template is subsampled and registered with the scan. Thereby, all prior matches are preserved and influence the registration procedure. Further, the original template is registered with the subsampled template (the result of the previous step) and all points of the subsampled template are taken as prior matches. In this way, a linear speedup is achieved. We adopt the subsampling strategy both for the global non-rigid and partial registrations. During the global registration, we also subsample the scan, since no decay in accuracy is observed, which results in an even higher speedup. The parameters of the proposed pipeline for human scan-template registration are listed in Table 8.3. We use the same parameter set in all experiments.

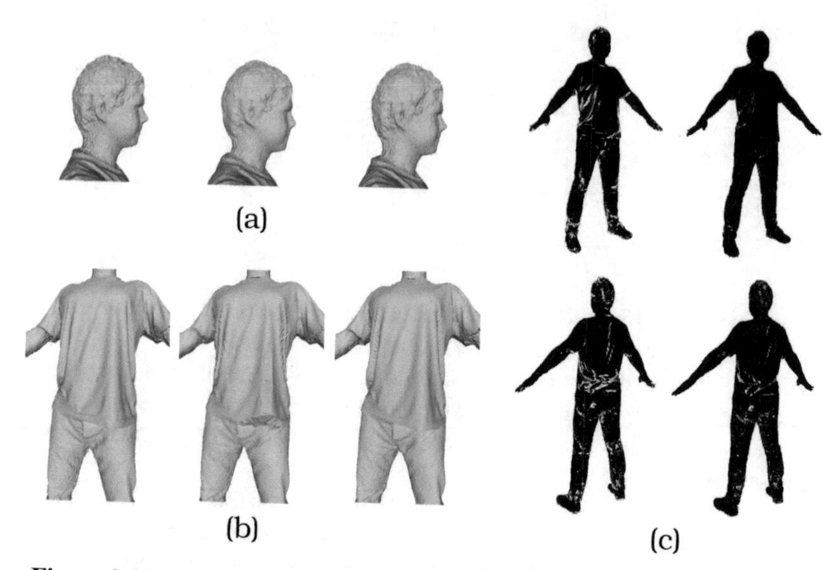

Figure 8.16: Accuracy evaluation of the proposed approach: (a) head of the original scan (left), registration result with the nearest-neighbour projection and result with the reverse nearest-neighbour projection (right); (b) torso of the original scan (left), registration result with the nearest-neighbour projection and result with the reverse nearest-neighbour projection (right); (c) detailed comparison of projection algorithms by means of Hausdorff distance (Red > Yellow > Green > Blue scale): nearest-neighbour (left), reverse nearest-neighbour (right). Better viewed with zoom.

8.3.3.1 Experiments with Real Data

We test our approach on over a hundred real-world scans with various appearance (clothes, hair, body metrics). Here, we show several results on the new data as well as on the FAUST dataset [45]. The scans are obtained from a multi-view system and reconstructed with an improved version of [99]. An exemplary scan with $4.66 \cdot 10^5$ points is shown in Fig. 8.12, on the left. All scans are captured in the pose similar to those of the human body template. While running, the template deforms, as shown in Fig. 8.12-(middle).

The result (Fig. 8.12, on the right) is very accurate. The appearance of the template is visually very close to the appearance of the original scan, despite that the template contains 2.65 times fewer points. Of course, not all details, especially in regions with a high total variation, can be captured by the template. Fig. 8.16 provides a closer look at the result. As expected, even with the reverse nearest-neighbour projection, there are areas with a lower resolution of the structure. We make several observations about the displacement fields. First, there are no obvious distortions in the face area, despite the fact that the template has not matched the head after rigid pre-alignment. The leg area demonstrates symmetric point drift, even though both parts are treated by the pipeline independently. The runtime for the processed scan from this experiment amounts to 2097 seconds. For other scans, the runtime lies in the interval $\{400; 2500\}$ seconds depending on the number of points in the scans.

We also run the pipeline on several scans from the FAUST dataset [45]. This dataset serves for evaluation purposes of mesh registration algorithms with an emphasis on varying poses, in inter- and intra-individual evaluation scenarios. Since we do not solve the problem of posing, we register several scans in similar poses to our segmented template. Results are shown in Fig. 8.17. The accuracy of results is in-line with the accuracy of other real-world scans. However, due to different hand poses (different from the pose of our template) and absence of prior matches, accuracy in the area of hands is lower.

We also tested the proposed pipeline with a lower-resolved 10k template for several scans. Exemplary results are shown in Fig. 8.18. This experiment demonstrates that the pipeline can work both with different scans and different templates, but, of course, the accuracy is lower than for the high-resolved template.

8.3.3.2 A System for Treatment of Social Pathologies

The proposed pipeline can be adopted for generation of animatable virtual avatars (AVA) resembling in appearance real persons. AVAs are widely used in the film industry, entertainment, augmented and virtual reality applications, and nowadays find their way into medicine and rehabilitation. Thus, our target application is an interactive system for curing pathologies such as schizophrenia or autism accompanied by social interaction burdens. The movement neuroscience and cognitive science suggest that it is easier for the respective patients to interact with subjects (real persons, virtual avatars or robots) looking similar to them. Thus, a system which allows the generation of AVAs and changing their appearance gradually in a virtual environment is a concept for the next-generation therapy of this type of social pathologies.

There are several key requirements for the target system. First, it should produce high-accuracy 3D scans. Second, to leverage real-time morphing, all scans must contain an equal number of points. We experienced that the accuracy of modern affordable RGB-D sensors is too low to achieve the desired clinical effect, and opted

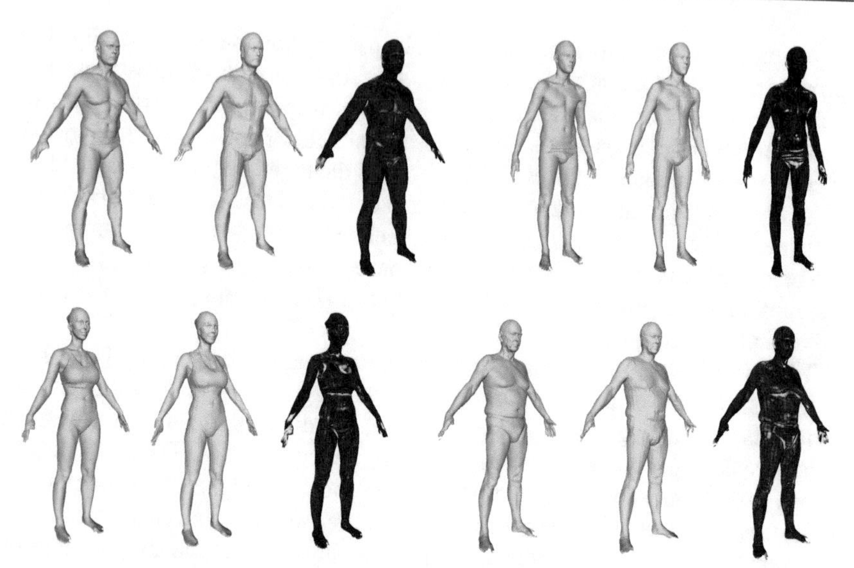

Figure 8.17: Results on the FAUST dataset [45] shown in triples. For every triple: original scan (left), registration of the scan with the 10k template (middle), Hausdorff distance (Red > Yellow > Green > Blue scale) between the scan and our result. Our solution operates on 3D points, results shown as meshes for visualisation purposes.

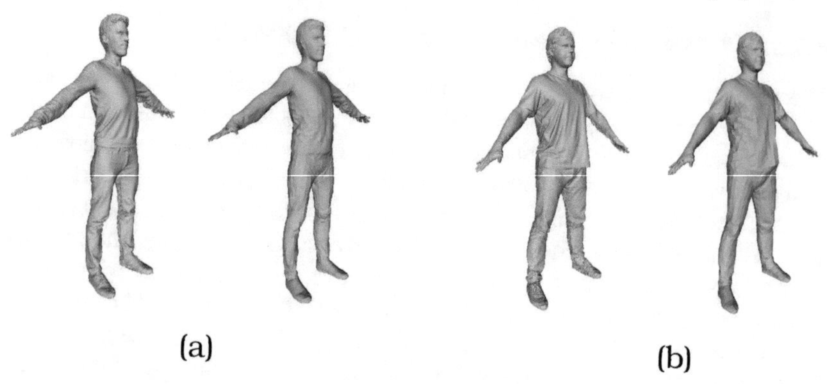

Figure 8.18: Results of the registration pipeline with a template containing 10^4 points. For both (a) and (b), the original scan is on the left and the registration result on the right, respectively.

for a multi-view reconstruction system consisting of multiple high-resolution RGB cameras. In contrast to modern RGB-D based systems which output a mesh from an implicit representation [207], multi-view systems output point clouds. Thus, the subsequent design decision — adaptation of the proposed framework — is caused by the necessity to mesh the input scan accurately. Due to the topology-preserving property of ECPD, the point topology of the template can be directly transferred to the registration result. On the other hand, the proposed pipeline gracefully solves two other tasks in one sweep — texturing of the AVA, co-registration of the joints for rigging (a core feature for animatibility of an AVA) and transfer of skinning weights. Since the original scan and the registered template are placed into the same coordinate system and are similar in appearance, the original high-resolution texture can be applied to the registered template. Moreover, the proposed framework is purely point-cloud based, and we are free to augment the template with the joints (a sparse point set).

Among over 140 generated AVA of the patients, only every twentieth AVA generation has required manual intervention (*e.g.*, a fine adjustment of elbow joints orientation or ECPD parameters). All in all, our template-based solution and scanning in a predefined pose contributed to a well-balanced trade-off between the generality of the pipeline and high-quality results.

In Fig. 8.19, prototypes of the multi-view system and AVAs obtained with the proposed pipeline are shown. The AVAs are placed in a virtual environment mimicking the real surgery and can be moved interactively (*e.g.*, by a pre-recorded motion sequence or a real-time capture of patient's movements).

8.4 Summary and Conclusion

In this chapter, we have proposed a probabilistic rigid and non-rigid point set registration approach with prior correspondences — ECPD. ECPD is the first probabilistic point set registration algorithm allowing embedding of prior correspondences in a closed form, to the best of our knowledge.

In the rigid case, a synergy of joint registration with the explicit embedding enhances the registration accuracy compared to the decoupled processing. As shown experimentally, in the scenarios with only one or two prior matches, rigid ECPD outperforms CPD with a separate pre-alignment step. Moreover, if the scans contain clustered outliers, rigid ECPD outperforms CPD with pre-alignment step in terms of cloud-to-cloud distance in ROI.

In the non-rigid case, the proposed coarse-to-fine acceleration scheme with correspondence preserving subsampling counterbalances the polynomial computational complexity. It splits the problem into two subproblems of smaller size and

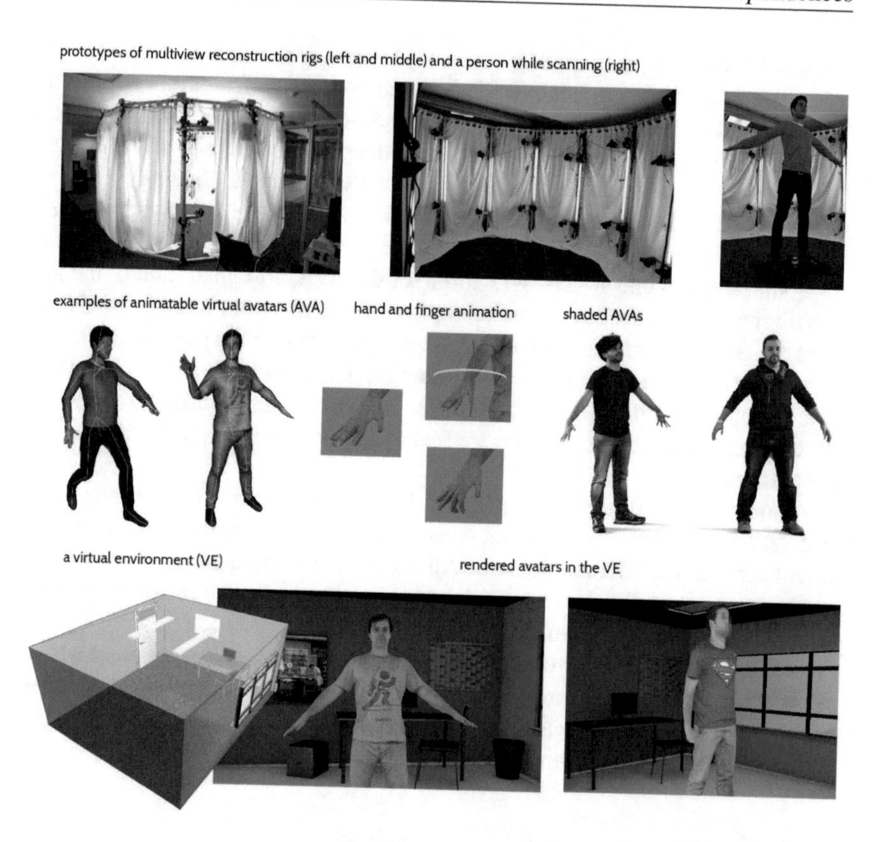

Figure 8.19: We adopt the proposed framework for registration of full 3D human body scans for the treatment of social pathologies. First, a patient is reconstructed on a high-resolution multi-view RGB system. Next, a unified template with augmented joints is aligned with the body scan using the proposed framework. Next, the result is textured, and the skinning weights are computed. Finally, the AVA is optionally shaded and placed in a virtual environment. In the course of the therapy, the patient interacts with the AVA (*e.g.*, repeats its movements) whose appearance changes gradually with time. The face of a real patient, bottom right, is pixelated.

reduces the number of operations by a linear factor. We have provided implementation details for the proposed method for a heterogeneous system with a multicore CPU and a GPU. Our experiments demonstrated that in applications where prior matches are available, non-rigid ECPD significantly improves the registration accuracy and enhances the robustness of point set registration under complex non-rigid deformations and in the presence of structured outliers compared to CPD.

Based on the rigid and non-rigid variants of ECPD, we have developed a pipeline for human appearance transfer and shown its application in the treatment of social pathologies. The proposed semi-automatic pipeline solely relies on point clouds and can handle large point sets in reasonable time due to the sophisticated sub-sampling strategy and a heterogeneous implementation of non-rigid ECPD. The achievable accuracy is in line with the current state-of-the-art competitor methods. A current limitation of the pipeline lies in the requirement of small pose differences between the reference and template, as it is not pose-invariant. Though, the introduction of partial registrations remedied the situation to a large extent.

In future work, rigid ECPD can be applied in further medical scenarios with only one or two available markers. For non-rigid ECPD, one of the next steps is to enable individual α weights for every correspondence. In both cases, a sensitivity analysis for prior correspondences will provide additional insights about the performance and accuracy of ECPD when correspondences are noisy or erroneous. The proposed framework for human appearance transfer can be combined with kinematic motion capture [268]. Moreover, generalising the pipeline for the pose-invariant operation is a useful direction as well.

9 Point Set Registration Relying on Principles of Particle Dynamics

A New class of approaches for point set registration based on particle dynamics is introduced. We formulate PSR as a modified N-body problem with additional constraints. In an N-body problem (see Sec. 2.3.3), positions and velocities of particles are updated under gravitational forces. Gravitational interaction induces the name of the new method class, *i.e.*, we refer to each of methods as Gravitational Approach (GA). In total, we discuss three techniques — two methods for rigid PSR (Secs. 9.1 and 9.2) and a scheme for non-rigid PSR (Sec. 9.3). Each of the methods evinces unique properties and allows to handle new challenging situations not well covered by the existing techniques. Among the strengths of the proposed methods is handling of missing data, clustered outliers and uniformly distributed noise. In the rigid case, GA shows a broader basin of convergence compared to several widely-used and state-of-the art methods. The accelerated rigid GA allows to efficiently handle large point sets with the multiply-linked character of interactions in the strong sense — the property which is highlighted and which is rarely intrinsic to other methods. In the non-rigid case, GA balances well surface smoothness, capture of fine local details and robustness to noise.

The proposed methods advance the field of point set registration and enable new applications. We discuss acceleration techniques and implementation details for every method. Each section contains a comprehensive evaluation section assessing different aspects of the proposed approaches.

9.1 Rigid Gravitational Approach (GA) with Second-Order ODEs

In this section, we introduce Gravitational Approach (GA) for rigid point set registration based on solving differential equations of second order for every particle. Inspired by astrodynamics, GA does not possess a direct ancestor. Many physical

V. Golyanik, *Robust Methods for Dense Monocular Non-Rigid 3D Reconstruction and Alignment of Point Clouds*,

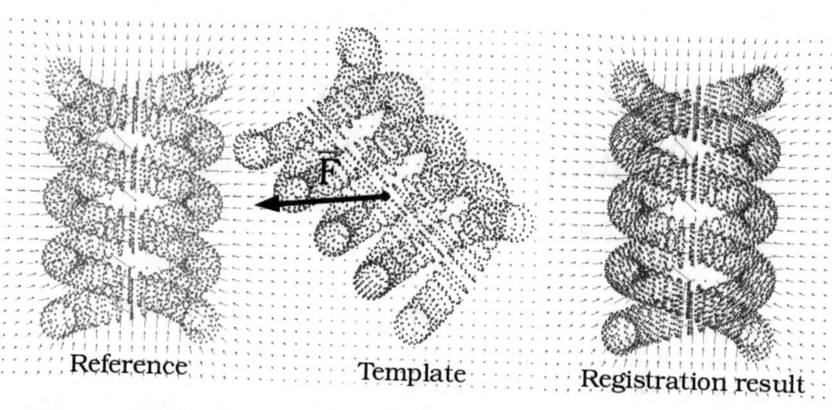

Figure 9.1: Point set registration with the gravitational approach: template moves in the gravitational field induced by the reference. Coordinates of the template points are individually updated by solving equations of particle motion in a viscous medium, whereupon rigidity constraints are applied. Left: initial misalignment of the helix point sets [59] and the induced gravitational field; right: registration result after 150 iterations.

phenomena involve an interaction of elements and allow a quantitative description of the system's state based on the configuration of their elements (gravitational or electrostatic potential energy, quantum state). We couple N-body simulation with rigid body dynamics. In an N-body simulation, future trajectories of n bodies with the specified initial coordinates, masses and velocities are estimated. Thus, the main idea of GA consists in the modelling of rigid systems of particles under gravitational forces in a viscous medium (see Fig. 9.1). Particle movements are expressed by differential equations of Newtonian mechanics. In our model, every point from both a reference and a template point set is treated as a particle with its own position, velocity and acceleration (in the text, the terms point and particle are used interchangeably). GA is a multiply-linked algorithm, as all template points are moving in the superimposed force field induced by the reference. To impose rigidity constraints, laws of rotational and translational motion of rigid bodies are employed. To resolve rotation, a formulation based on singular value decomposition (SVD) for finding an optimal rotation matrix is used [158, 202].

This method was extensively applied to recover rotation matrixes in computer vision applications [50, 203, 218, 327]. Similar to ICP and CPD, GA can operate in multiple dimensions. GA also supports a basic form of including prior correspondences into the registration procedure by assigning different masses to regular and

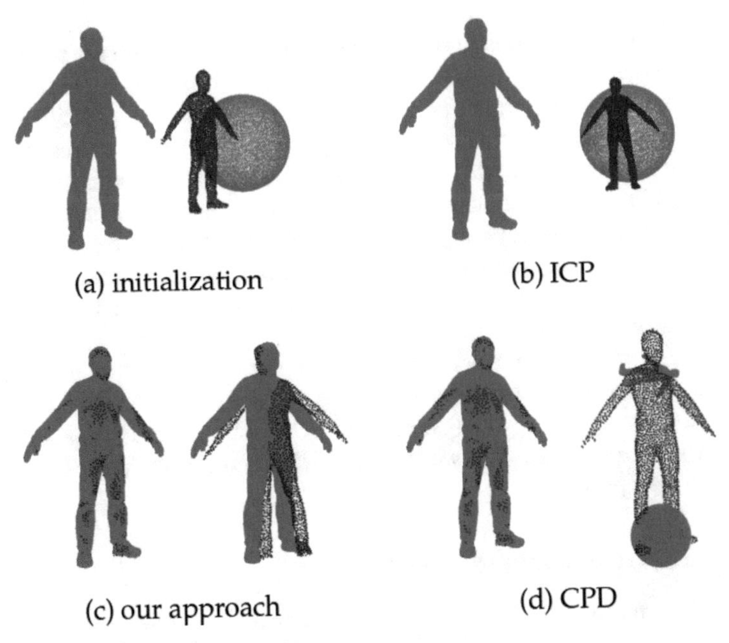

Figure 9.2: Registration results of ICP [40], CPD [203] and our approach on real data with introduced clustered outliers. (a) Initialisation; template (shown in blue) is located between the reference human scan and outliers arranged as a sphere (shown in red). (b) ICP registration result — the algorithm is trapped into a local minimum; (c) GA registration results (left: an optimal parameter; right: a suboptimal parameter); (d) CPD registration results (left: an optimal parameter; right: a suboptimal parameter).

matching points. Embedding the prior correspondences is not trivial, especially in the case of probabilistic approaches (see chapter 8).

The motivation for a conceptually new registration algorithm is manifold. Firstly, because of a new formulation, an algorithm with unique properties among point set registration algorithms is obtained. Thus, GA can take an initial velocity of the template as a parameter. There are application scenarios where such velocity can be estimated (*e.g.*, from an optic flow between frame pairs in simultaneous localisation and mapping (SLAM) systems). Secondly, registration algorithms are sensitive to noise and improving the performance of the rigid point set registration is a fundamental task in computer vision. GA can perform better than the state-of-the-art CPD in some scenarios with structured outliers and in the presence of noise (see Fig. 9.2 and Sec. 9.1.2 for a detailed description). Thirdly, we place parallelizability of operations to the foreground, because many existing methods (e.g CPD and KC) contain a significant portion of serial code. Fourthly, in GA point set registration is formulated as energy minimisation. We show one possible minimisation through the forward integration but many other minimisation techniques can be tried out (non-linear optimisation of the gravitational potential energy function, simulated annealing for a globally optimal solution). Besides, a new algorithm is, of course, interesting both from the theoretical and practical point of views and can encourage further ideas in the area.

To the best of our knowledge, formulating point set registration problem as a modified N-body simulation was not shown in the literature so far. We also did not find any evidence for interpreting the problem as an object moving in a force field, considering early and pre-ICP works in the field (as a starting point for pre-ICP works we used [40]).

9.1.1 Gravitational Approach

In a point set registration problem, two D-dimensional point sets are given, *i.e.*, a reference $\mathbf{X}_{N\times D} = (X_1,\ldots,X_N)^T$ and a template $\mathbf{Y}_{M\times D} = (Y_1,\ldots,Y_M)^T$. We search parameters of the rigid transformation, i.e a tuple $(\mathbf{R},\mathbf{t},\mathbf{s})$ which optimally aligns the template point set to the reference point set.

Since we target an efficient point set registration, we adopt the N-body problem (see Sec. 2.3.3) while abstracting from a realistic physical model and alter the simulation objective. Specifically, the following assumptions and modifications are made:

i. Every point represents a particle with a mass condensed in an infinitely small area of space,

ii. A reference $\mathbf{X}$ induces a constant inhomogeneous gravitational field,

iii. Particles Y_i move in the gravitational field induced by the reference and do not affect each other,

iv. $\mathbf{Y}$ moves rigidly, *i.e.*, the transformation of the template particle system is described by the tuple $(\mathbf{R},\mathbf{t},\mathbf{s})$,

v. A collisionless N-body simulation is performed, since the number of particles cannot be changed according to the problem definition,

vi. Astrophysical constants (*e.g.*, G) and units are considered as algorithm parameters, and

vii. A portion of kinetic energy is dissipated and drained from the system — the physical system is not isolated.

Modification (ii) reflects that the reference point set remains idle. Physically, it is said to be fixed by an external force. Modification (vii) — the introduction of *energy dissipation* or *viscosity* term — arises through the physical analogy of movement with friction in a viscous medium (gas, fluid), whereby a part of the kinetic energy transforms to heat.

In GA, potential and kinetic energy are continuously redistributed. Second-order ODEs in Eq. (2.79) without an external stimulus describe endless oscillatory phenomena. If a part of the kinetic energy which has been converted from the potential energy under the influence of the gravitational field is dissipated, the system gradually converges to its most stable state with locally minimal potential energy. This state corresponds to a locally optimal solution to the point set registration problem. Moreover, the viscosity term is necessary to assure the algorithm's convergence. Without viscosity, it would be difficult to refine the solution, as **Y** can have a high speed close to a local minimum.

We find the force exerted on a particle Y_i by all particles of the reference **X**:

$$\vec{F}_{Yi} = -Gm^{Yi} \sum_{j=1}^{N} \frac{m^{Xj}}{\|r^{Yi} - r^{Xj}\|^2} \hat{\mathbf{n}}_{ij} \,, \tag{9.1}$$

where m^{Yi} (m^{Xi}) and r^{Yi} (r^{Xi}) denote mass and absolute coordinates of a particle Y_i (X_i), respectively, and $\hat{\mathbf{n}}_{ij} = \frac{(r^{Yi} - r^{Xj})}{\|r^{Yi} - r^{Xj}\|}$ is a unit vector in the direction of the force. Note that we depart from the notations used in Sec. 2.3.3, as we deal with two non-overlapping particle sets. Besides, instead of position vectors absolute point coordinates are used. The gravitational force in Eq. (9.1) can lead to a singularity during a collisionless simulation, since two or more particles can be pushed infinitely close to each other. The singularity can be avoided by revising gravitational interaction at small scales. Thus, we introduce the softening length ε — a threshold distance, below which gravitational interaction does not increase severely. The force acting on a particle takes the form of a cubic spline [4, 284]:

$$\vec{F}_{Yi} = -Gm^{Yi} \sum_{j=1}^{N} \frac{m^{Xj}}{(\|r^{Yi} - r^{Xj}\|^2 + \varepsilon^2)^{3/2}} \hat{\mathbf{n}}_{ij} \,. \tag{9.2}$$

The dissipation term is expressed by a drag force acting in the opposite to the particle's velocity direction with a magnitude proportional to its speed:

$$\vec{F}^{d}_{Yi} = -\eta \, v^{Yi}, \tag{9.3}$$

where the dimensionless constant parameter η jointly reflects properties of the particle Y_i and the viscous medium. Thus, the resultant force exerted on a particle Y_i reads

$$\vec{\mathbf{f}}_{Yi} = \vec{F}_{Yi} + \vec{F}^d_{Yi}\,. \tag{9.4}$$

Using Euler's method for second order ODEs we perform forward integration, *i.e.*, solve the system in Eq. (2.79) and get updates for an *unconstrained* velocity and displacement of every particle Y_i:

$$\vec{v}^{t+1}_{Yi} = \vec{v}^{t}_{Yi} + \Delta t\, \frac{\vec{\mathbf{f}}_{Yi}}{m^{Y_i}} \text{ (velocity)}, \tag{9.5}$$

$$\vec{d}^{t+1}_{Yi} = \Delta t\, \vec{v}^{t}_{Yi} \text{ (displacement)}. \tag{9.6}$$

Unconstrained velocities and displacements can be combined into the velocity and displacement field matrices $\mathbf{V}_{M\times D}$ and $\mathbf{D}_{M\times D}$:

$$\mathbf{V} = \begin{bmatrix} \vec{v}^{t+1}_{Y1} & \vec{v}^{t+1}_{Y2} & \dots & \vec{v}^{t+1}_{Ym} \end{bmatrix}^T, \tag{9.7}$$

$$\mathbf{D} = \begin{bmatrix} \vec{d}^{t+1}_{Y1} & \vec{d}^{t+1}_{Y2} & \vdots & \vec{d}^{t+1}_{Ym} \end{bmatrix}^T. \tag{9.8}$$

$\mathbf{V}$ and $\mathbf{D}$ are subjects to further regularisation which depends on the type of point set registration.

9.1.1.1 Gravitational Potential Energy

Between two bodies of masses m_1 and m_2 the gravitational potential energy (GPE) U_p emerges:

$$U_p = -G\frac{m_1 m_2}{r}, \tag{9.9}$$

where G is the gravitational constant and r is the distance between centers of mass of the bodies. GPE is defined as the work done by the gravitational force to displace the particle from one to another position. Accordingly, the energy function for two interacting rigid systems of particles can be defined as

$$E(\mathbf{R},\mathbf{t},\mathbf{s}) = -G\sum_{i,j} \frac{m^{Yi}\, m^{Xj}}{\|\mathbf{R}\, r^{Yi}\, \mathbf{s} + \mathbf{t} - r^{Xj}\| + \varepsilon}. \tag{9.10}$$

At a local minimiser ($\mathbf{R}_{opt}, \mathbf{t}_{opt}, \mathbf{s}_{opt}$) *the total GPE of the system and the value of the energy function E are locally minimal.* Thus, it is possible to express the GA stopping criterion by a difference of GPEs in several consecutive iterations.

9.1.1.2 Rigidity Constraints

In the rigid case, rigidity constraints on the displacement field $\mathbf{D}$ must be imposed and rigid body physics takes effect.

Resolving Translation

Since distances between points are preserved, several simplifications can be carried out. First, the resultant force exerted on a rigid body is equal to the sum of the forces exerted on individual particles:

$$\vec{F}_{\mathbf{Y}}^{t+1} = \sum_{i=1}^{M} \vec{\mathbf{f}}_{Yi}^{t}. \tag{9.11}$$

Second, the resultant velocity changes depending on the action of the resultant force on the total mass of the template $m^{\mathbf{Y}}$ per unit of time as

$$\vec{v}_{\mathbf{Y}}^{t+1} = \vec{v}_{\mathbf{Y}}^{t} + \Delta t \, \frac{\vec{F}_{\mathbf{Y}}^{t+1}}{m^{\mathbf{Y}}}. \tag{9.12}$$

From the resultant velocity $\vec{v}_{\mathbf{Y}}^{t+1}$, the resultant translation can be computed as

$$\vec{\mathbf{t}}^{t+1} = \Delta t \, \vec{v}_{\mathbf{Y}}^{t+1}. \tag{9.13}$$

Resolving Scale[1]

We find a scale $\mathbf{s}$ in the least-square sense. $\mathbf{s}$ relates the current position of a template $\mathbf{Y}_t$ with the predicted position $\mathbf{Y}_{t+1} = \mathbf{Y}_t + \mathbf{D}$ as

$$\mathbf{Y}_{t+1} = \mathbf{Y}_t \mathbf{s}. \tag{9.14}$$

Suppose $\hat{\mathbf{Y}}_t$ and $\hat{\mathbf{Y}}_{t+1}$ are column vectors of length DM with vertically stacked entries of $\mathbf{Y}_t$ and $\mathbf{Y}_{t+1}$, respectively. In that case, the following proposition holds.

Proposition 2. *In Eq.* (9.14) *the optimal scaling factor* $\mathbf{s}$ *in the least-squares sense is equal to the ratio of two vector dot products* $\frac{\hat{\mathbf{Y}}_t^T \cdot \hat{\mathbf{Y}}_{t+1}}{\hat{\mathbf{Y}}_t^T \cdot \hat{\mathbf{Y}}_t}$.

[1] this ansatz requires prior matches or local point interaction; see [124] for the latest insights on scale resolution in GA

Proof. Eq. (9.14) has no exact solution, unless $\hat{\mathbf{Y}}_{t+1}$ lies in the column space of $\hat{\mathbf{Y}}_t$. Nevertheless, we can solve

$$\Upsilon_{t+1} = \hat{\mathbf{Y}}_t s, \tag{9.15}$$

where Υ_{t+1} is a projection of $\hat{\mathbf{Y}}_{t+1}$ to the column space of $\hat{\mathbf{Y}}_t$ by normal equations. After rewriting Eq. (9.15) in terms of the known variables we obtain:

$$s = (\hat{\mathbf{Y}}_t^T \hat{\mathbf{Y}}_t)^{-1} \hat{\mathbf{Y}}_t^T \hat{\mathbf{Y}}_{t+1} = \frac{\hat{\mathbf{Y}}_t^T \hat{\mathbf{Y}}_{t+1}}{\hat{\mathbf{Y}}_t^T \hat{\mathbf{Y}}_t} \tag{9.16}$$

which corresponds to the optimal in the least squares sense solution to scaling. □

Resolving Rotation

Rigorously, rotation can be inferred from a *torque* acting on a rigid body. A torque (a moment of force) is a physical quantity reflecting the tendency of the force to change the angular momentum of the system, *i.e.*, to rotate an object. Resolving rotation rigorously applying physics of rotational motion introduces to GA an additional parameter ω (angular velocity) and generates a bunch of new intermediate quantities.
Thus, computing R using torque requires several non-trivial steps. Instead, a different method is used in this section. To recap the initial conditions, in every iteration the starting and final position vectors of M points are given and the task is to find a rotation matrix which optimally aligns both vectors. This can be efficiently addressed by solving a corresponding *Generalised Orthogonal Procrustes Problem*. The disadvantage is the loss of ω, since no angular acceleration from previous iterations is considered. The solution in a closed-form is given in Lemma 1. It resembles the Kabsch algorithm [158] and is provided without proof.

Lemma 1. *Given are point matrices* $\mathbf{Y}$ *and* $\mathbf{Y}_D = \mathbf{Y} + \mathbf{D}$. *Let* μ_Y *and* μ_{YD} *be the mean vectors of* $\mathbf{Y}$ *and* $\mathbf{Y}_D$, *respectively,* $\hat{\mathbf{Y}} = \mathbf{Y} - \mathbf{1}\mu_Y^T$ *and* $\hat{\mathbf{Y}}_D = \mathbf{Y}_D - \mathbf{1}\mu_{YD}^T$ *point matrices centred at the origin of the coordinate system and* $\mathbf{C} = \hat{\mathbf{Y}_D}^T \hat{\mathbf{Y}}$ *a covariance matrix. Let* $\mathbf{US}\hat{\mathbf{U}}^T$ *be SVD of* $\mathbf{C}$. *Then the optimal rotation matrix* $\mathbf{R}$ *reads*

$$\mathbf{R} = \mathbf{U}\Sigma\hat{\mathbf{U}}^T, \tag{9.17}$$

where $\Sigma = diag(1, \ldots, sgn(|\mathbf{U}\hat{\mathbf{U}}^T|))$.

The covariance matrix C has dimensions 3×3. Thus, the sequential code portion dedicated to the (parts of) SVD computation is negligible.

Algorithm 9 Gravitational Approach for Rigid Point Set Registration

Input: a reference $\mathbf{X}_{N\times D}$ and a template $\mathbf{Y}_{M\times D}$
Output: parameters $(\mathbf{R},\mathbf{t},\mathbf{s})$ aligning $\mathbf{Y}$ to $\mathbf{X}$ optimally
Parameters: $G \in (0,1]$, $\varepsilon \in (0;0.5)$, $\eta \in (0;1]$, m^{Xj}, m^{Yi}, $j \in \{1,\dots,N\}$, $i \in \{1,\dots,M\}$, $\vec{v}^{\,0}_{\mathbf{Y}}$, ρ_E, Δt, K

1: **Initialisation:** $\mathbf{R}=\mathbf{I}$, $\mathbf{t}=\mathbf{0}$, $\mathbf{s}=\mathbf{1}$, $E_{curr}=E(\mathbf{R},\mathbf{t},\mathbf{s})$, $E_{prev}=0$
2: **while** $|E_{curr}-E_{prev}|>\rho_E$ **do**
3: update force $\vec{\mathbf{f}}_{Yi}$ for every particle Y_i (Eqs. (9.2)–(9.4))
4: update velocity and displacement matrices $\mathbf{V}$ and $\mathbf{D}$ (Eqs. (9.5)–(9.8))
5: compute translation $\mathbf{t}_k$ according to Eqs. (9.11)–(9.13)
6: compute scale $\mathbf{s}_k$ as stated in Proposition (2)
7: update rotation $\mathbf{R}_k$ (Eq. (9.17))
8: $\mathbf{Y}_{t+1}=\mathbf{s}_k\mathbf{Y}_t\mathbf{R}_k+\mathbf{t}_k$ (Eq. (9.18))
9: $\mathbf{R}=\mathbf{R}\mathbf{R}_k$, $\mathbf{t}=\mathbf{t}+\mathbf{t}_k$, $\mathbf{s}=\mathbf{s}\mathbf{s}_k$ (optimal parameters)
10: $E_{prev}=E_{curr}$, update E_{curr} according to Eq. (9.10)
11: **if** the current iteration number k exceeds K **then**
12: break
13: **end if**
14: **end while**

Finally, having resolved the translation, scale and rotation it is possible to update the template's pose as

$$\mathbf{Y}_{t+1}=\mathbf{s}\mathbf{Y}_t\mathbf{R}+\mathbf{t}. \tag{9.18}$$

The centre of mass of $\mathbf{Y}_t$ must coincide with the origin of the coordinate system for the rotational update.

GA is summarised in Alg. 9. As an optional parameter, a non-zero template's velocity $\vec{v}^{\,0}_{\mathbf{Y}}$ can be provided. Following the rigorous approach to resolving rotation would require an additional parameter ω as well as a modification of line 7 in Alg. 9. As stated so far, GA has complexity $\mathcal{O}(MN)$, since every particle of the reference perturbs every template's particle. The stopping criterion is formulated in terms of the difference in the gravitational potential energy (GPE) (see Eq. (9.10)).

9.1.1.3 Acceleration Techniques

Acceleration techniques from both areas of N-body simulations and point set registration can be adopted for GA. They enable a drop in computational complexity to at least $\mathcal{O}(N \log M)$ as well as a speedup in a corresponding complexity class in terms of the number of operations.

Various techniques to accelerate N-body simulations were developed in the past three decades [284]. Ahmad-Cohen (AC) neighbour scheme employs two time scales for each particle [18]. Thereby, force evaluations for neighbouring particles occur more frequently than for distant particles. For smaller time steps, contributions from the distant points are approximated. Various strategies for neighbourhood selection were proposed [6]. Though the AC scheme allows achieving a speedup, the complexity class remains $\mathcal{O}(n^2)$. Barnes and Hut [31] introduced a recursive scheme for the force computation based on the space subdivision and particle grouping in an octree. The algorithm achieves $\mathcal{O}(n \log n)$ for an N-body simulation. Adopting it for GA will decrease the complexity to $\mathcal{O}(M \log N)$. Fast multipole methods (FMM) [130] also employ hierarchical space decomposition but additionally take advantage of multipole expansions. Thus, adjacent particles in the near-field tend to accelerate similarly under forces exerted by particles in the far field. This class of algorithms exploits the idea of rank-deficiency of the $n \times n$ interaction matrix considering the nature of the far-field interactions. This results in an $\mathcal{O}(n \log n)$ algorithm, whereby a multiplicative constant depends on the approximation accuracy. An $\mathcal{O}(n)$ algorithm is also possible but requires a lot of additional effort accompanied by an increase in the multiplicative constant. When adopting FMM for GA, the complexity can theoretically drop to $\mathcal{O}(M \log N)$ and $\mathcal{O}(M+N)$, respectively. Interestingly, CPD employs the Fast Gauss Transform (FGT) to approximate sums of Gaussian kernels in $\mathcal{O}(M+N)$ time. FGT was developed by Greengard and Strain following the principles of FMM [131].

Common acceleration techniques from the area of point set registration can be applied to GA. Firstly, both the reference and the template can be subsampled. Assume s_X and s_Y are corresponding subsampling factors. The speedup amounts to $\sim s_X s_Y$ is this case. Subsampling has to be used with caution since it can cause loss of information. Secondly, a coarse-to-fine strategy can be applied — starting on a rough scale, the solution is refined while involving more and more points for the registration. Thirdly, dedicated data structures such as kd-tree for nearest neighbour search can be used. Such data structures, hierarchically reordering the points according to their spatial positions also find their application in tree codes and FMM.

Furthermore, parallel hardware can be used to speedup GA, since the algorithm is inherently data- and task-parallel with the portion of parallel code $> 99\%$. Though, with the decrease in computational complexity of a GA variant (*e.g.*, from $\mathcal{O}(MN)$ to $\mathcal{O}(M \log N)$), memory complexity and the effort to parallelise the algorithm may increase. In 2007, Nyland *et al.* reported a fiftyfold speedup of the GPU implementation of an all-pairs N-body simulation compared to a tuned serial implementation [211]. Later, a first GPU implementation of the Barnes-Hut octree was presented. It allows simulating interactions of $5 \cdot 10^6$ particles with 5.2 seconds per time step [60]. A recent tendency is to unify tree codes and FMM with automatic parameter tuning for heterogeneous platforms [318]. Using an efficient implementation, it should be possible to run GA for point sets with 10^7 points on a single GPU in a reasonable time.

Apart from the abovementioned techniques, a further one is conceivable for GA. Since only the template point set is moving and its particles do not affect each other, the force field induced by the reference can be precomputed once in a grid. Thereby, the gravitational force field can be sampled with a higher density in the proximity to the reference points. This can be especially advantageous when many templates are registered with the same reference and there is no memory restriction to achieve the desired accuracy. This technique exhibits a resemblance with the particle-mesh class of methods in the area of N-body simulations (see *e.g.*, [6]). In the particle-mesh methods, particles interact with each other through a mean force field changing over time. The methods achieve complexity $\mathcal{O}(n \log n)$. In the case of GA, the technique raises the algorithm's memory complexity and preprocessing time, but reduces computational complexity to $\mathcal{O}(M)$, since the gravitational force field remains constant and does not need to be recomputed.

9.1.2 Evaluation

In this section, we focus on the qualitative evaluation of the first GA implementation and compare it with ICP and CPD in synthetic and real-world scenarios. The Matlab implementations of the ICP and CPD algorithms are taken from [190] and [201], respectively. GA is implemented in C++ and runs on a system with 3.5 GHz Intel Xeon E5-1620 processor and 32 GB RAM.

9.1.2.1 Experiments on Synthetic Data

In the first experiment, we compare ICP, CPD and GA in a registration scenario with the Bunny point cloud from the Stanford 3D Scanning Repository [278]. We copy the downsampled version of the Bunny (1889 points), translate it and change its orientation (angle $\phi \in [0; \frac{\pi}{2}]$). The resulting point set serves as a template and the original one as a reference. We also introduce uniformly and Gaussian distributed noise so that for each noise type 5%, 10%, 20%, 40% or 50% of points in the resulting point cloud represent noise. In the case of the uniform distribution, noise is added to the bounding box of the 3D scan. In the case of Gaussian noise, the mean value in the centre of the bounding box is taken. For every noise combination out of ten, 500 random transformations are applied resulting in 500

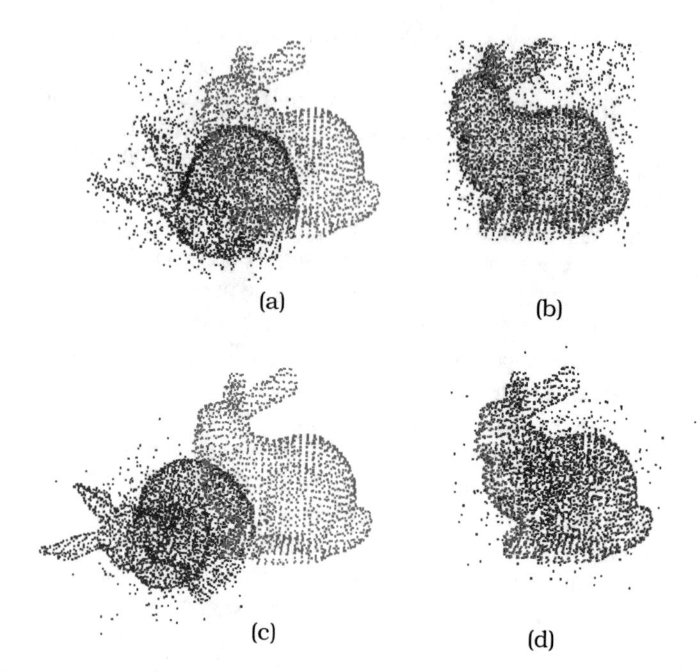

Figure 9.3: Registration results from the experiment on synthetic data (Stanford bunny [278]). The reference is shown in red, the noisy template in blue. (a) Initialisation, the template contains 40% of uniformly distributed noise; (b) result with a uniformly distributed noise; (c) initialisation, the template contains 40% of Gaussian noise; (d) result with Gaussian noise. GA is more robust to Gaussian noise in terms of the mean distance and RMSE, but it resolves rotation more often under uniformly distributed noise.

random initial misalignments. For every initial misalignment and noise combination, rigid registrations with ICP, CPD and GA are performed. To assure the highest accuracy, a CPD version without FGT is used. CPD takes one parameter, i.e the estimated amount of noise in a dataset which is set to the corresponding noise level in every run. For GA, $G = 6.67 \cdot 10^{-5}$ is set. We measure mean distance and root-mean-square error (RMSE) between the reference and a registered template. The noise is removed while computing the metrics since we are interested in the quality of data alignment and exact correspondences are known in advance. For all algorithms, failed registrations are not considered in the computation of the

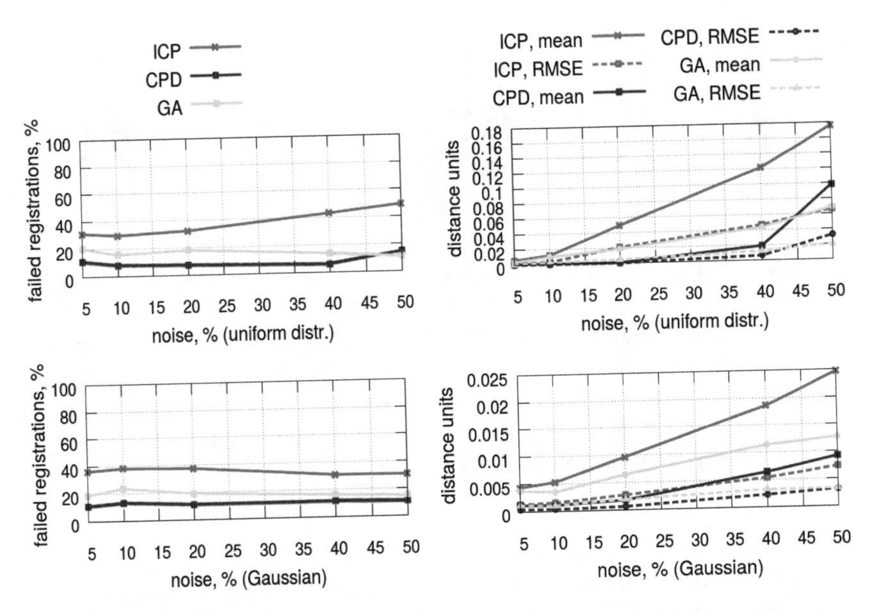

Figure 9.4: Results of the experiment on synthetic data: the reference is the Stanford Bunny [278], the template is a randomly transformed copy of it with 5%, 10%, 20%, 40% and 50% added uniformly distributed and Gaussian noise. Metrics are calculated over 500 runs for every noise level and type.

metrics. Instead, the amount of failures is reported separately. A criterion for registration failure is defined in terms of a threshold on RMSE. We observe that in the experiment with the 3D Bunny, RMSE is either < 0.2 or > 0.4. In the former case, point clouds always appear to be registered correctly, at least approximately well, whereas in the latter case they are never registered correctly. Thus, we set the failure threshold to 0.3. In Fig. 9.3, exemplary results of GA are shown (40% noise level). Running time of GA ranges from 1.5 to 10 minutes per run depending on the noise level. The algorithm converges at most after 100 iterations when possible oscillations around the local minimum attenuate. Results are summarised in Fig. 9.4.

GA shows intermediate performance between ICP and CPD. In average, it fails rarer than ICP and more often than CPD when resolving rotation. The angle of initial misalignment causing GA to fail lies in the range $[\frac{\pi}{4}; \frac{\pi}{2}]$, whereby the higher the angle, the smaller the probability to resolve rotation correctly. CPD starts to fail when the angle of initial misalignment exceeds $65°$. Results of the

Figure 9.5: Results of the experiment with structured outliers and missing parts. (a) the processed reference 3D model, $1.45 \cdot 10^5$ points; (b) reconstruction with removed 7% of the points, $1.34 \cdot 10^5$ points. (c) initialisation; (d) GA registration result; (e) cloud-to-cloud distance visualised with a Blue<Green<Yellow<Red colour scale; the mean distance amounts to 0.109, the RMSE to 0.63; outliers and missing parts explain the high RMSE.

experiment confirm the tendency — since the set of initial misalignments is equal for all algorithms, direct angle comparison in failure cases is performed. All three algorithms are stable against Gaussian noise while resolving rotation (the number of failures does not correlate with the level of Gaussian noise). CPD and GA are also stable to uniformly distributed noise. In the case of 50% of uniformly distributed noise, GA outperforms CPD both in terms of the mean distance, RMSE and amount of correct registrations. Here, the difference between a probabilistic approach and our method comes to light: in the case of GA, more distant points contribute more significantly (hyperbolic expression) than in the case of CPD (Gaussian vicinity) allowing for more robust cumulative compensation.

9.1.2.2 Experiments on Real Data

We evaluate GA with several experiments on real data. Fig. 9.2 depicts the course of the first one. We take a human body scan ($4.2 \cdot 10^4$ points) and a template ($3 \cdot 10^3$ points) reconstructed on a multi-view system with an algorithm described in [99].

We add synthetic clustered outliers (forming a sphere) to the reference. Initially, the template is located exactly between the human scan and the sphere (Fig. 9.2-a). Both point sets are registered with ICP, CPD and GA. ICP fails to associate the human scan with the human template. Being influenced by direct nearest neighbours, it converges to the sphere (if the template is located closer to the human scan, ICP resolves rotation less accurate than CPD and GA) (Fig. 9.2-b). GA and CPD, provided appropriate parameters are chosen, resolve registration correctly (Fig. 9.2-c,-d). If the corresponding parameter is set suboptimally (either G for GA or the weight w for CPD), the result of GA is not as accurate, whereas CPD can fail to resolve the example. The experiment demonstrates the gravitational nature of GA — subspaces with a higher total mass win against outliers, even if they are clustered.

The second experiment on real data is selected to demonstrate the performance of GA in the presence of structured outliers and with missing parts. We use two scans of the guardian lions reconstructed with the multi-view algorithm [99]. The reference (Fig. 9.5-a) represents a processed 3D model. We register it with a rough reconstruction of the lion with 7% of contiguous points removed from the area of the head (Fig. 9.5-b). Fig. 9.5-b,-c show initialisation and the registration result, respectively. During the registration, a 30x subsampling of both point clouds is used and the recovered transformation is applied to the initial template. Results are highly accurate despite of outliers and missing parts.

In the third experiment on real data, the influence of different particle masses is evaluated. The point sets are obtained from two different images of the Orion constellation. For every point, a weight according to the grayscale pixel intensity is set. Fig. 9.6 depicts the course of the experiment. The template and the reference contain 324 and $\sim 10^4$ points, respectively, whereby the template contains $\sim$95% noise including clustered outliers. GA downweights darker points corresponding to the noise and emphasizes star clusters with the higher weights. It successfully accomplishes the task and the corresponding star clusters are aligned correctly, as can be observed in Fig. 9.6-f. CPD is not able to incorporate weighting information and fails, although the noise weight w is set to 0.95. This example shows an advantage of GA against CPD — incorporating weights — which can influence the registration procedure in a favourable way. Different masses can be also assigned to particles if prior macthes between point sets are known in advance.

In the experiments $G \in [6.67 \cdot 10^{-6}; 6.67 \cdot 10^{-5}]$ and $\eta \in [0.2; 0.9]$ were chosen and the step size Δt was fixed to 1 so far. G and η counterbalance each other and should be set depending on the scale and the total mass of the points involved in the registration. Higher values of η can lead to faster convergence but can also hinder the algorithm from finding a solution.

9.1.2.3 Experiments on SLAM Datasets

In this section, we describe experiments on data from simultaneous localisation and mapping (SLAM) datasets. We use point clouds from two RGB-D datasets, *i.e.*, the Stanford 3D Scene Dataset [329] (captured by a pattern projection sensor) and CoRBS [303] (captured by a time-of-flight camera). One of the goals of SLAM is to reconstruct a scene given multiple depth maps (which can be unambiguously converted into point clouds) and colour images captured by an RGB-D sensor

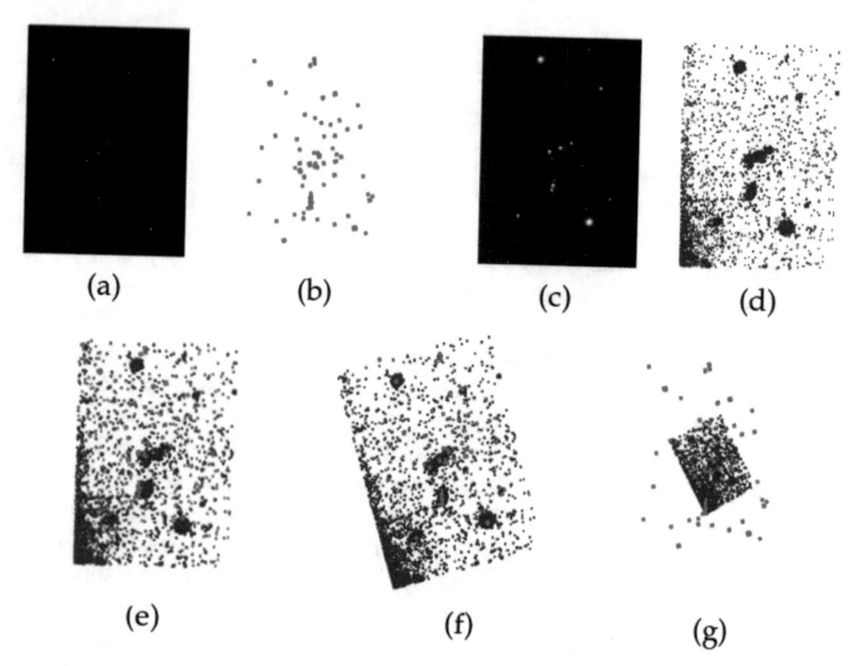

Figure 9.6: Experiment with prior correspondences as applied to image registration. Pictures are converted into 2D point sets, whereby pixel intensities determine point masses. (a) Reference image [279] and (b) the corresponding point set; (c) template image [255] and (d) the corresponding point set; (e) initial alignment of the point sets; (f) recovered transformation by GA; (g) registration result of CPD.

such as Kinect. An RGB-D sensor outputs RGB images as well as depth maps for discrete moments of time.

The course of the experiment is equal for both datasets. The difference concerns resolution of the depth maps and, as a consequence, the number of points in the point clouds. In Fig. 9.9 depth maps involved in the experiments are shown. Every depth map corresponds to a frame in a recorded RGB-D sequence. The resolutions of the depth maps are 640×480 and 512×424 pixels for the *copyroom* (from the Stanford Scene Dataset) and the *electrical cabinet* dataset (from CoRBS), respectively. The depth maps are converted to the corresponding point clouds using the parameters provided by the authors of the datasets, *i.e.,* intrinsic camera parameters and scaling factors for the depth maps. In Fig. 9.7, several examples of point clouds are shown.

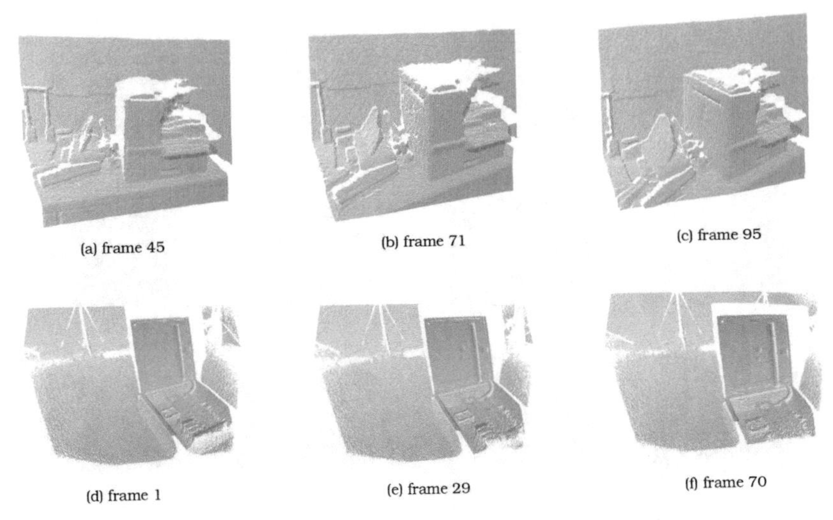

Figure 9.7: Selected point clouds converted from the depth maps corresponding to the frames from the Stanford 3D Scene Dataset (a)-(c) and CoRBS (d)-(f).

For the registration, we choose several frame combinations with different frame intervals between references and templates. If a frame interval is reasonably large, rigid registration can become challenging because of clustered outliers, missing parts, effects of the depth sensor distortion and noise presented in both point sets. The goal of the experiment is to demonstrate that GA can potentially be used in a SLAM system to register point clouds. Therefore, we compare cloud-to-cloud distances between point clouds on the initialisation step and after the registration is finished. Thus, we can qualitatively observe if GA can improve the template's pose based at least on a single criterion.

For several challenging frame pairs, rigid registration with GA is performed. Point clouds contain $\approx 2.67 \cdot 10^5$ and $\approx 2 \cdot 10^5$ points for the Stanford 3D Scene Dataset and CoRBS, respectively. All point clouds are subsampled so that each of them contains 2000 points. The algorithm converges at the latest after 100 iterations when oscillations attenuate. The runtime per iteration amounts to ≈ 0.5 sec. $G = 1.27 \cdot 10^{-4}$ for the Stanford 3D Scene Dataset and $G = 8.27 \cdot 10^{-4}$ for CoRBS is set. In all experiments, $\eta = 0.2, \varepsilon = 0.1$ are set and the scaling is fixed since the point clouds do not differ in scale significantly in this experiment. Note that the colour information presented in the colour images is not used — all points are of equal masses. This makes the scenario more challenging for GA. Colour

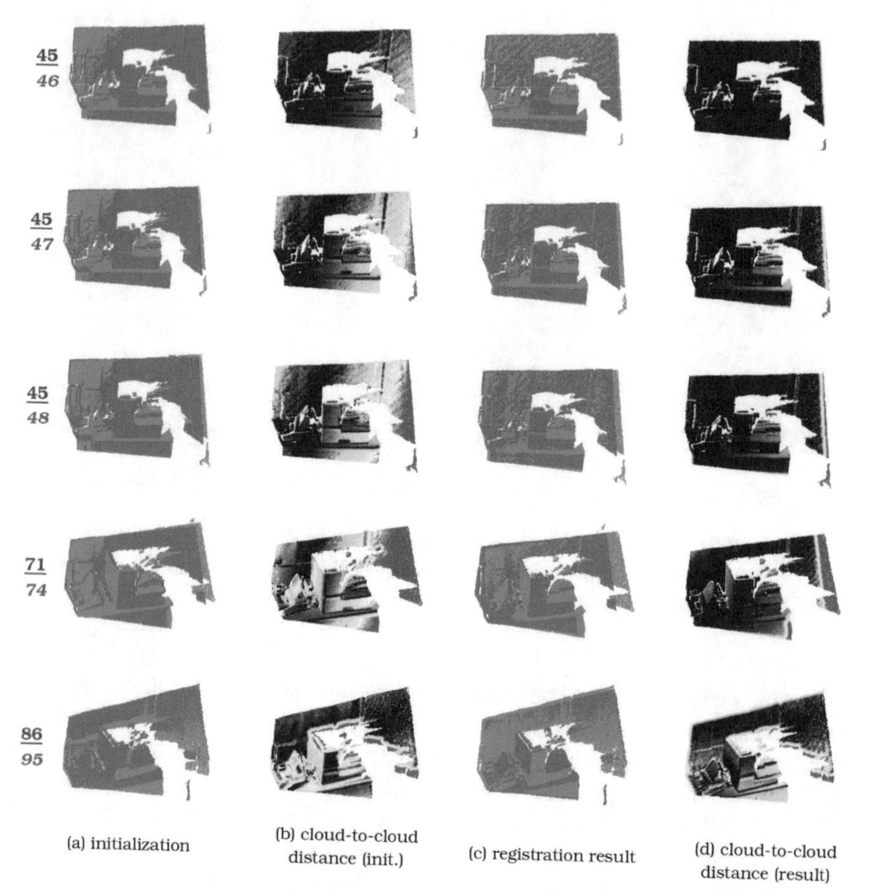

Figure 9.8: Results of the experiment on the Stanford 3D datasets [329]: (a) initial misalignments (initialisation); references are shown in cyan and templates in orange; on the left, corresponding frame numbers are provided (in cyan for the reference frames and in orange for the template frames). (b) the cloud-to-cloud distance corresponding to the initial misalignment; (c) GA registration results; (d) the cloud-to-cloud distance corresponding to the registration result; the cloud-to-cloud distances are coded with a Blue<Green<Yellow<Red colour scale; the saturation point (red value) amounts to 0.08 distance units.

information provided by the RGB images can be used to assign different point

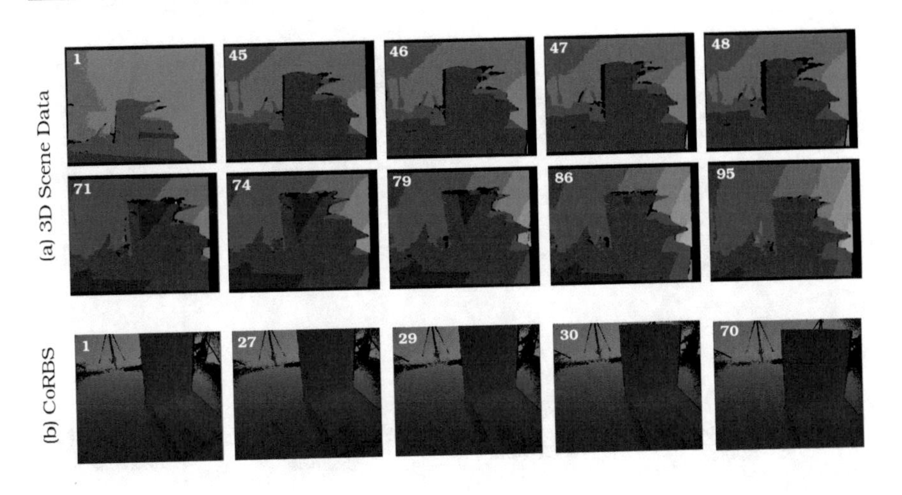

Figure 9.9: Depth maps involved in the experiment: (a) selected frames from the *copyroom* dataset [329]; (b) selected frames from the *electrical cabinet* dataset [303]; the contrast of the depth maps is enhanced for better perceptibility; corresponding frame numbers are given in the top left corners of the images.

weights both in a reference and in a template, similar to the Orion experiment (see Fig. 9.6, Sec. 9.1.2).

In Fig. 9.8, results on the Stanford 3D Scene Dataset are summarised. In the first column (Fig. 9.8-a), initialisations are shown — the point clouds are taken directly after conversion from the depth maps. The references are shown in cyan and the templates are shown in orange. On the left, frame numbers of the reference frames (cyan) and the template frames (orange) are provided. The second column (Fig. 9.8-b) contains colour-coded cloud-to-cloud distances between the templates and references. In the third column (Fig. 9.8-c), the GA registration results and, similarly, in the fourth column (Fig. 9.8-d) the corresponding colour-coded cloud-to-cloud distances between the registered templates and references are provided. Analogously to the Stanford 3D Scene Dataset, the results on the CoRBS Dataset are summarised in Fig. 9.10. Note that every cloud-to-cloud distance plot evinces the same colour saturation for a given dataset. This simplifies the visual comparability of the results.

In all attempts, GA is able to improve cloud-to-cloud distances between the point clouds significantly. This demonstrates the robustness of GA against the

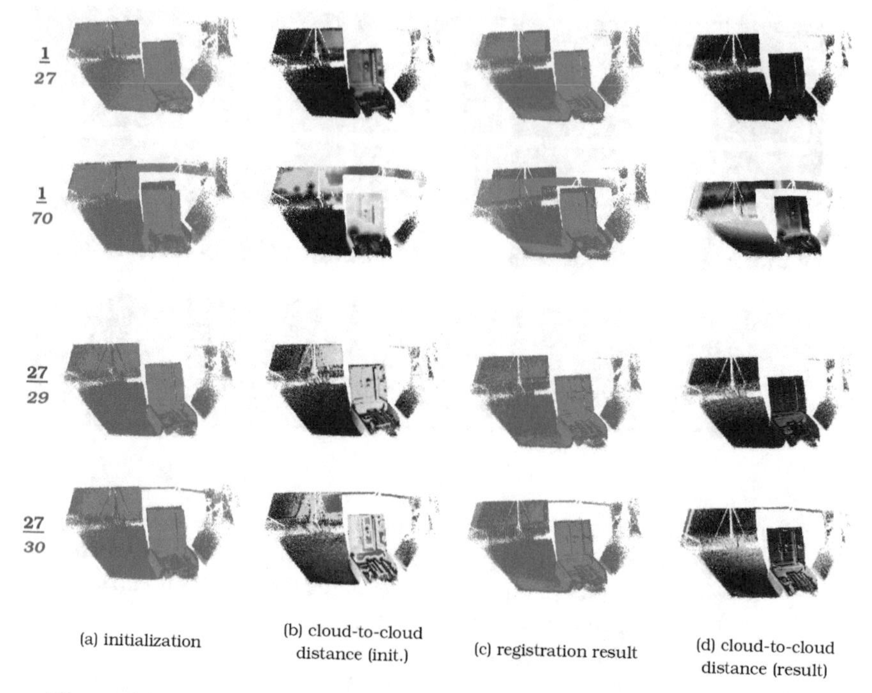

Figure 9.10: Results of the experiment on the CoRBS dataset [303]: (a) initial misalignments (initialisation); (b) cloud-to-cloud distance corresponding to the initial misalignment; references are shown in cyan and templates in orange; on the left, corresponding frame numbers are provided (in cyan for the reference frames and in orange for the template frames). (c) GA registration results; (d) cloud-to-cloud distance corresponding to the registration result; cloud-to-cloud distances are coded with a Blue<Green<Yellow<Red colour scale; the saturation point (red value) amounts to 0.1 distance units.

typical real-world point cloud artefacts occurring in different combinations. Thus, the experiment shows that GA can potentially be used in challenging real-world scenarios such as scene completion for SLAM. Accordingly, an accelerated GA will be tested in the SLAM scenario in future work. Moreover, additional information such as initial velocity and colours will be used in future experiments.

9.1.2.4 Discussion

The experiments confirm that it is possible to register point sets through modelling a rigid system of particles in a force field. The results evince suitability of the proposed method to cope with real-world scenarios. In the above experiments, GA performs robustly in the presence of large amounts of noise, especially uniformly distributed noise. We believe that the unique properties such as embedding of prior correspondences through different point masses and outlier suppression need to be further investigated.

The current limitation of the proposed approach consists in its limited capability to resolve scale which requires a special parameter tuning. If the parameters are set suboptimally, the template can shrink to a single point. Also in the current implementation, GA can handle large point sets through subsampling.

9.2 Accelerated Gravitational Approach with Altered Laws of Physics and Non-Linear Least Squares

The previous section has shown that Newtonian particle dynamics can be adapted for point set registration. The proposed GA outperformed competing methods in scenarios with substantial portions of noise and moderate angles of initial misalign-mentbetween a reference and a template. The solution with ODE's has several weaknesses such as sensitivity to the parameter choice, slow convergence and a large number of iterations until convergence. Recall that behind the scenes, GA finds transformations corresponding to the locally minimal gravitational potential energy (GPE). The physically accurate GPE is not solvable by general-purpose iterative optimisation techniques (*e.g.*, Gauss-Newton algorithm).

In this section, we propose an enhanced method which overcomes the limitations of GA, see Fig. 9.11 for an overview of the method. We apply a negative element-wise reciprocal transform to the GPE and enable efficient solution to rigid point set registration in the framework of non-linear least squares (NLLS). The proposed transform changes the form of the energy functional to a weighted sum of squared residuals and inverts physics while preserving the advantages of multiply-linked particle dynamics (*e.g.*, notion of a particle's mass).

We introduce a new acceleration technique based on the idea of Barnes and Hut initially developed for N-body simulations [31]. Applied to RA, a Barnes-Hut (BH) tree allows to efficiently handle large point sets with hundreds of thousands

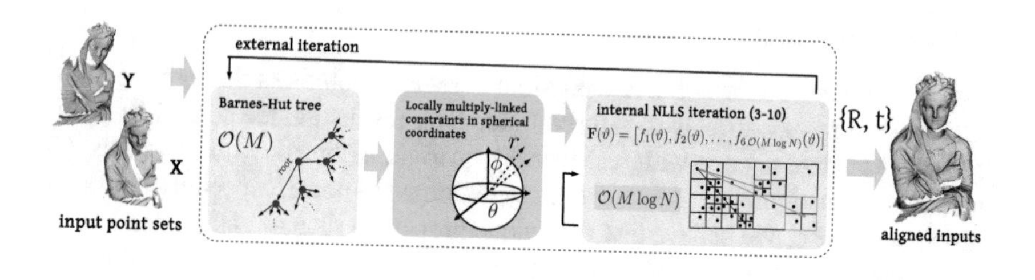

Figure 9.11: An overview of the proposed BH-RGA approach. In every external iteration, we compute a Barnes-Hut tree for the reference and template jointly and specify an optimisation problem either with or without additional constraints in spherical coordinates. Next, NLLS solver runs for few internal iterations while requiring $\mathcal{O}(M \log N)$ time for approximating gravitational potential energy of all-to-all particle interactions.

of points while preserving the globally multiply-linked nature of interactions. In other words, this happens not at the expense of constraining interactions to local vicinities but rather by accumulating contributions of individual reference points into clusters. A BH tree encompasses multiple extractable representations of **A** at a low computational cost with spatially varying cluster configurations. Next, we show that the accuracy of our method can be enhanced by a local multiply-linked policy. We formulate a new alignment criterion in spherical coordinates, in terms of vector lengths and angles between vectors. We observe an additional improvement in the registration accuracy in cases with missing data and when point sets exhibit moderate structure variations (Sec. 9.2.1.2).

We call the new method *Barnes-Hut Rigid Gravitational Approach* (BH-RGA). Among further advantages of BH-RGA is acceptance of anchor points and colours. We design a set of systematic tests for RA with a heightened complexity and test the performance of BH-RGA on it. Systematic experiments demonstrate that BH-RGA surpasses performances of baseline methods in terms of the convergence basin and accuracy when handling incomplete, noisy and perturbed data. As real-world examples, we align partially overlapping point clouds originating from cultural heritage and lidar measurements.

9.2.1 The Enhanced Particle Dynamics Based Gravitational Approach

Consider *negative elementwise reciprocal transform* $\xi^{-}(\cdot)$ of Eq. (9.10):

$$\xi^{-}(\mathbf{E}(\mathbf{R},\mathbf{t})) = \sum_{i,j} \frac{1}{Gm_i m_j} \left\|\mathbf{R}\mathbf{x}_i + \mathbf{t} - \mathbf{y}_j\right\|_2 + \varepsilon. \quad (9.19)$$

Eq. (9.19) changes the model of the simulated world. In the inverse world, the potential between two particles is inversely proportional to the product of their masses and directly proportional to the distance between the points. The meaning of a mass has changed — now, the mass is comprehended as a property of the matter so that the less its value, the more significant is the interaction.

It is noteworthy that in both cases, *i.e.*, $\mathbf{E}(\mathbf{R},\mathbf{t})$ and $\xi^{-}(\mathbf{E}(\mathbf{R},\mathbf{t}))$, *the further are two particles apart from each other, the higher is the GPE between them.* Note that

$$\lim_{\|\mathbf{RY}+\mathbf{T}\|_{\mathscr{F}} \to \|\mathbf{X}\|_{\mathscr{F}}} \mathbf{E}(\mathbf{R},\mathbf{t}) = -\infty, \text{ and} \quad (9.20)$$

$$\lim_{\|\mathbf{RY}+\mathbf{T}\|_{\mathscr{F}} \to \|\mathbf{X}\|_{\mathscr{F}}} \xi^{-}(\mathbf{E}(\mathbf{R},\mathbf{t})) = 0 \quad (9.21)$$

(here, we use a shorthand $\mathbf{T}_{M\times D} = \begin{pmatrix}\mathbf{t} & \mathbf{t} & \dots & \mathbf{t}\end{pmatrix}^{\mathsf{T}}$). Thus, $\xi^{-}(\cdot)$ preserves the optimality condition which was postulated in GA, *i.e.*, an optimal alignment of $\mathbf{X}$ and $\mathbf{Y}$ will still be reached when the GPE of the $\{\mathbf{X},\mathbf{Y}\}$ system is locally minimal. In the context of RA, (9.19) has several advantages versus (9.10).

The multiplicative and additive factors G and ε do not influence the optimal configuration; ε is redundant also because gravitational collapse results in a zero GPE. Moreover, $\xi^{-}(\cdot)$ does not distort multiply-linked nature of interactions, and the meaning of the distance is preserved. Last but not least, Eq. (9.19) can be minimised by Gauss–Newton algorithm (see Sec. 9.2.1.4).

9.2.1.1 Acceleration with a Barnes-Hut Tree

Our main idea is to cluster the particles dependent on the distance, accumulate their contributions to GPE and surpass the quadratic computational complexity. Whereas previous approximations [203, 286] *de facto* restrict interactions to local neighbourhoods, we preserve global multiply-linking in our approximation.

Suppose $\mathbf{x}_j$ interacts with K sufficiently distant particles $\mathbf{y}_k$, $k \in \{1,\dots,K\}$ and $\mathbf{y}_k$ are sufficiently close to each other. Let $\hat{\mathbf{x}}_j = \mathbf{R}\mathbf{x}_j + \mathbf{t}$ for briefness. Consider that variance of the distances from $\mathbf{x}_j$ to every $\mathbf{y}_k$ is below some small ζ. *If located sufficiently far, the total impact of all elements* $\mathbf{y}_k$ *in a volume of space* V *to a particle* $\mathbf{x}_j$ *can be well approximated by the impact of the combined single particle*

$\tilde{\mathbf{y}}$. *The mass of* $\tilde{\mathbf{y}}$ *equals to the mass integral of* $\mathbf{y}_k$ *over* V, *and the position of* $\tilde{\mathbf{y}}$ *equals to the centre of mass of the elements* $\mathbf{y}_k$ *in* V:

$$\sum_{k=1}^{K} m'_j m'_k \left\| \hat{\mathbf{x}}_j - \mathbf{y}_k \right\| \approx m'_j \left(\sum_{k=1}^{K} m'_k \right) \left\| \hat{\mathbf{x}}_j - \tilde{\mathbf{y}} \right\|, \tag{9.22}$$

with m'_i and m'_j denoting in the following the inverse masses. Next, we recall how a BH tree is computed in the general case and used for GPE calculation.

Building BH tree

A BH tree is initialised as a root node with 2^D empty external nodes. Always starting from the root, a new particle p_i is added to the tree following several rules, and every insertion results in a new leaf (an occupied external node). The insertion rules are: 1) if an external node v is empty, add p_i to v and done; 2) if an internal (= non-empty) node v_i is encountered, update its centre of mass and the mass; update all internal nodes hierarchically descending until an external node is reached (either occupied or an empty one); 3) if an external node v'_e is occupied, treat v'_e as an internal node, *i.e.*, introduce centre of mass and a mass of the node, and then split v'_e into 2^D regions; after splitting, there will be one occupied external node and $2^D - 1$ empty external nodes; add p_i to one of 2^D external nodes following either the rules for inserting into an empty or an occupied external node [31].

To compute GPE at every p_i location, naïvely we would need to compute m BH trees, each containing $\mathbf{A} \cup p_i$. After all, it is very costly to build an individual BH tree for every template point. Instead, we build a joint BH tree on the point set union $\mathbf{A} \cup \mathbf{B}$. Since template points per definition are not influenced by each other, we set all masses of $\mathbf{B}$ to zeroes. This allows us to include all points of the template into the tree (consider their positions) but exclude their effect (mass) to other points after this step, *i.e.*, while calculating GPE at location p_i, no template point distorts the potential between p_i and $\mathbf{A}$. On the other hand, to obtain a non-zero potential, we set the mass of p_i (a leaf in the BH tree) to its original value. This mass does not affect the force acting on p_i, as p_i never interacts with any cluster containing it; the BH tree does not need to be updated. After the potential is computed, we set the mass of p_i again to zero in order to avoid interaction between template points.

Using BH tree

The potential at position p_i is calculated by traversing the BH tree starting from the root towards the leaves and requires a single parameter — the distance threshold γ. It determines whether traversing further downwards the current internal node is

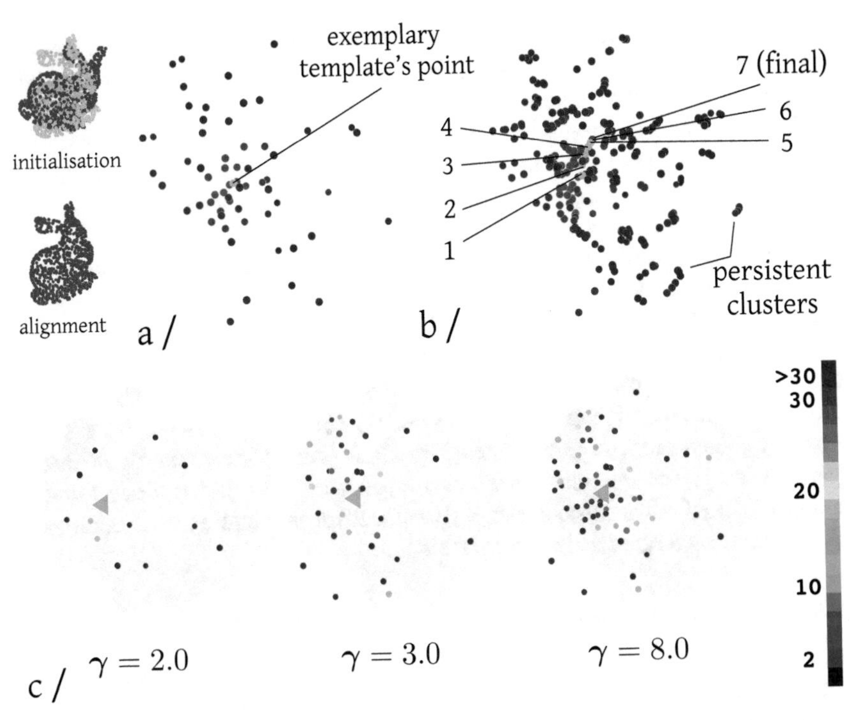

Figure 9.12: Clusters fetched during alignment (*clean-500* experiment, see Sec. 9.2.2.1) from a BH-octree: a) cluster configuration in the beginning; b) overlayed cluster configurations for all seven iterations. The colour encodes the distance to the exemplary template's particle (the darker, the more distant), and massless clusters are shown in black; c) cluster configuration in the first iteration, for a single template's point (shown as a triangle) and three different γ values. The colour scheme is given on the right. Each colour encodes a mass range of the cluster. Initially, all points were assigned unit masses.

finished (in this case, v_i contributes to GPE by a collective potential), or all subcells of v_i must be visited until a sufficiently far internal node or a leaf is reached.

If **A** and **B** contain duplicated points, the subdivision in BH tree will continue infinitely long. In practice, the depth of a BH tree is restricted to avoid infinite splits. The earlier the node is used for the force evaluation, the higher is the amount of individual particle-particle interactions which are bypassed. The time and memory requirements of computing a BH tree for a given union $\mathbf{A} \cup \mathbf{B}$ are constant, and the speed of GPE calculation depends on γ.

The NLLS optimisation step in Fig. 9.11 features a schematic visualisation of a multiply-linked particle interaction with a BH quadtree in 2D — the further are the particles from the target particle, the wider are the regions which can be built for the accumulative interaction (marked by blue, orange and green squares). Fig. 9.12 visualises cluster configurations fetched during a registration for a single tempalte's point, for one of the instances from the *clean-500* experiment (see Sec. 9.2.2.1) with $\gamma = 5.0$ (fixed for all iterations). The input point clouds contain 817 points each, and the cardinality of the configurations fetched from the BH-octree lies in the range $[49;62]$ points. Some of the clusters are massless and can be removed from the respective configuration as they comprise only massless templates' points. Distinct is the trajectory of the examplary point over all iterations. Some distant particles are joined into clusters which are persistent over all iterations, whereas compositions of other clusters change dynamically from iteration to iteration while the template undergoes a rigid transformation.

9.2.1.2 Local Enhancement with Spherical Coordinates

In RA, the most challenging subtask is rotation resolution. We propose to impose additional constraints in spherical coordinates in terms of the radial distance r, the azimuthal angle $\theta \in [-\pi;\pi)$ and polar angle $\phi \in [0;\pi]$. In a metric space, rotation alters all three coordinates. Rotation in spherical coordinates around the origin of the coordinate system does not affect r, and differences in θ and ϕ will always be constant for rigid bodies. Besides, local structures comprise a stronger cue when it comes to the local refinement especially when the inputs do not completely overlap. We can now think of rigid point set registration as a process of finding an optimal state where radial distances, as well as the angles θ and ϕ, coincide altogether as far as possible. The local multiply-linked enhancement policy is imposed by thresholds on r, θ and ϕ. *i.e.,* we add additional residuals for every spherical coordinate to Eq. (9.19). In the out-of-the-range cases, the residuals are set to constants so that within the range, the residuals are always smaller.

In the enhancement policy with spherical coordinates, we perform pre-processing of the input point sets in Euclidean space. First, we register both point sets to the origin of the coordinate system. Second, we normalise the coordinates so that all points reside in the space bounded by $\{x,y,z\} \in [-1;1]$. Upon the *mathematical convention*, conversion of a point $\mathbf{p} = \mathbf{p}(x,y,z)$ from Cartesian to spherical representation (r, θ, ϕ) is performed as follows:

$$r = \sqrt{x^2 + y^2 + z^2}, \quad \theta = \tan^{-1}\left(\frac{y}{x}\right), \tag{9.23}$$

$$\phi = \tan^{-1}\left(\frac{\sqrt{x^2+y^2}}{z}\right). \tag{9.24}$$

tan is a periodic function defined in the intervals $\left(-\frac{\pi}{2} + k\pi; \frac{\pi}{2} + k\pi\right)$, for all $k \in \mathbb{Z}$, and $\tan^{-1}$ returns an angle in the interval $\left(-\frac{\pi}{2}; \frac{\pi}{2}\right)$. To determine θ and ϕ in the ranges upon the used convention, we perform quadrant-dependent angular corrections. The reverse transformation from the spherical to Cartesian coordinates is performed by

$$x = r\sin(\phi)\cos(\theta), \;\; y = r\sin(\phi)\sin(\theta), \;\; z = r\cos(\phi). \tag{9.25}$$

9.2.1.3 Handling Varying Point Densities

Handling of point sets of varying densities has been indicated as challenging for many approaches, with specialised techniques and pre-processing steps (such as re-sampling) required to handle the setting efficiently [151]. For BH-RGA, we propose to normalise total point mass per unit volume in linear time. In *volumetric mass normalisation* (VMN), the bounding box of a point set is split in multiple non-overlapping voxels, and in every voxel, the pre-defined mass is uniformly distributed among all points in the voxel.

9.2.1.4 Energy Minimisation

BH-RGA minimises a sum of squared residuals $f_r(\vartheta) = c_r \left\|\mathbf{R}\mathbf{x}_i + \mathbf{t} - \mathbf{y}_j\right\|$ weighted by the product of inverse masses $c_r = m'_j m'_k$. Let $\mathbf{F}(\vartheta) : \mathbb{R}^6 \rightarrow \mathbb{R}^{6\mathcal{O}(M \log N)}$ denote a vector-valued function with $f_r(\vartheta)$, *i.e.*, $\mathbf{F}(\vartheta) = \left[f_1(\vartheta), f_2(\vartheta), \ldots, f_{6\mathcal{O}(M \log N)}(\vartheta)\right]^{\mathsf{T}}$, and the exact number of residuals depends on the BH tree configuration and γ. The optimisation objective can be compactly written as

$$\vartheta' = \arg\min_{\vartheta} \|\mathbf{F}(\vartheta)\|_2^2, \tag{9.26}$$

with $\mathbf{R}$ parametrised by the axis-angle representation. Since minimising the GPE in Eq. (9.26) is an overconstrained non-linear optimisation problem, we solve it in the least squares sense with a Levenberg-Marquardt (LM) algorithm [175, 189] by iteratively linearising $\mathbf{F}(\vartheta)$ in the vicinity of the current solution. For the higher resistance to outliers, we apply Huber loss to every $f_r(\vartheta)$ defined as:

$$\|a\|_\varepsilon = \begin{cases} \frac{1}{2}a^2, & \text{if } |a| \leq \varepsilon \\ \varepsilon(|a| - \frac{1}{2}\varepsilon), & \text{otherwise.} \end{cases} \tag{9.27}$$

The BH tree is renewed after every successful $\Delta\vartheta$ update and the objective (9.26) is formulated for the new set of $f_r(\vartheta)$ in every external iteration. The number of corresponding internal LM iterations is kept in the range $[3;10]$, see Fig. 9.11.

9.2.2 Experimental Evaluation

In this section, we describe evaluation methodology and results. BH-RGA is implemented in C++. The experiments are performed on a system with 32 Gb RAM and quadcore Intel *i7-6700K* processor running at 4GHz. For NLLS, we use *ceres* solver [242]. We compare our BH-RGA with Iterative Closest Point (ICP) [40], Coherent Point Drift (CPD) [203], GMM Registration (GMR) [157] and GA proposed in Sec. 9.1 [115]. CPD and GMR are evaluated both as 6 DoF ($\mathbf{R}, \mathbf{t}$) and 7 DoF ($\mathbf{R}, \mathbf{t}$ and scale s) variants. To evaluate the influence of prior matches, we also compare BH-RGA with rigid extended CPD (E-CPD) guided by prior matches proposed in Sec. 8.1.2 [123]. ICP is taken from the *Matlab*'s repository [311], and implementations of other methods are publicly available. In every experiment, we report mean root mean square error (RMSE) and standard deviation of the RMSE denoted by σ.

9.2.2.1 Quantitative Evaluation

Main Tests

In the tests with synthetic data, we use a subsampled Stanford *bunny* [278]. In the first *clean-500* test, we sample the 3D rotation space by the angular displacement of $\frac{2\pi}{10}$ radians and synthesise 500 different initial configurations (the duplicated states are avoided). In this experiment, we test the methods for their ability to resolve rotation in scenarios with noiseless data. In contrast, the second test is designed to evaluate the ability of RA to converge under severe amounts of noise. We add 100% of uniform noise to the *clean-500* and obtain *N500-U100* dataset. The range of the noise generating function is a bounding sphere of the template. Several methods demonstrate poor performance on *N500-U100*, and we also generate *N500-U50* dataset with 50% of uniform noise according to the same rule. Next, we choose a single initial configuration which can be successfully resolved by all methods and add three different noise patterns to the template — 100% of uniform noise, 100%

Table 9.1: Summary of the qualitative evaluation of the compared methods with *clean-500*, *N500-U100*, *U100*, *G100* and *GS100* datasets.

methods and metrics		*clean-500*	*N500-U50*	*N500-U100*	*U100*	*G100*	*GS100*
ICP [40]	succes. (in %)	62 (12.4%)	36 (7.2%)	19 (3.8%)	33 (66%)	**50 (100%)**	**50 (100%)**
	RMSE (σ)	0.005 (0.016)	0.022 (0.3)	0.042 (0.031)	0.091 (0.081)	**0.007 (0.002)**	**0.002 (5*E*-4)**
CPD (7 DoF) [203]	succes. (in %)	130 (26%)	**128 (25.6%)**	**109 (21.8%)**	**48 (96%)**	**50 (100%)**	**50 (100%)**
	RMSE (σ)	0.04 (6*E*-5)	**0.064 (0.003)**	**0.088 (0.003)**	0.098 (0.066)	**0.027 (1.7*E*-3)**	**0.046 (1.7*E*-3)**
CPD (6 DoF) [203]	succes. (in %)	**143 (28.6%)**	98 (19.6%)	62 (12.4)	**48 (96%)**	**50 (100%)**	**50 (100%)**
	RMSE (σ)	**0.006 (0.009)**	0.025 (0.014)	0.034 (0.017)	0.061 (0.148)	**0.01 (0.003)**	**7.5*E*-3 (2.3*E*-3)**
GMR (7 DoF) [157]	succes. (in %)	126 (25.2%)	113 (22.6%)	0 (0%)	0 (0%)	**50 (100%)**	**50 (100%)**
	RMSE (σ)	7*E*-5 (8*E*-5)	0.084 (0.005)	n / a	n / a	**0.04 (7.5*E*-3)**	**1.5*E*-3 (4.7*E*-4)**
GMR (6 DoF) [157]	succes. (in %)	131 (26.2%)	79 (15.8%)	87 (17.4%)	36 (72%)	26 (52%)	25 (25%)
	RMSE (σ)	8*E*-5 (2*E*-4)	5*E*-4 (2*E*-4)	6*E*-4 (2*E*-4)	0.16 (0.077)	0.37 (0.4)	0.37 (0.38)
GA [115]	succes. (in %)	21 (4.2%)	6 (1.2%)	3 (0.6%)	19 (38%)	1 (2%)	10 (20%)
	RMSE (σ)	0.029 (0.021)	0.049 (0.025)	0.03 (0.012)	0.164 (0.082)	0.289 (0.087)	0.163 (0.072)
BH-RGA (ours)	succes. (in %)	132 (26.4%)	**132 (26.4%)**	**100 (20%)**	**50 (100%)**	**50 (100%)**	**50 (100%)**
	RMSE (σ)	0.009 (0.004)	**0.032 (0.013)**	**0.059 (0.021)**	**0.056 (0.017)**	**0.04 (0.01)**	**0.022 (0.006)**

of Gaussian noise (the range volume is the same as in *N500-U100*) and 100% of the Gaussian noise with the mean values coinciding with the template point locations. The resulting datasets are abbreviated as *100U*, *100G* and *100GS*, respectively. Table 9.1 summarises the outcomes of the experiments with the new datasets. If RMSE is below 0.1, the registration succeeds. This value is not set arbitrarily — in the case of successful alignments RMSE was in most cases much lower than 0.1. Otherwise, it was in most cases considerably larger than 0.1.

Upon our expectation, BH-RGA is not the most accurate method on the *clean-500* dataset (it is the second best regarding the success rate). Perfectly matching data is, however, not common in practical applications. With an increasing noise, the relative accuracy and performance of BH-RGA compared to all other methods increases. While CPD (7 DoF) achieves a lower RMSE than BH-RGA on *clean-500*, the situation inverts on *N500-U50* and *N500-U100*, with the success rates of BH-RGA close to those of CPD (and the lowest RMSE compared to CPD). At the same time, ICP and GMR considerably lower their success rates on *N500-U50* and *N500-U100*. BH-RGA was the only successful method on *U100* — it resolves all cases with the lowest RMSE. The second best method is CPD (it resolves 48 cases out of 50), while demonstrating a 50% to 66% higher RMSE on *N500-U100* and *U100*, respectively. GMR is able to hold on par with ICP only in the 6DoF mode. On *N500-U100* and *U100*, GMR with 7 DoF is not able to align any data with RMSE lower than 0.1 (our success criterion). On *G100* and *GS100*, most methods successfully recover transformations for all cases, except of GMR (6 DoF) and GA which recover 52%, 25%, 1% and 10% of transformations, respectively; here, ICP, CPD and GMR (7 DoF) are mostly more accurate than BH-RGA (with the exception of CPD with 7 DoF on *GS100*).

The performance of baseline GA in our tests is in most cases inferior to the results of other methods. Our experimental setting is different from the one proposed in

Table 9.2: RMSE and σ (in parentheses) in *U256* and *G256* experiments.

method	*U256*	*G256*
ICP	**9*E*-3 (7*E*-3)**	**0.015 (0.012)**
CPD (7 DoF)	0.051 (9E-3)	0.064 (0.021)
CPD (6 DoF)	0.016 (0.014)	0.045 (0.058)
GMR (7 DoF)	0.019 (0.028)	0.065 (0.051)
GMR (6 DoF)	0.853 (1.16)	1.027 (1.226)
GA	0.149 (0.143)	0.207 (0.158)
BH-RGA (ours)	**0.015 (4*E*-3)**	**0.019 (9.5*E*-3)**

Table 9.3: Comparison of E-CPD [123] and BH-RGA with prior matches.

configuration		*RMSE*	σ	*success rate*
E-CPD	no priors	0.08	**0.013**	85 (17%)
	1 prior	0.084	**0.012**	89 (18%)
	2 priors	0.076	**0.014**	**376 (75%)**
	3 priors	0.056	0.018	**488 (97%)**
BH-RGA	no priors	**0.05**	0.018	**124 (25%)**
	1 prior	**0.043**	0.013	**199 (40%)**
	2 priors	**0.011**	0.017	191 (38%)
	3 priors	**0.0086**	**0.005**	165 (33%)

Sec. 9.1 where the angle of initial misalignment is randomly selected in the range $[0; \frac{\pi}{2}]$. GA was starting to fail in the angular misalignment range $[\frac{\pi}{4}, \frac{\pi}{2}]$. In our *clean-500* and *N-500* experiments, the angles are sampled with the 36° step implying that starting from the second value (out of ten), the angle is always $\geq 72°$. Thus, it is a suboptimal evaluation setting for GA. A similar situation is with the $U100$, $G100$ and $GS100$ experiments. Though the angle of initial misalignment is in the optimal range, the excessive amount of noise hinders GA to converge.

Recall that GA takes more than 300 iterations to converge and requires an energy dissipation term and multiple parameters. BH-RGA can handle scenarios which GA cannot handle well, due to the altering the laws of physics and the energy optimisation with NLLS. Thanks to NLLS, the number of iterations required by BH-RGA to converge is two orders of magnitude lower compared to GA, and all that with much fewer parameters that we need to specify.

Deteriorated Data

Next, we propose a new test for evaluating how robust are the methods to the corrupted data — the *U256* and *G256* experiments. We perturb template points with uniform and Gaussian noise. In total, there are 256 magnitude indexes generating states with the increasing degree of distortion (the maximum displacement or variance of the Gaussian in every state are equal to the scaled magnitude index times the length of the bounding box of the object in the x direction). The specifics of both experiments is that at the optimum, no point of the template physically coincides with any reference point. Thus, we report the mean RMSE between the reference and the template transformed by the recovered $\mathbf{R}, \mathbf{t}$, see Table 9.2. CPD and GMR have higher RMSE than ICP and BH-RGA, *i.e.*, both probabilistic methods are sensitive to point perturbations which agrees with our expectations. ICP shows the lowest RMSE. Especially for small perturbations, conditions for ICP are optimal. BH-RGA slightly concedes to ICP as its RMSE is a little higher

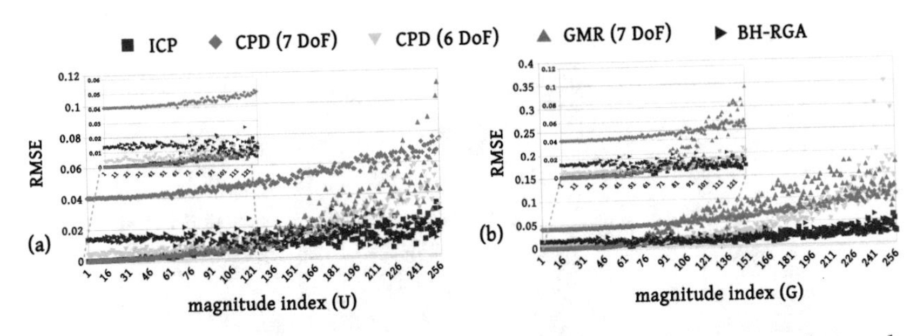

Figure 9.13: M. RMSE between the transformed unperturbed template and the reference as a function of the point perturbation magnitude index, in the *U256* (a) and *G256* (b) experiments.

for small perturbation magnitudes. On the other side, BH-RGA has the smallest σ. Additionally, the results are visualised in in Fig. 9.13 as scatter plots. Noticeable is the variance of the RMSE for different tested methods, *i.e.,* ICP [40], CPD [203], GMR [157] and our BH-RGA. The graphs allow localised interpretation of the results, in addition to RMSE averaged over all magnitude indices.

Prior Matches

Different masses in BH-RGA can serve as a weak form of prior correspondences since the optimal alignment (a locally minimal GPE) is more likely when points with larger masses are close to each other. We call unordered sets of relevant points with an unknown relation *anchor points*. If anchor points are available, masses of corresponding points can be set *two or three* orders of magnitude larger than the other masses, and the initial conditions will guide the registration process. A method which can perform rigid prealignment conditioned upon prior correspondences is a hybrid transformation estimation and correspondence estimation approach E-CPD [123]. Especially in the cases with only one or two reliable prior matches, this technique brings advantages. In contrast, the effect of anchor points in BH-RGA is closer to *registration in registration*. We have repeated the *N500-U50* test with up to three prior correspondences. CPD with no prior correspondences resolves 17% of the cases and its success rate increases to 18% and 75% for the case of one and two prior matches. BH-RGA resolves 25% with no prior matches and 39% with one or two available prior matches, see Table. 9.3.

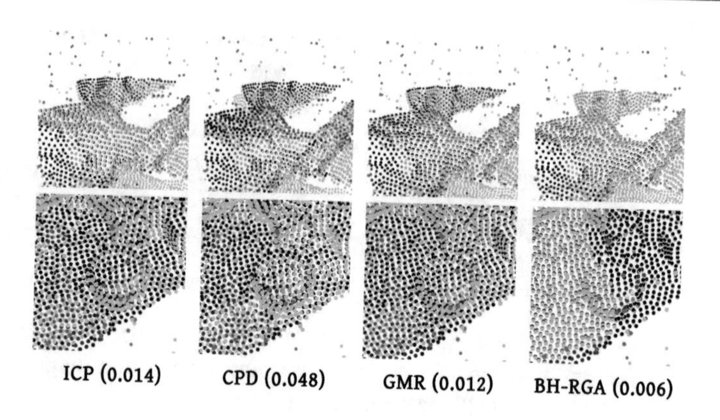

Figure 9.14: Fragment comparison after the alignment of two frames of the *sleeping2*.

Spherical Enhancement

Spherical enhancement is a new type of locally multiply-linked constraints effective for point sets with structured outliers and noise. In the next experiment, the reference and the template originate from the same SINTEL *sleeping_2* scene [61]; they overlap by 70% and contain 7*k* points including 30% of added uniformly distributed noise each. Best results obtained by ICP, GMR, CPD and BH-RGA as shown in Fig. 9.14. BH-RGA reduces RMSE of GMR by the factor of two.

Runtime and Convergence

We evaluate the runtime of BH-RGA for the inputs of different sizes with different thresholds γ. For each approximation level of the BH-tree, we report the achieved accuracy in the well-defined setting. We take a frame from the SINTEL *sleeping2* sequence as a reference and its translated and rotated clone as a template. The translation roughly amounts to one-third of the point cloud extent in the x-direction, and the rotation is either 5° or 24° around all axes. In total, there are seven versions of the point cloud obtained by subsampling, with the cardinality of ca. 5*k*, 12*k*, 25*k*, 50*k*, 105*k*, 205*k* and 446*k* (the original resolution). For each combination of the resolution and template transformation, we test multiple $\gamma \in [0.0625; 64.0]$. The runtime evaluation metrics for 5° rotations are summarised in Fig. 9.15, *i.e.*, runtimes (full and per iteration), runtime ratio of the BH-tree generation, RMSE and the number of residuals.

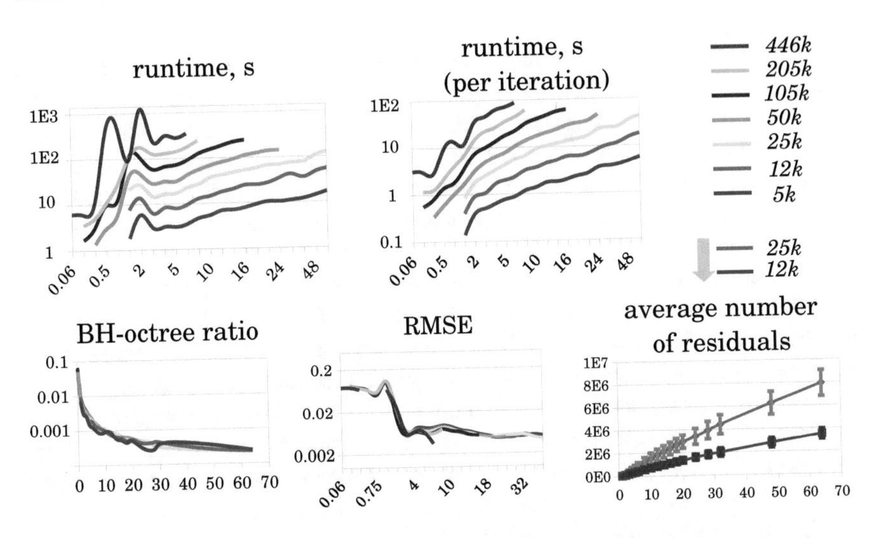

Figure 9.15: Graphs of different runtime evaluation metrics as the functions of the BH-tree threshold γ (SINTEL *sleeping2* sequence, different subsampling rates and a different number of points in the point clouds).

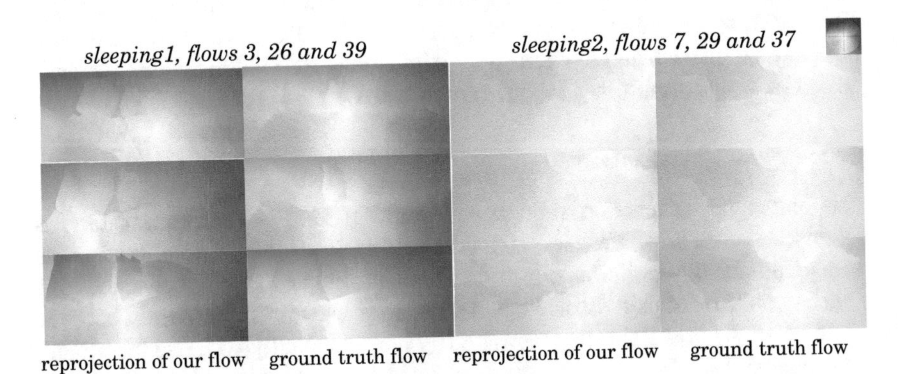

Figure 9.16: Examples of reprojected 3D flows obtained by BH-RGA, compared to ground truth optical flows, for the *sleeping1* and *sleeping2* sequences from the SINTEL collection [61].

Several patterns can be noticed in the runtime statistics. When γ is fixed, RMSE is similar for different numbers of points. Conversely, with the increasing γ, the alignment accuracy increases for all resolutions, and smaller and smaller runtime fraction is spent on building BH-tree while not being influenced much by the number of points. For $\gamma > 2.0$, BH-tree runtime ratio is $\leq 0.5\%$. In comparison, the solver runtime ratio is always around 80% suggesting that an up to fivefold acceleration with a GPU Gauss-Newton solver is possible. Next, as the γ increases, the number of iterations decreases and stabilises. For $\gamma \geq 6$, the number of iterations is ≤ 4. The average number of residuals (number of point-to-cluster interactions) reaches $\approx 1.8E7$ for $\gamma = 6.0$ and $446k$ points. The graph metrics for the case of $24°$ rotations coarsely follow the shapes of the graph metrics for the visualised case of $5°$ rotations. Larger misalignments require larger γ for a successful registration and the overall runtime increases by the factor of two (note that in this experiment, we do not perform a translational pre-alignment). The per-iteration runtime increases by $\approx 40\%$, and the number of iterations for $\gamma \geq 6$ increases by one-two.

The runtime evaluation shows that BH-RGA can register two point sets of $446k$ points each in 11 seconds, with the RMSE of 0.068 ($\gamma = 0.25$). The RMSE of 0.0065 requires 264 seconds ($\gamma = 3.0$). The comparable RMSE of 0.0067 for the point sets with $25k$ points requires 41 seconds ($\gamma = 12.0$). BH-RGA registers very large point sets in a globally multiply-linked manner in tolerable time, whereas tested implementations of competing methods were not able to cope with the data. Thus, [203] reports 10 minutes for two point sets of $\approx 35k$ points each when using fast Gauss transform and low-rank approximation of the Gaussian affinity matrix. BH-RGA would require ≈ 1 second for this task with $\gamma = 0.25$.

Scene Flow Visualisation

For point cloud sequences of rigid scenes, results of BH-RGA can be visualised as reprojected scene flow. We take 50 frames long *sleeping_1* and *sleeping_2* sequences with predominantly rigid frame-to-frame motion from the SINTEL collection [61], register point clouds pairwise and report the average endpoint error (AEPE) between reprojected 3D flow fields parameterised by the recovered $\{\mathbf{R}, \mathbf{t}\}$, and the ground truth optical flow. The AEPE for *sleeping_1* and *sleeping_2* amounted to 2.242 and 0.914, respectively, see Fig. 9.16 for visualisations. Naïve GPE calculations would require $2 \cdot 10^{11}$ evaluations in every iteration, and running a method with quadratic complexity becomes prohibitive for point sets containing more than $10k$ points. No other method was able to execute on such massive data (unordered point sets).

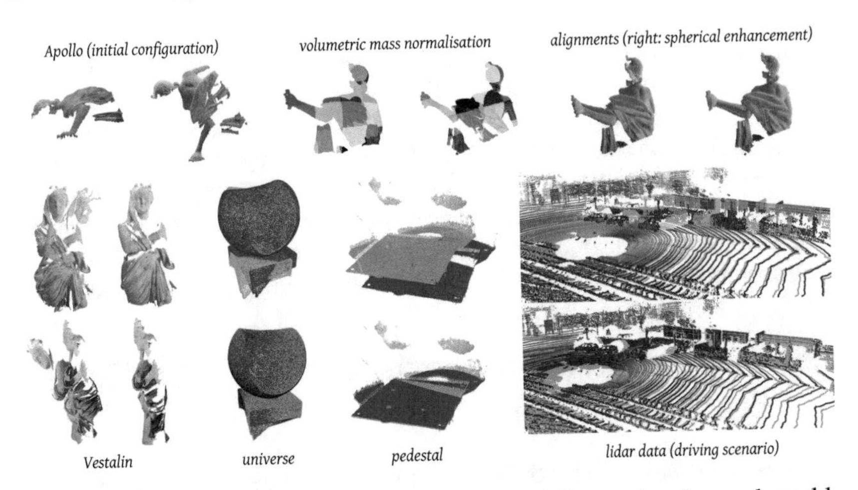

Figure 9.17: BH-RGA accurately registers partially overlapping real-world data. *Vestalin* is aligned with the colour cue and three anchor points, and for *pedestal* we use two anchor points. For *Apollo* and the lidar data, we apply volumetric mass normalisation (VMN) helping to counterbalance varying point densities (each colour corresponds to a voxel in VMN). For the *Vestalin*, the initialisations are shown on the left, and the alignments are shown on the right. For the *universe*, *pedestal* and *driving scenario*, the initialisations are on the top, and the alignments are on the bottom.

9.2.2.2 Evaluation with Real-World Data

BH-RGA is well suitable for registration of real-world data including large partially overlapping point sets and lidar data. Fig. 9.17 comprises several such scenarios. The *Apollo* and *Vestalin* datasets are partial 3D reconstructions of statues acquired with a structured light technique [233]. Due to the scanning process, they are represented in different reference frames. The partial scans of the *Apollo* dataset contain $\approx 30\%$ of clustered outliers. Due to the varying point densities, we apply volumetric mass normalisation (VMN). Though not a part of the BH-RGA algorithm itself, VMN is often useful in practice — it allows initialising masses uniformly per volume unit (see Sec. 9.2.1.3). *Vestalin* and *Pedestal* datasets are processed with three and two anchor points, respectively. For *Vestalin*, we additionally vary masses based on point intensities. In *Pedestal*, we use two anchor points in the lower left and right corners of the plates. *Universe* dataset represents two 3D reconstructions of a sculpture (the shapes differ in the lower parts).

We also test BH-RGA on lidar data from the KITTI dataset [109]. Fig. 9.17-(bottom right) shows an excerpt from the *2011_09_26_drive_0001* sequence. We take the first and the third frames of the sequence so that displacements are moderate and the outlier ratio amounts to ca. 20%. Due to the highly varying point densities in the scans, we apply VMN. It can be well noticed on buildings and cars that the scans are accurately registered. In this experiment, no anchor points or further prior knowledge were used. Both scans contain ca. 120*k* points and the registration is performed in ≈ 90 seconds on a CPU without subsampling or parallelisation.

9.3 Gravitational Approach for Non-Rigid Point Set Registration

In this section, we introduce the extension of GA (see Sec. 9.1) to the non-rigid case — Non-Rigid Gravitational Approach (NRGA). Similar to the rigid case, the optimal alignment of a deformable point set is attained in the state of a locally minimal gravitational potential energy of two particle swarms. We show that NRGA handles challenging scenarios with noisy and partially overlapping data and overcomes several limitations of previous approaches. NRGA accurately aligns a facial template with a known topology to a highly noisy reference point cloud obtained from a multi-view system. Our result captures the facial appearance while preserving the initial topology and suppressing the uniformly distributed noise component of the reference.

NRGA is the first method for non-rigid point set registration in the gravitational class proposed in the literature, to the best of our knowledge. In NRGA, the template and the reference are split into as many overlapping subsets as there are points in the template and a series of locally multiply-linked N-body problems is solved on the corresponding subsets. At the same time, the unconstrained position updates are combined into individual per-point displacements with *Coherent Collective Motion* (CCM) regulariser [294].

To maintain the topology, some methods rely on global motion coherency constraint [203, 300] or local shape signatures [186]. In contrast, the CCM regulariser in NRGA can be viewed as a locally-aware global topology preservation operator. Further, we embed curvature in Eq. (9.2). The curvature is an inherent metric of the space-time in theory of General Relativity (see Sec. 2.3.3). In gravitational point set alignment, the curvature can serve as a shape matching descriptor enhancing the registration accuracy. Though the curvature was used in shape and image registration before [92, 102, 265, 269, 274], its application in point set registration is not straightforward. One reason is that for an arbitrary point set, the curvature is not

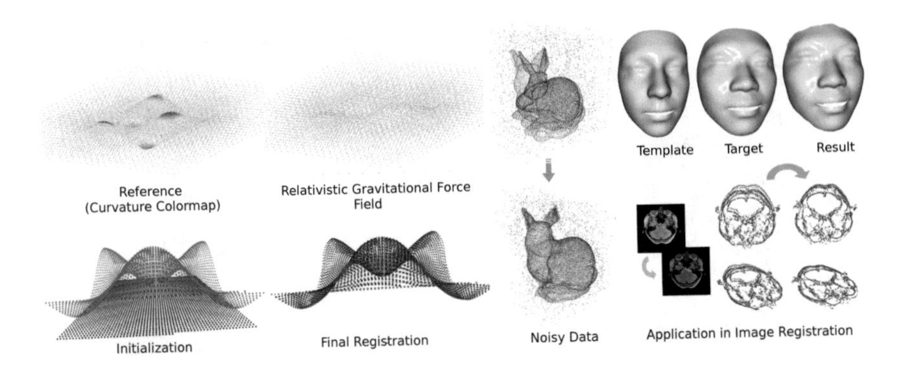

Figure 9.18: Similar to GA (see Sec. 9.1), NRGA uses the law of universal gravitation for non-rigid alignment of point sets. On the left, the *cosine surface* $f(x,y) = \cos(x)\sin(y)$ induces *relativistic gravitational force* field to attract the plane $f(x,y) = -1$, where $x,y \in [-\pi,\pi]$. On the right, several registration results are demonstrated, *i.e.*, noisy *bunny* [278], human faces of different expressions (from [55] on BU-3DFE [317]), and image registration of computer tomography of a human brain [297].

everywhere defined. Fig. 10.8 visualises the main principle of NRGA and shows some registration results in different scenarios.

9.3.1 Non-Rigid Gravitational Approach (NRGA)

Given two point sets, a *reference* or target $\mathbf{X}_{N\times D} = (\mathbf{X}_1,\dots,\mathbf{X}_N)^T$, and a *template* or source $\mathbf{Y}_{M\times D} = (\mathbf{Y}_1,\dots,\mathbf{Y}_M)^T$, with N and M D-dimensional points, respectively, our objective is to find a mapping $\mathscr{T}: \mathbb{R}^D \to \mathbb{R}^D$ which minimises a distance error function between two sets. In NRGA, this error function is the total GPE of $\mathbf{Y}$ w.r.t $\mathbf{X}$, *i.e.*, a sum of weighted inverse values of the Euclidean distances:

$$\phi_a = \arg\min_{\mathbf{T}} \sum_{i=1}^{M} \sum_{j\in\mathcal{N}(i)} \omega_{ij} \left(d\left(\mathscr{T}(\mathbf{Y}_i,\mathbf{T}_i) - \mathbf{X}_j\right)\right)^{-1}, \tag{9.28}$$

where $\omega_{ij} = -Gm_i m_j$ denotes the weight of GPE of point i amongst its nearest neighbours $j \in \mathcal{N}(i)$, and $d(\mathbf{Y}_i,\mathbf{X}_j) = \left\|\mathbf{Y}_i - \mathbf{X}_j\right\| + \varepsilon$ is the numerically stabilised Euclidean distance. The transformation $\mathscr{T}(\mathbf{Y},\mathbf{T})$ registers $\mathbf{Y}$ to $\mathbf{X}$ using the set of optimal transformation tuples $\mathbf{T} = \{\mathbf{T}_1,\dots,\mathbf{T}_M\}$. We employ a collision-less

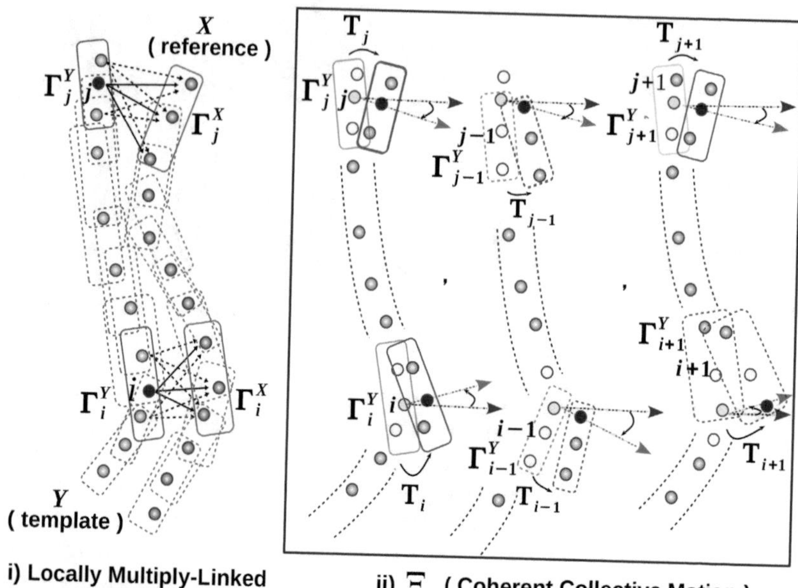

Figure 9.19: Two main steps of NRGA. *(left)* Solving locally multiply-linked N-body problems for the corresponding regions. These local gravitational interactions estimate updated unconstrained point positions. Points i and j select few nearest points from Y and X and induce the regions $\Gamma_i^{\mathbf{Y}}$ and $\Gamma_i^{\mathbf{X}}$ bounded by solid blue and red colour lines from $\mathbf{Y}$ and $\mathbf{X}$, respectively. Rigid transformation parameters $\mathbf{T}_i$ are obtained for all regions with Procrustes alignment between consecutive states. *(right)* Applying coherent collective motion regulaliser on final position updates, as a consensus filter on all $\mathbf{T}_i$.

N-body problem in a curved space (Eq. (2.80)) and iteratively find the optimal transformation parameters $\mathbf{T}_i$. In NRGA, the points are interpreted as particles with individual masses. The overall registration includes the following steps:

- *first*, we build k-d trees on $\mathbf{X}$ and $\mathbf{Y}$ separately. Since $\mathbf{X}$ is always static and remains unchanged, we computer the k-d tree on $\mathbf{X}$ only once and rebuild the k-d tree on $\mathbf{Y}$ in every iteration as $\mathbf{Y}$ deforms.

- *second*, in every iteration, each $\mathbf{Y}_i$ selects a number of nearest neighbours (see Sec. 9.3.1.2) from $\mathbf{X}$ and $\mathbf{Y}$ to define zones of influence $\Gamma_i^{\mathbf{X}}$ and $\Gamma_i^{\mathbf{Y}}$ so that the solution space for every $\mathbf{Y}_i$ is narrowed down. We use the *k*-d trees to obtain the nearest neighbours. For every $\mathbf{Y}_i$, $\Gamma_i^{\mathbf{X}}$ attracts the corresponding zone $\Gamma_i^{\mathbf{Y}}$ employing *N*-body problem parameterised by the Gaussian curvature of $\mathbf{X}$. This results in a locally multiply-linked point-to-point interaction. Once the force fields are computed, we independently update the template points in $\Gamma_i^{\mathbf{Y}}$. From the previous and current states of $\Gamma_i^{\mathbf{Y}}$, unconstrained transformation tuple $\mathbf{T}_i$ is obtained using Procrustes alignment [158]. We serially estimate $\mathbf{T}$ for all $\Gamma_i^{\mathbf{Y}}, i \in \{1,\ldots,M\}$. Every tuple $\mathbf{T}_i$ represents a transformation for the group of points in $\Gamma_i^{\mathbf{Y}}$. Conversely, a template point $\mathbf{Y}_i$ can have multiple transformation representatives as the adjacent regions overlap. In every iteration, only the velocity of representative points from each group is updated.
- *third*, we apply coherent collective motion (CCM) operator Ξ to consolidate multiple unconstrained transformation fields and velocities for every $\mathbf{Y}_i$ into an optimal per-point update. The locally optimal solution is achieved when GPE of the system reaches its local minimum.

The two iterative steps of NRGA are shown in Fig. 9.19.

9.3.1.1 Modified *N*-Body Problem

We follow the modifications of the *N*-body problem proposed in Sec. 9.1.1. The forces are computed on demand, whereas the curvature of $\mathbf{X}$ is precomputed following [213]. Only $\mathbf{X}$ induces gravitational field and remains fixed. We avoid gravitational interactions between the particles from the same set because, otherwise, the template and reference would deform at rest. Next, the external potential energy in Eq. (2.78) is considered as a dissipative loss, similar to Eq. (9.3). We add an external drain force $\mathbf{F}_i^{ext}$ to $\mathbf{F}_i^{a}$ as a force proportional to a small fraction η of the particle's velocity acting in the opposite direction:

$$\mathbf{F}_i^{ext} = -\eta\, \dot{\mathbf{r}}_i. \tag{9.29}$$

As a result, a part of the system's kinetic energy is dissipated in every iteration.

9.3.1.2 Distributed Locally Multiply-Linked Policy

NRGA is a locally multiply-linked method which iteratively registers $\mathbf{Y}$ to $\mathbf{X}$. The groups of nearest neighbour multi-sets from $\mathbf{Y}$ and $\mathbf{X}$ — further the metasets of point indices — are denoted by $\mathscr{R}_{\mathbf{Y}} \equiv \{\Gamma_1^{\mathbf{Y}},\ldots,\Gamma_K^{\mathbf{Y}}\}$ and $\mathscr{R}_{\mathbf{X}} \equiv \{\Gamma_1^{\mathbf{X}},\ldots,\Gamma_K^{\mathbf{X}}\}$, respectively. In total, there are $K = M$ elements in $\mathscr{R}_{\mathbf{Y}}$ and $\mathscr{R}_{\mathbf{X}}$ each. The number of elements in the metasets is picked as a proportion $\rho_{\mathbf{Y}}$ and $\rho_{\mathbf{X}}$ of the total number of points in $\mathbf{Y}$ and $\mathbf{X}$, respectively. Variation in the values of the proportion factors decides the magnitude of overlaps between the adjacent zones. $\rho_{\mathbf{Y}}$ and $\rho_{\mathbf{X}}$ can be different depending on the dissimilarity factors of input point sets (*e.g.*, densities and noise ratios). It is possible to set $\rho_{\mathbf{X}} = \rho_{\mathbf{Y}} = \rho$ when the point densities or the number of points are roughly equal in $\mathbf{X}$ and $\mathbf{Y}$. We choose $\rho_{\mathbf{X}}$ and $\rho_{\mathbf{Y}}$ as a function of N and M, respectively, and typically $\rho \in [0.02, 0.1]$. Hence, $P = \lfloor \rho M \rfloor$ (or $\lfloor \rho_{\mathbf{X}} N \rfloor$ and $\lfloor \rho_{\mathbf{Y}} M \rfloor$) nearest neighbours define each subset of the regions.

Locally multiply-linked N-body problem is solved on the multi-sets of corresponding regions. The particle-particle interactions are restricted between the regions — the goal is to refine the local transformation rather than to recover a single global transformation. Henceforth, local rigid transformation tuple $\mathbf{T}_k = \{\mathbf{R}_k, \mathbf{s}_k, \mathbf{t}_k\}$ for every region $\Gamma_k^{\mathbf{Y}}$ is estimated with rotation $\mathbf{R}_k$, scale $\mathbf{s}_k$ and translation $\mathbf{t}_k$ ($\mathbf{T}_k$ has 7 DoF). The force exerted by $\Gamma_k^{\mathbf{X}}$ on $\mathbf{Y}_p$ is:

$$\mathbf{F}^a_{p \in \Gamma_k^{\mathbf{Y}}} = -G m_p \sum_{q \in \Gamma_k^{\mathbf{X}}} \frac{m_q \left(\mathbf{r}_p \left(1 - \frac{\kappa_q \mathbf{r}_{pq}^2}{2} \right) - \mathbf{r}_q \right)}{\left(\left\| \mathbf{r}_p - \mathbf{r}_q \right\|^2 + \varepsilon^2 \right)^{\frac{3}{2}} \left(1 - \frac{\kappa_q \mathbf{r}_{pq}^2}{4} \right)^{\frac{3}{2}}}, \tag{9.30}$$

where $p \in \Gamma_k^{\mathbf{Y}}$. The notations in Eq. (9.30) are akin to the notations used in Eq. (2.80), *i.e.*, $\mathbf{r}_p$ and $\mathbf{r}_q$ (m_p and m_q) are coordinates (masses) of the points from the template and reference, respectively; $r_{pq} = \left\| r_p - r_q \right\|$, and κ_q stands for the Gaussian curvature of X_q. We apply a coarse-to-fine policy for softening radius parameter ε which results in a sharp directional displacement towards true correspondences, as it falls exponentially in every iteration t:

$$\varepsilon(t) = \varepsilon \exp\left(1 - \frac{t}{\xi} \right), \tag{9.31}$$

where $\xi \leq 150$ is the maximum number of iterations. In every pairwise gravitational interaction, we stack the forces exerted to the points $p \in \Gamma_k^{\mathbf{Y}}$ into a force matrix $\mathscr{F}_k$:

$$\mathscr{F}_k^t = \left[\ldots, \mathbf{F}_p^a - \eta v_p^t, \ldots \right]^T, \tag{9.32}$$

and point velocities in the region:

$$\mathscr{V}_k^t = [\ldots, v_p^t, \ldots]. \tag{9.33}$$

Next, we obtain the unconstrained velocity updates

$$\mathscr{V}_k^{t+1} = \mathscr{V}_k^t + \Delta t \, \mathscr{F}_k^t \circ [\ldots, m_p^{-1}, \ldots]^T, \tag{9.34}$$

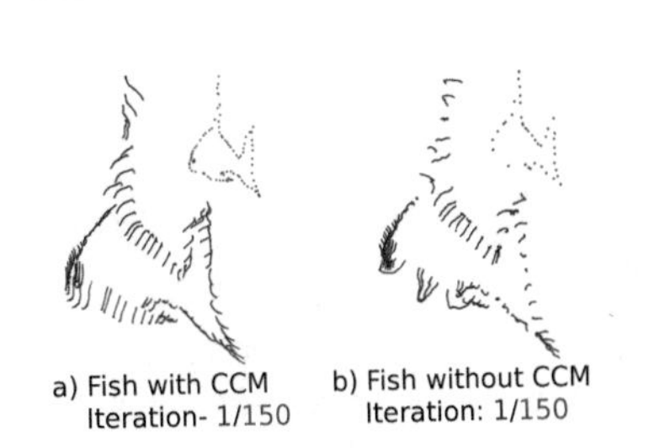

Figure 9.20: Point trajectories of *fish* and *bunny* datasets regularised by the proposed CCM – (a) and (c). The same trajectories collapse or cross over each other in the absence of the CCM regulariser – (b) and (d). The shown trajectories are complete path integrals of the particles leading from the initial misalignment (blue) to the registration result (red). Each superscripted figure is the final aligned template.

where $\circ$ denotes elementwise matrix multiplication, and unconstrained displacement updates:

$$\mathscr{D}_k^{t+1} = \Delta t\, \mathscr{V}_k^{t+1}. \tag{9.35}$$

Taking the current coordinates of P points $(\mathbf{Y}_p)^t$ and the new unconstrained updates of the state space as $(\mathbf{Y}_p)^t + \mathscr{D}_k^{t+1}$, $\forall p \in \Gamma_k^{\mathbf{Y}}$, we first resolve $\mathbf{s}_k$ as a curl-free component of the displacement field $\mathscr{D}_k^{t+1}$, and then solve the absolute orientation problem [144, 158] to obtain $\mathbf{R}_k$. Finally, $\mathbf{t}_k$ is obtained as the mean of $\mathscr{D}_k^{t+1}$. We serially estimate $\mathbf{T}_k$ for all $k \in \{1, \ldots, K\}$ regions and forward them to the CCM step (Sec. 9.3.1.3) for the final position updates. Note that only the velocity of the main representative point per region is updated.

9.3.1.3 Coherent Collective Motion Regulariser

In Sec. 9.3.1.2, we have shown how locally multiply-linked N-body problem is solved separately for individual overlapping regions. $\mathbf{Y}_i$ can appear in several regions. Consider a set $\mathbf{\Psi}_i = \left\{\forall k : \mathbf{Y}_i \in \Gamma_k^{\mathbf{Y}}\right\}$. Let $(i, \mathbf{\Psi}_i)$ be a mapping between a point index and its shared region indices (*e.g.*,, the Fig. 9.19 illustrates that point i is shared by three regions: $\Gamma_i^{\mathbf{Y}}$, $\Gamma_{i-1}^{\mathbf{Y}}$ and $\Gamma_{i+1}^{\mathbf{Y}}$). The CCM operator Ξ is defined for

$(i, \boldsymbol{\Psi}_i)$ and it regularises the velocity v_i^t of $\mathbf{Y}_i$:

$$\Xi(v_i^t) = |v_i^t|(\vartheta)\left(\sum_{k\in\boldsymbol{\Psi}_i} v_k^t\right), \tag{9.36}$$

where $(\vartheta)(.)$ is the normalisation operator. Ξ preserves the velocity magnitude of the representative point $\mathbf{Y}_i$ and replaces its direction by the mean of normalised force directions of all $\mathbf{Y}_k$, $\forall k \in \boldsymbol{\Psi}_i$ from its shared groups. The CCM operator ensures the point trajectories do not cross over and relative point positions are preserved. Originally, the Ξ operator was proposed by Vicsek *et al.* [294] in the context of interactions encountered in biological systems. We apply Ξ operator to the transformation tuple $\mathbf{T}_i$:

$$\Xi(\mathbf{T}_i) = \left\{(\vartheta)\left(\sum_{k\in\boldsymbol{\Psi}_i} \mathbf{R}_k\right), (\vartheta)\left(\sum_{k\in\boldsymbol{\Psi}_i} \mathbf{t}_k\right), (\vartheta)\left(\sum_{k\in\boldsymbol{\Psi}_i} \mathbf{s}_k\right)\right\}. \tag{9.37}$$

For a given point, this consensus filter averages out several transformation tuples. The nature of collective particle dynamics in NRGA is similar to the *smoothed particle hydrodynamics* [197]. Fig. 9.20 shows the impact of the CCM regulariser on point trajectories.

9.3.1.4 Algorithm and Complexity Analysis

A breeze over NRGA summarised in Alg. 10 allows estimation of its complexity:

$$\mathcal{O}(\underbrace{\xi}_{\text{iterations}}\ (\underbrace{M\log M}_{\text{k-d tree}} + \underbrace{M\rho M\rho N}_{\text{NRGA: N-body}} + \underbrace{M}_{\text{NRGA: CCM}})) = \mathcal{O}\left(M^2N\right). \tag{9.38}$$

The coarse analysis is not taking into account that usually $\rho \in [0.02; 0.1]$. Suppose $\rho = 0.05$ and $M \leq 5\cdot 10^4$. In this case, $\rho^2 M^2 \leq M^{1.45}$. Thus, the revised complexity will be at most $\mathcal{O}(M^{\beta}N)$, with $\beta = 1.45$. With smaller M, β will drop. The core implementation of NRGA requires only several hundred lines of an unoptimised C++ code.

9.3.2 Experimental Evaluation

We run experiments on a platform with 32 GB RAM and Intel i7-6700K CPU running at 4.0GHz. We set $\eta = 0.05$, $G = 1.67$, $\rho = 0.05$, $\Delta t = 0.006$, $\xi = 100$ and $\varepsilon = 0.1$. Other methods [24, 70, 203] are running entirely in Matlab, except GMMReg [156] which calls a Linux executable compiled from a C++ source.

Algorithm 10 Non-Rigid Gravitational Approach

Input: a reference $\mathbf{X}_{N \times D}$ and a template $\mathbf{Y}_{M \times D}$
Output: a displacement field $\mathscr{T}(\mathbf{Y}, \mathbf{T})$ registering $\mathbf{Y}$ to $\mathbf{X}$
Parameters: $\varepsilon \in (0,1]$, $\eta \in (0,1]$, G, $m(Y)$, $m(X)$, Δt, $\rho \in [0.02, 0.1]$, $\varepsilon_E = 10^{-4}$, ξ

1: **Initialisation:** $\mathbf{T} = \mathbf{0}$
2: build a k-d tree on $\mathbf{X}$
3: **while** $|\phi_g^{\text{curr}} - \phi_g^{\text{prev}}| > \varepsilon_E$ *or* $t \leq \xi$ **do**
4: build a k-d tree on $(\mathbf{Y})^t$
5: build multi-set of regions $\mathscr{R}_\mathbf{X}$ (Sec. 9.3.1.2)
6: build multi-set of regions $\mathscr{R}_\mathbf{Y}$ (Sec. 9.3.1.2)
7: **for all** $k \in \{1, \ldots, M\}$ **do**
8: select the regions $\Gamma_k^\mathbf{Y}$ and $\Gamma_k^\mathbf{X}$ from $\mathscr{R}_\mathbf{Y}$ and $\mathscr{R}_\mathbf{X}$
9: compute $\mathscr{F}_k^t$ using Eqs. (9.29) – (9.32)
10: compute $\mathscr{V}_k^{t+1}$ and $\mathscr{D}_k^{t+1}$ using Eqs. (9.34) – (9.35)
11: compute $\mathbf{T}_k = \{\mathbf{R}_k, \mathbf{s}_k, \mathbf{t}_k\}$ for $\Gamma_k^\mathbf{Y}$:
12: ▷ solve for $\mathbf{s}_k$ from the curl-free component of $\mathscr{D}_k^{t+1}$
13: ▷ solve absolute orientation problem for $\mathbf{R}_k$ [144, 158]
14: ▷ $\mathbf{t}_k$ = mean($\mathscr{D}_k^{t+1}$)
15: **end for**
16: **for all** $\mathbf{Y}_k$, $k \in \{1, \ldots, M\}$ **do**
17: update velocities $v_k^{t+1} = \Xi\left(v_k^{t+1}\right)$ using Eq. (9.36)
18: update transformation $\mathbf{T}_k^{t+1} = \Xi\left(\mathbf{T}_k^{t+1}\right)$ using Eq. (9.37)
19: **end for**
20: update $\mathbf{Y}^{t+1} = \mathscr{T}(\mathbf{Y}^t, \mathbf{T})$
21: **end while**

9.3.2.1 Evaluation Methodology and Datasets

A perform quantitative evaluation on synthetic benchmark datasets *fish*, *line* (2D) and *bunny* (3D) with known ground truth correspondences. The input template is modified in four degradation scenarios and 100 experiments are run for each of the scenarios. We report the root-mean-squared error (RMSE) for every run defined as:

$$E_{rmse} = \sqrt{\frac{1}{N} \sum_{j=1}^{N} \left(r_{ij} - E_{mean}\right)^2}, \tag{9.39}$$

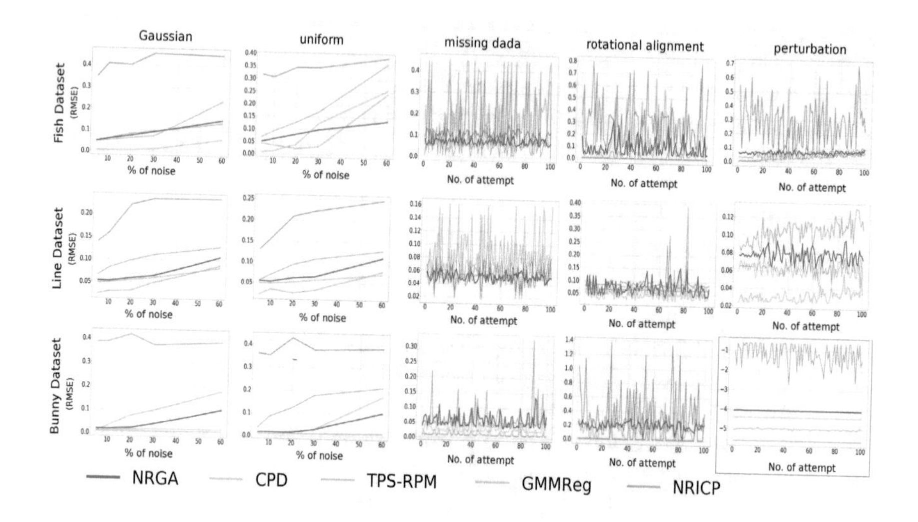

Figure 9.21: Quantitative results on *fish, line* and *bunny* show the error statistics for the cases with Gaussian noise, uniform noise, missing data patterns, different initial misalignments and point perturbations. NRGA is the most stable method under uniform noise and missing data, with lowest RMSE. The high peaks of GMMReg [156] and NRICP [24] are due to the inherent instability. The bottom right graph shows $\log_e(E_{rmse})$ values.

where E_{mean} denotes the mean error. *First*, we introduce uniformly and Gaussian distributed noise into the templates, in the amounts ranging from 5% to 60% of the total points, in every instance of the experiment. Additional input noises increase the number of uncertain correspondences and make the problem more ill-posed. *Second*, we randomly perturb the initial point positions in the template. The direction and magnitude of perturbation are obtained as the realisations of a Gaussian distribution. The magnitude of perturbation increases linearly with every experiment. *Third*, we randomly delete 20% of the original data in a chunk from the *templates*. In the *fourth* scenario, we vary the degree of the initial misalignment of the template so that the axis-angle orientation values $(\theta_x, \theta_y, \theta_z)$ are the realisations from a uniform distribution $\mathscr{U}(-\frac{\pi}{4}, \frac{\pi}{4})$. Although an optimal rigid pre-alignment is crucial for an accurate non-rigid registration, we show that NRGA can still cope with fairly misaligned point sets.

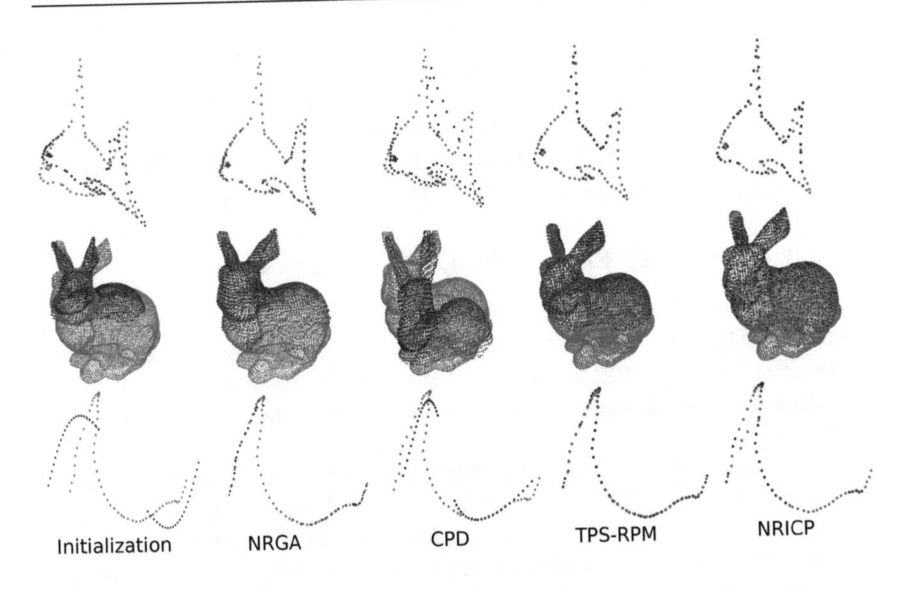

Figure 9.22: NRGA outperforms other evaluated methods in the handling of missing data. In this picture, results of evaluated methods on *fish*, *bunny* and *line* datasets are shown. In this experiment, more than 30% of points have been removed from all templates.

Compared methods

We compare several widely-used and state-of-the-art methods for non-rigid point set registration, *i.e.*, non-rigid ICP [24], TPS-RPM [70], CPD [203] and GMM-Reg [156]. We choose optimal settings for every method — this allows us to compare the best possible performances in every scenario.

9.3.2.2 Experimental Results on Synthetic Data

Quantitative results shown in Fig. 9.21 reveal NRGA's strength against uniform noise — with the increasing noise, the relative accuracy of NRGA increases (for the case of 60% of noise, NRGA shows the lowest RMSE on two datasets). For the Gaussian noise, the winner is CPD — this approach explicitly models Gaussian noise, and we were always choosing the respective optimal parameter. Nevertheless, NRGA is close to GMMReg and TPS-RPM which do not make such assumption. NR-ICP fails on the experiments with noise. In the case of missing data, NRGA consistently outperforms all compared methods by a significant margin. NRGA shows stability as the error does not vary much across the different scenarios and datasets. In Fig. 9.22, selected results from the experiments with missing data are shown. Note how NR-ICP, TPS-RPM and CPD either stretch or dilate the template. In contrast, NRGA displaces points towards the appropriate regions with the attractive force function parameterised by the shape curvature.

9.3.2.3 Experimental Results with Qualitative Interpretation

In Fig. 10.8, we demonstrate the performance of NRGA in image registration scenario. 2D point sets originate from computer tomography (CT) of a human brain [297]. Both images represent the same physical state and differ by a rectification applied on the reference image. 2D point sets are obtained by the sampling of image contours.

Regarding the practical applications of NRGA, qualitative registration results on human faces (BU-3DFE datasets [317]) using different methods are shown in Fig. 9.23. The *template* and *target* faces differ by the facial expression and scaling factor. NRGA achieves high correspondence accuracy and geometric consistency thanks to the locally aware global topology preserving CCM operator. Recall that NRGA uses point positions exclusively (and not other geometric attributes like faces or vertex normals).

Next, we register a coarse synthetic face template containing $4k$ points with two real human head scans obtained from a multi-view system. The results are shown in Fig. 9.24. The first scan is a part of the full-body 3D reconstruction from [120] (see Sec. 8). It contains complex details such as curly hair and rough skin membranes, and some parts are missing. The second scan is reconstructed from multiple views and is used raw, *i.e.*, without noise filtering. The numbers of points in the template and the reference are roughly equal. A large portion of the points represents low-amplitude noises. Fig. 9.24 shows that NRGA accurately fits the template onto the real scans in these challenging scenarios. Especially the second scan is fit much more accurately compared to what is attainable by other non-rigid point set registration methods. The pronounced noise component strongly influences the accuracy of other methods like CPD or ECPD (even if some accurate prior correspondences are given). The consequence is the deteriorated topology and, overall, an unsuccessful appearance transfer.

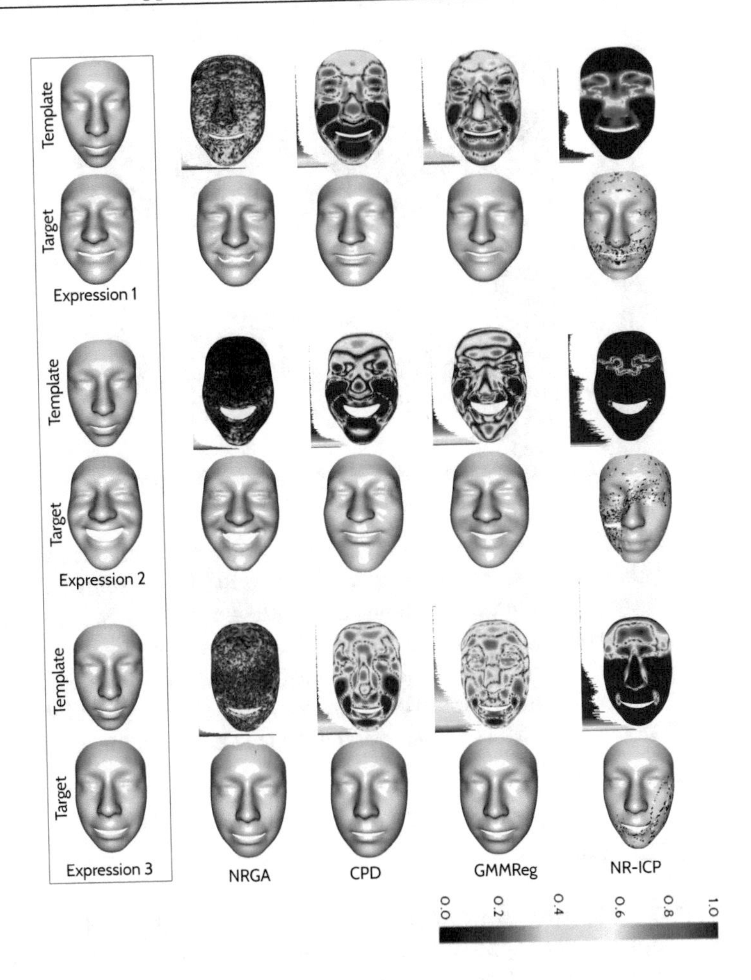

Figure 9.23: Qualitative results of non-rigid point set registration methods: NRGA, CPD [203], GMMReg [156], NR-ICP [24] when applied to human faces from the BU-3DFE dataset [55, 317] differing in the expressions and scaling. The upper and lower rows in all pairs illustrate the Hausdorff distance between the target and the deformed template as a colour-map (with a histogram for every registration; the colour coding in distance units is on the bottom).

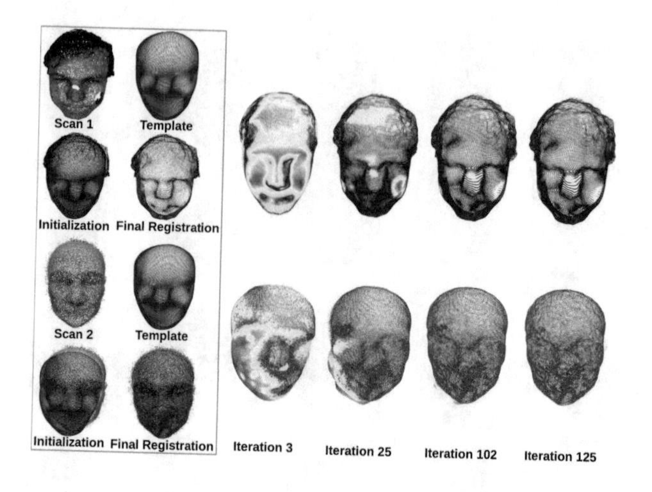

Figure 9.24: The experiment with real data: a coarse synthetic face template is registered to two scans from multi-view systems. The first scan is detailed and contains only little noise. The second scan is a raw noisy 3D reconstruction of a head statue with low-amplitude noises. In both cases, NRGA robustly registers *template* to the *scans*. Results after several iterations are shown on the right. The colour represents the per-point distance error with the same coding as in Fig. 9.23.

9.4 Conclusion

In this chapter, we introduced a new physically-based class of point set registration methods — gravitational approaches. The new methods are based on particle dynamics in a force field induced by the reference point set. The rigid GA, as well as BH-RGA, are globally multiply-linked methods.

All GA approaches are well parallelizable and allow for embedding of prior matches through point weights. Various acceleration techniques can be adopted for GA reducing the computational complexity or providing a speedup in a respective complexity class. Experiments on synthetic and real data show that GA, BH-RGA and NRGA are robust against clustered outliers and large amounts of uniformly distributed noise. Several experiments with real data demonstrated the maturity of the new method class for practical applications (*e.g.*,, medical image registration and facial appearance transfer).

We have also shown that particle-dynamics based RA can be accelerated using a BH tree, stabilised by inverting the laws of physics and adding additional constraints in spherical coordinates. Our enhanced formulation significantly reduces the number of parameters and enables RA which is well-posed w.r.t. parameters. Even for large point sets of dimensionality going beyond the extents feasible for other general-purpose RA methods, the proposed BH-RGA performs global multiply-linked updates. This contrasts with other known techniques which de facto only perform locally multiply-linked updates. As a result, BH-RGA is especially robust to noise and achieves state-of-the-art performance whenever existing techniques do. Furthermore, BH-RGA accepts anchor points of different origins, *i.e.*, prior matches, key points and even point intensities, and handles varying point densities with volumetric mass normalisation.

NRGA interprets alignment problem as a series of altered relativistic N-body problems in a space with constant Gaussian curvature. The curvature of the reference operates as a local shape descriptor and influences gravitational attraction. The core CCM regularisation merges multiple update proposals for the same point from several overlapping regions.

There are multiple venues for further investigation of the framework of particle dynamics based RA. The most promising ones are multibody registration and parallelisation of the proposed methods.

10 Application of Point Set Registration and Monocular Non-Rigid 3D Reconstruction to Scene Flow Estimation

IN this chapter we will show how methods for point set registration and monocular non-rigid reconstruction can be adopted for scene flow estimation from RGB-D and monocular image sequences.

The problem of scene flow estimation consists in the recovery of 3D flow fields between the underlying geometry states observed by an RGB-D or a monocular camera. Not only the camera pose but also the observed states can differ from frame to frame if the geometry is non-rigid or comprises several related or independent rigid motions. In other words, scene flow refers to a 3D point velocity field of an observed scene. It has been a key element in various computer vision applications such as 3D reconstruction [206, 334], motion analysis and prediction [163, 305] as well as stereo matching [147, 193].

Both proposed methods — NRSfM-Flow (Sec. 10.1) and Multiframe Scene Flow (Sec. 10.2) — are tested on a variety of datasets including human and animal faces, synthetic SINTEL dataset [61], a dataset for autonomous driving vKITTI [101], as well as office dataset with piece-wise rigid motion [261].

10.1 Scene Flow from Monocular Image Sequences

In this Section, we show that the developed formulation of monocular surface recovery allows to reconstruct scene flow from monocular image sequences. Since only points observed in the reference frame are tracked and reconstructed throughout an image sequence, the correspondences between the points are always established. We use this property to efficiently compute the derivative of the geometry *w.r.t* time.

V. Golyanik, *Robust Methods for Dense Monocular Non-Rigid 3D Reconstruction and Alignment of Point Clouds*,

Scene flow recovery from monocular image sequences is an emerging field in computer vision. While existing Monocular Scene Flow (MSF) methods extend the classical optical flow formulation to estimate depths, disparities and 3D motion, we propose in this Section a framework based on Non-Rigid Structure from Motion (NRSfM) technique — NRSfM-Flow. Therefore, both problems are formulated in the continuous domain and relation between them is established. To cope with real data, we propose two preprocessing steps for image sequences — redundancy removal and translation resolution — which increase the quality of reconstructions and speedup computations. In contrast to the existing MSF methods which can cope with non-rigid deformations, our solution makes no strong assumptions about a scene such as known camera motion or camera velocity constancy and can handle occlusions. NRSfM-Flow is qualitatively evaluated on challenging real-world data. Experiments provide evidence that the proposed approach achieves high accuracy and outperforms state of the art in terms of the ability to reconstruct MSF with less prior knowledge about a scene.

10.1.1 Monocular Scene Flow as an Emerging Field

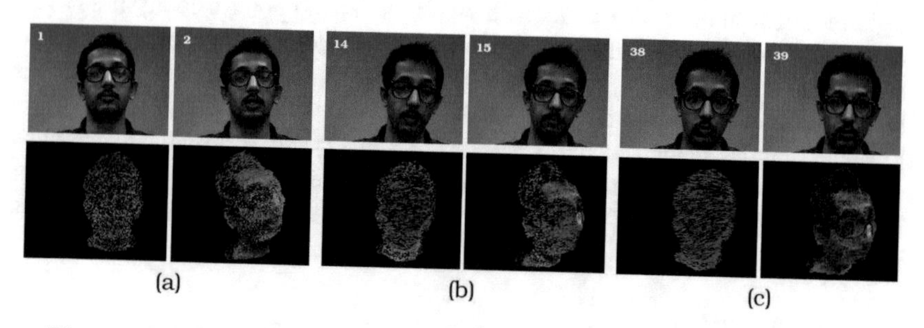

Figure 10.1: Results of the proposed NRSfM-Flow framework on the *human face* sequence, for several pairs of frames. For every pair of frames: input frames (top row), recovered scene flow (down left) and geometry with overlayed 3D motion fields from a new viewpoint. Better viewed in colour. Our approach is able to recover scene flow from monocular image sequences depicting non-rigid scenes and does not make any strong assumptions on the scene or type of camera motion. It is robust to occlusions.

Scene flow recovery from monocular image sequences is an emerging field in computer vision. As of today, this topic was sparsely discussed in the literature and only a few works exist. Scene flow refers to a dense 3D velocity vector field of a

moving and possibly non-rigidly deforming scene, see Fig. 10.1 for an example. The concept is similar to optical flow, i.e. a dense 2D velocity vector field in an image plane. Scene flow finds applications in autonomous driving, motion segmentation, motion capture, egomotion, 4D reconstruction, scientific visualisation and other domains. In real applications, scene flow computation is mostly based on stereo/multi-view camera settings or sensors directly outputting depth.
The earliest work on monocular scene flow (MSF) recovery was carried out by Birkbeck *et al.*. The variational method proposed in [42] supports short image sequences and can handle rigid, articulated and non-rigid motion. Several aspects can restrict the applicability of the approach in practice, *i.e.,* camera motion is assumed to be known in advance and constant in a short temporal frame. Another limitation is a high sensitivity to occlusions. In [41], the same authors proposed a solution without the requirement of known camera motion, but with a known rigidly moving base-mesh geometry approximation of the scene. Mitiche *et al.* recently proposed a variational method for the concurrent recovery of structure and scene flow [195]. As shown experimentally, the algorithm can handle noiseless scenarios with rigid motion including scenes with few moving objects. Tikhonov regularisation used in the algorithm tends to oversmooth depth and motion discontinuities in the recovered 3D flow fields. Further studies are required to make the technique applicable in more complex and realistic scenarios. Motivated by remote patient monitoring and driver assistance systems, Xiao *et al.* proposed an MSF estimation algorithm based on energy functional minimisation [312]. Along with brightness and gradient constancy assumptions, velocity constancy over a short period of time is reflected in the energy functional. The algorithm can use at least three frames and does not require optical flow as an input. In experiments, an application of the proposed technique in a challenging real-world driving scenario was demonstrated which is commonly tackled by stereo based methods. Support of scenes exhibiting non-rigid deformations is limited.

The discussed methods share several common attributes. Firstly, they are formulated as energy minimisation problems solved by an Euler-Lagrange differential equation. They extend a classic optical flow formulation[1] by estimating depths/disparities and a 3D motion field instead of image motion and estimate correspondences and geometry simultaneously. Secondly, intrinsic camera parameters are assumed to be known. Thirdly, the reviewed methods operate in a batched manner, i.e. they compute scene flow after the complete image sequence is acquired. Fourthly, support of non-rigidly deforming structures is limited. Processing of non-rigid scenes was demonstrated in the papers by Birkbeck *et al.* [41,42], but

[1] which has its roots in the seminal work by Horn and Schunck [145]

assumptions which need to be satisfied limit their applicability in practice considerably. Multiple common aspects encourage us to classify the reviewed MSF methods into a separate class which we refer to as *direct* MSF methods.

Extension of optical flow to three dimensions is one possible approach to the problem of MSF recovery. Another one is to adopt Non-Rigid Structure from Motion (NRSfM) techniques. NRSfM allows reconstruction of non-rigidly deforming and moving 3D surfaces from monocular image sequences given coordinates of tracked points for every frame (combined in a measurement matrix). Birkbeck *et al.* [41,42] mention NRSfM as a class of techniques that can be potentially used to recover geometry of deformable scenes — a scenario similar to MSF recovery. However, an explicit step for 3D motion field estimation is not present in NRSfM methods. Another possible limiting factor for adopting NRSfM for scene flow estimation at that time was the lack of dense techniques.

NRSfM methods take advantage of motion and non-rigid deformation as cues to infer geometry. They are based on the factorisation of a measurement matrix with coordinates of the tracked points into non-rigid shape and motion for every frame — an inverse problem inherently ill-posed in the sense of Hadamard. Additional constraints are required to obtain a unique and reasonable solution. For an orthographic camera, the factorisation idea was first proposed by Tomasi and Kanade for the rigid case [281] and later extended to the non-rigid case by Bregler *et al.* [53]. In [53], every non-rigid shape is represented by a linear combination of basis shapes, wherein the basis shapes and the weights are unknowns. This idea was further improved in successor methods proposing different types of constraints and optimisation methods for higher stability and reconstruction accuracy [126,127,216,226,282] and for sequential operation [9,214]. Akhter *et al.* proposed the employment of a trajectory basis [22] instead of the metric one. It allows reduction of the number of unknowns in NRSfM, since the trajectory basis is fixed in advance. As a result, NRSfM in the trajectory space can lead to more stable reconstructions. Several papers investigated ways to improve this class of methods and eliminate weaknesses such as ambiguity in trajectory bases [290] or to model point trajectories more realistically [333]. The first dense NRSfM method was shown in [238] followed by [104].

In contrast to direct MSF methods, NRSfM relies on correspondences obtained in a separate step. The demand for sufficient motion implies a certain length of the image sequence. Another assumption commonly made in NRSfM is that a scene is centred throughout the whole image sequence (similar to the direct method of [42]), which can limit the application of NRSfM in real-world scenarios.

Several steps are required to adopt current NRSfM methods to MSF. Accordingly, the following contributions are made in this section: 1) relation between NRSfM and MSF is established. A novel analytical framework is introduced which allows to analyse and relate both problems in the continuous domain on a high level of abstraction; 2) a solution to MSF recovery based on the extensively studied NRSfM under orthography is proposed — the NRSfM-Flow; 3) two preprocessing steps are proposed — to resolve translation in a scene and to compress the input image sequence by eliminating redundant frames (frames with low variance in scene appearance) — so as to reduce the runtime and enhance the overall accuracy of the approach; 4) NRSfM-Flow is designed and implemented as a framework combining state-of-the-art methods for correspondence computation [105, 267], non-rigid geometry reconstruction [104] and proposed preprocessing steps; 5) results on several real-world image sequences are shown and performance of the proposed approach is evaluated qualitatively. We consider MSF estimation as a standalone field in computer vision. To the best of our knowledge, we are the first to propose an NRSfM-based MSF method, formulate NRSfM in the continuous domain and propose explicit translation resolution and redundancy removal for NRSfM. Our approach can handle challenging scenarios with non-rigid motion which cannot be handled by the current direct monocular scene flow methods.

The rest of this section is organised as follows. In Sec. 10.1.2, the formulation of NRSfM in the continuous domain as well as the relation between NRSfM and scene flow is derived. In Sec. 10.1.3, the NRSfM-Flow framework is described together with preprocessing steps broadening the scope of NRSfM, followed by experiments in Sec. 10.1.4 and conclusions in Sec. 6.3.

10.1.2 MSF and NRSfM in the Continuous Domain

In this section, relation between NRSfM and MSF is established. Therefore, both problems are formulated in the continuous domain. Continuous representation often allows one to analyse problems and reveal their properties on a high level of abstraction. Moreover, multiple variants of discretisation, numerical and optimisation methods are possible in combination with it. In other words, the problem statement and its implementation is kept separate.

Assume an orthographic camera observes a 3D non-rigidly deforming scene $\mathbf{S}(\mathbf{p},t)$ consisting of 3D points $\mathbf{p}$ from the continuous point space domain $\Omega \subset \mathbb{R}^{3+1}$. Every point possesses a colour, hence an additional space dimension denoted by $+1$. The observed scene is different at each time $t \in \mathrm{T} \subset \mathbb{R}$:

$$\mathbf{S}(\mathbf{p},t) : \Omega \times \mathrm{T} \to \mathbb{R}^{3+1}. \tag{10.1}$$

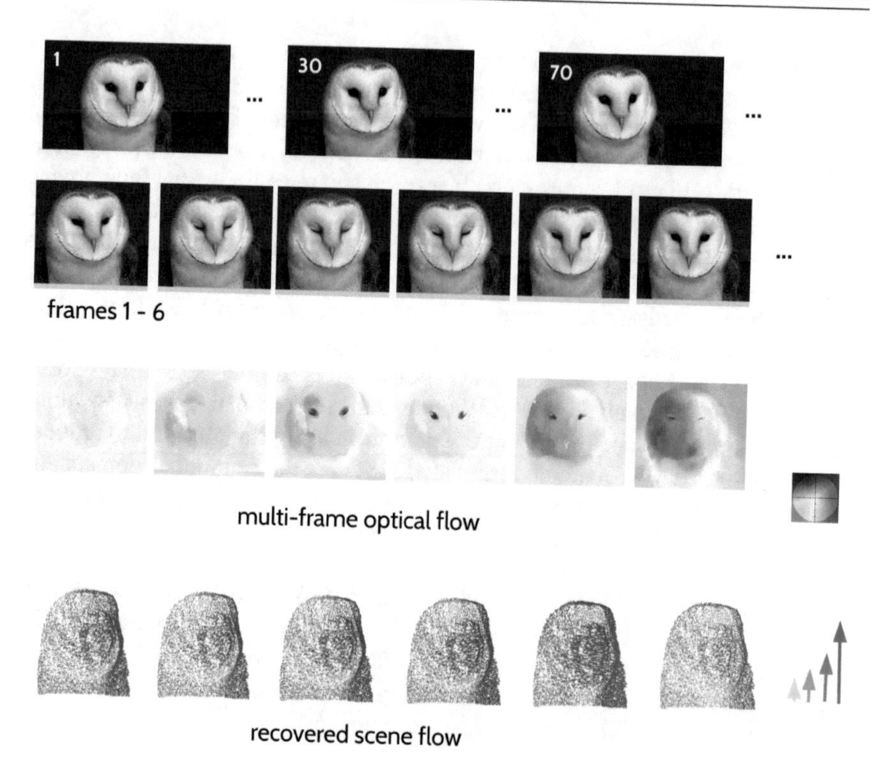

Figure 10.2: Overview of NRSfM-Flow. In the preprocessing steps, the scene is centred and redundant frames are removed. Next, optical flow between the reference frame and other frames in the sequence is calculated. Finally, the non-rigid states are reconstructed with NRSfM and the scene flow is obtained by taking the derivative of the geometry *w.r.t* time. For optical flow visualisation, we use Middlebury colour scheme [29]. The colour scheme for scene flow is given bottom right (colours encode vector lengths).

The scene $\mathbf{S}(\mathbf{p},t)$ continuously produces 2D projections (images) on the camera sensor containing 2D points $\mathbf{v}$ from the image domain $\Psi \subset \mathbb{R}^{2+1}$:

$$\mathbf{I}(\mathbf{v},t) : \Psi \times \mathrm{T} \rightarrow \mathbb{R}^{2+1}. \tag{10.2}$$

We assume that the scene is registered to the origin of the coordinate system and the camera translation $T(t)$ is always $\mathbf{0}$. The scene and its image is related as

$$\mathbf{W}(\hat{\mathbf{v}},t) = \mathbf{W}_\tau(\hat{\mathbf{v}},t) + \mathbf{C}(\hat{\mathbf{v}}) = \mathbf{R}(t)\, \mathbf{S}(\mathbf{p},t), \tag{10.3}$$

where $\mathbf{R}(t) : \mathrm{T} \to SO(3)$ is the camera pose, $\mathbf{W}(\hat{\mathbf{v}},t)$ is a measurement function (image coordinates of the tracked points); the correspondence (2D motion field) function $\mathbf{W}_\tau(\hat{\mathbf{v}},t)$ outputs a 2D displacement field relative to a reference time τ and $\mathbf{C}(\hat{\mathbf{v}})$ is the point displacement function in image coordinates relative to the origin of the coordinate system of the image. Note that $\hat{\mathbf{v}} \in \hat{\Psi} \subset \mathbb{R}^2$ are 2D colourless points and $\hat{\mathbf{v}} \subset \mathbf{v} : \mathbf{v}$ are visible at time τ; in $\mathbf{I}(\mathbf{v},t)$, point displacements are given relative to the changing reference time τ, whereas in the case of $\mathbf{W}_\tau(\hat{\mathbf{v}},t)$ the reference time τ is fixed (see Fig. 10.3 for geometric interpretations). An infinitesimal change in camera pose and 3D scene structure $\Theta(\mathbf{p},t) : \Omega \times \mathrm{T} \to \mathbb{R}^3$ can be described by a derivative of the right side of Eq. (10.3):

$$\Theta(\mathbf{p},t) = \frac{\partial \mathbf{R}(t)}{\partial t} \mathbf{S}(\mathbf{p},t) + \mathbf{R}(t) \frac{\partial \mathbf{S}(\mathbf{p},t)}{\partial t}, \tag{10.4}$$

where $\frac{\partial}{\partial t}$ denotes a partial derivative with respect to time t. Note that unlike $\mathbf{S}(\mathbf{p},t)$, $\Theta(\mathbf{p},t)$ represents a continuous 3D vector field, i.e. the 3D output encodes relative displacements of points $\mathbf{p}$, or *scene flow*. The scene flow is composed of a *rotational component* $\rho(t) = \frac{d\mathbf{R}(t)}{dt} \in SO(3)$ and a *deformational component* $\frac{\partial \mathbf{S}(\mathbf{p},t)}{\partial t}$.

Recall that in NRSfM, there is an inherent rotational ambiguity, i.e. the rotational component can be explained either by a camera or an object movement or a combination of both; without prior knowledge it is not possible to determine a cause of the observed rotation. This ambiguity means that all combinations are possible and lead to equivalent observations. Assume that the camera is fixed, namely $\forall t : \mathbf{R}(t) = \mathbf{I}$ and the object rotates. In this case $\mathbf{S}(\mathbf{p},t)$ also covers observed rotations in the scene. We can exploit the rotational ambiguity to simplify Eq. (10.4) — the rotational component is equal to zero and only the right term remains in the expression:

$$\Theta(\mathbf{p},t) = \frac{\partial \mathbf{S}(\mathbf{p},t)}{\partial t}. \tag{10.5}$$

To ensure $\forall t : \mathbf{R}_{2\times 3}(t) = \mathbf{I}_{2\times 3}$, the recovered rotation must be applied to the observed non-rigidly deforming structure in 3D space. Therefore, we obtain the $\mathbf{R}_{3\times 3}(t)$ matrix by extending $\mathbf{R}_{2\times 3}(t)$ with a third row. The third row is equal to a cross product of the first two rows, which guarantees orthonormality of the $\mathbf{R}_{3\times 3}(t)$ matrix. Likewise, the deformational component in the image plane can be computed from the image function $\mathbf{I}(\mathbf{v},t)$ as a continuous 2D vector field referred to as *optical flow* $\Xi(\mathbf{v},t)$ (algorithms for computing optical flow from images are *e.g.*, [145, 323]). The relation between the optical flow Ξ and the measurement

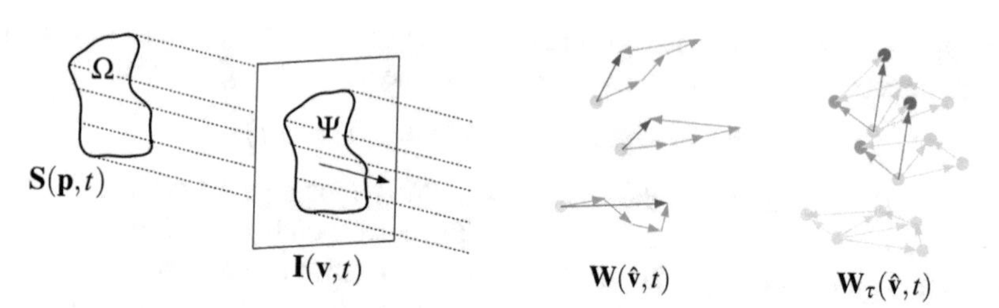

Figure 10.3: Left: orthographic projection of a 3D object to a 2D image plane — the projection lines are parallel and intersect at infinity. Middle: function $\mathbf{W}(\hat{\mathbf{v}},t) = \mathbf{W}_\tau(\hat{\mathbf{v}},t) + \mathbf{C}(\hat{\mathbf{v}})$ as in Eq. (10.3). $\mathbf{W}(\hat{\mathbf{v}},t)$ outputs *absolute* coordinates of the tracked points. Right: function $\mathbf{W}_\tau(\hat{\mathbf{v}},t)$ visualised, see Eq. (10.6) — it outputs 2D displacements of the points visible at the reference time τ for every time t.

Table 10.1: Core equations of the proposed theoretical framework relating NRSfM and MSF.

domain	meaning	defined notions	equations
$\mathbf{p} \in \Omega \subset \mathbb{R}^{3+1}$	all 3D points of a scene	3D scene $\mathbf{S}(\mathbf{p},t)$, scene flow $\Theta(\mathbf{p},t)$	(10.3)–(10.5), (10.9)
$\hat{\mathbf{p}} \in \hat{\Omega} \subset \mathbb{R}^3$	reconstructed 3D points	reconstructed 3D surface $\mathbf{S}(\hat{\mathbf{p}},t)$	(10.8)
$\mathbf{v} \in \Psi \subset \mathbb{R}^{2+1}$	all observed 2D points	images $\mathbf{I}(\mathbf{v},t)$, optical flow $\Xi(\mathbf{v},t)$	(10.6)–(10.8)
$\hat{\mathbf{v}} \in \hat{\Psi} \subset \mathbb{R}^2$	2D points visible at time τ	measurement function $\mathbf{W}_\tau(\hat{\mathbf{v}},t)$	(10.3), (10.6)

function $\mathbf{W}_\tau$ reads:

$$\mathbf{W}_\tau(\hat{\mathbf{v}},t) = \int_\tau^t \Xi(\hat{\mathbf{v}},t)\,dt. \tag{10.6}$$

Change of τ causes changes in $\hat{\Psi}$, since the set of visible points is different each time. From Eqs. (10.3), (10.4) and (10.6), the relation between the infinitesimal optical flow and scene flow can be established as

$$\int_\tau^t \Xi(\hat{\mathbf{v}},t)\,dt + \mathbf{C}(\hat{\mathbf{v}}) = \mathbf{R}(t)\,\mathbf{S}(\hat{\mathbf{p}},t), \quad \text{or} \tag{10.7}$$

$$\Xi(\hat{\mathbf{v}},t) = \mathbf{R}_{2\times 3}(t)\frac{\partial \mathbf{S}(\hat{\mathbf{p}},t)}{\partial t}, \tag{10.8}$$

where points $\hat{\mathbf{p}}$ and $\hat{\mathbf{v}}$ are the reconstructed 3D points and their projections into the image plane, respectively. A summary of the domains and defined notions is given in Table 10.1. Similar to Eq. (10.3) which relates geometry of a non-rigid scene with its projection into an image plane, Eg. (10.8) relates changes in the geometry with changes in the projection. An accumulated scene flow (or equivalently, 3D point trajectories which can be also expressed as a set of 3D line integrals) in time interval $[t_1;t_2]$ reads as an integral

$$\int_{t_1}^{t_2} \Theta(\mathbf{p},t)\,dt. \tag{10.9}$$

Using the introduced space-time structures and relations, it is possible now to give the formal definitions of NRSfM and MSF recovery problems.

Definition of NRSfM. *Given the displacements of the projected points* $W_\tau(\hat{\mathbf{v}},t)$ *relative to time* τ *, the objective of an NRSfM problem is to recover the underlying non-rigidly deforming scene function* $S(\mathbf{p},t)$.

Definition of MSF. *Given projections of the observed scene* $\mathbf{I}(\mathbf{v},t)$*, the objective of an MSF problem is to reconstruct the scene flow function* $\Theta(\mathbf{p},t)$.

As follows from the definitions, the inputs and objectives of NRSfM and MSF are different, but related with Eqs. (10.6) and (10.4). Thus, it is possible to adopt NRSfM for estimation of scene flow from monocular image sequences using the proposed analytical framework.

10.1.3 The NRSfM-Flow Framework

In this section, the NRSfM-Flow framework for MSF recovery is introduced. It encompasses several steps including correspondence computation, geometry reconstruction as well as preprocessing steps for input image sequences. Though NRSfM-Flow is designed with batch processing in mind, it can be adopted for sequential processing.

To compute the measurement function $\mathbf{W}(\hat{\mathbf{v}},t)$, we adopt the state-of-the-art Multi-Frame Optical Flow (MFOF) method of Garg *et al.* [105]. In the case of severe occlusions presented in a scene, we also use the occlusion-aware MFOF of Taetz *et al.* [267]. To recover non-rigid geometry and camera pose, we choose the variational approach [104] combined with the GrabCut algorithm [234] for foreground-background segmentation. Thus, the methods aiming at high-quality reconstructions are combined in NRSfM-Flow.

Preprocessing steps. NRSfM methods require sufficient diversity in non-rigid deformations and camera motion as reconstruction cues. We propose to compress an input image sequence so that it fulfils temporal and spatial assumptions of NRSfM in an optimal way. We call this preprocessing step *redundancy removal*. Suppose at time t_a an instantaneous image is considered for further processing. The next instantaneous image will be taken at time t_b for which the inequality holds:

$$\left\| \int_{\hat{\Psi}} \int_{t_a}^{t_b} \Xi(\mathbf{v},t)\, dt\, d\hat{\mathbf{v}} \right\|_2 \geq \varepsilon, \tag{10.10}$$

where $\|.\|_2$ denotes a 2-norm and ε is a scalar threshold. In other words, if total flow (2-norm of the integrated flow field) in a time interval $[t_a; t_b]$ is above a threshold ε, then a view at time t_b exhibits sufficient diversity relative to the view at time t_a. Otherwise, another time interval $[t_a; t_b = t_b + dt]$ should be evaluated. The optimality criterion proposed in Eq. (10.10) can detect duplicate frames, small motions as well as oscillatory effects. Moreover, it can also serve as a discretisation criterion, since the regularisation parameter ε determines whether the observed motion provides a sufficient reconstruction cue. Though it is possible to resolve translation in a scene directly by registering the measurement matrix to the mean coordinates of the structure, we notice that resolving it before computing correspondences can increase the accuracy of reconstructions. Therefore, we propose an explicit *translation resolution* step. Assuming that an object is entirely visible at the reference time τ, we segment the scene into foreground and background and track the ROI throughout the image sequence using Kanade-Lucas-Tomasi feature tracker [183]. The output of the translation resolution is a frame size and the corresponding translation function $T(t)$. After applying redundancy removal and translation resolution, correspondences are computed faster. If required, reverse transformations can be applied in the postprocessing step.

The NRSfM-Flow framework is summarised in an algorithmic fashion in Alg. 11. Note that the framework does not prescribe particular algorithms for the steps 2–5. They can be chosen or tuned dependent on requirements (*e.g.*, perspective or orthographic camera model, *etc.*).

Implementation. Our test platform has 128 GB RAM, an NVIDIA GeForce TITAN Z GPU and an Intel Xeon E5-1650 v3 CPU. We use our own C++/CUDA C implementations of the methods [267] and [104] as well as a publicly available Matlab [277] version [235] of the method [105]. The preprocessing steps were implemented in C++ using the OpenCV 3.0 library [2]. The implementation of [104] supports heterogeneous platforms with a multi-core CPU and a CUDA-capable GPU. Since the framework is formulated in the continuous domain following Sec. 10.1.2, several discretisation aspects shall be mentioned here. In the beginning, we choose discretisation points to coincide with image frames, wherein for computing derivatives (step 7) forward finite differences between consecutive frames are

Algorithm 11 NRSfM-Flow Framework for MSF Recovery

Input: monocular image sequence $\mathbf{I}(\mathbf{v},t) : \Psi \times \mathrm{T} \rightarrow \mathbb{R}^{2+1}$.
Output: Scene flow $\Theta(\mathbf{p},t) : \Omega \times \mathrm{T} \rightarrow \mathbb{R}^3$
1: **Initialisation:** *depends on the underlying algorithms*
2: Resolve translation, find translation function $T(t)$
3: Compress image sequence (eliminate redundant frames) according to Eq. (10.10)
4: Compute measurement function $\mathbf{W}_\tau(\hat{\mathbf{v}},t)$
5: Factorize $\mathbf{W}_\tau(\hat{\mathbf{v}},t) + \mathbf{C}(\hat{\mathbf{v}})$ into non-rigid shapes $\mathbf{S}(\mathbf{p},t)$ and motion $\mathbf{R}(t)$
6: Apply $\mathbf{R}(t)$ to $\mathbf{S}(\mathbf{p},t)$
7: Compute scene flow according to Eq. (10.5)
8: Apply reverse transformations ($-T(t)$ and geometry duplication) if required
9: Save the final recovered scene flow in $\Theta(\mathbf{p},t)$

used. In this case, the accumulated scene flow between two consecutive frames is computed, as defined in Eq. (10.9). We can also estimate geometry between frames by interpolating the structure along the accumulated 3D motion fields.

10.1.4 Evaluation

In this section, the proposed NRSfM-Flow framework is qualitatively evaluated on several challenging real-world image sequences depicting non-rigid scenes. The recovered 3D flow fields and reconstructions are visualised. Additionally, projections of the scene flow into the image plane and optical flow between consecutive frames are compared. We follow colour schemes for optical and scene flow fields proposed in [29] and [288], respectively. Scenes which can be handled by our method should fulfil the requirements, i.e. provide sufficient reconstruction cues and preferably consist of a single non-rigidly deforming and moving (possibly translating) object. We are not aware of scene flow benchmark datasets fulfilling the aforementioned requirements. Therefore, we opt for several real-world image sequences. All existing works on MSF recovery [41, 42, 195, 312] employ a similar evaluation methodology.

The human face sequence. The face sequence was acquired with a Flea FL2-03S2C camera. It depicts a speaking person; arbitrary translations, facial expressions (non-rigid deformations) and self-occlusions are present in the sequence which makes it challenging for MSF recovery. Resolution of the images is 486×366. Translation resolution is applied in the preprocessing step. Exemplary results are shown in Fig. 10.1. Every recovered dense surface contains $5.6 \cdot 10^4$ points. Results

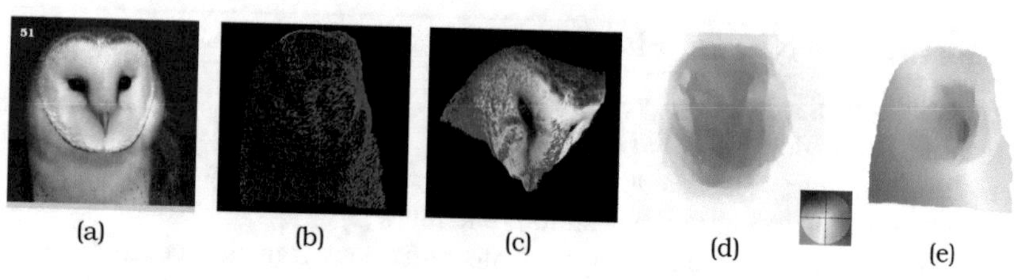

Figure 10.4: Experimental results on the *barn owl* sequence [82]: (a) input frame number 51; (b) result of the MSF recovery by NRSfM-Flow (between frames 51 and 52), the colour scheme is the same as in [288]; (c) geometry with an overlayed scene flow for the frame 51; (d) result of the optical flow between frames 51 and 52 by the TV-L1 method [323]; (e) projection of the recovered 3D motion field into the image plane. In (d) and (e) the colour scheme is replicated according to [29].

Figure 10.5: Examples of Poisson reconstructions: (a) shaded geometry from novel viewpoints from the *face* sequence (frame 1); (b) textured and shaded geometry from novel viewpoints from the *barn owl* sequence (frame 1).

of this experiment are qualitatively similar to the results on the *mouth* sequence shown in [42]. Though, several differences can be noticed. First, our results are less accurate in the areas of the forehead and mouth. The forehead is inherently poorly textured and correspondences, as well as reconstructions, are less accurate in this area. In the area of the mouth, the points are interpolated building a smooth surface so that the opening is less recognisable in the shaded Poisson surface. However, both reconstructions exhibit artefacts associated with correspondences (there are convexities and concavities). Thereby, our reconstructions are more accurate in the cheek and side areas, see Fig. 10.5-a. Recall that our method does not rely on a known camera motion. The length of the sequence is 80 frames. Reducing the length to 40 frames does not result in decay of reconstruction accuracy. The

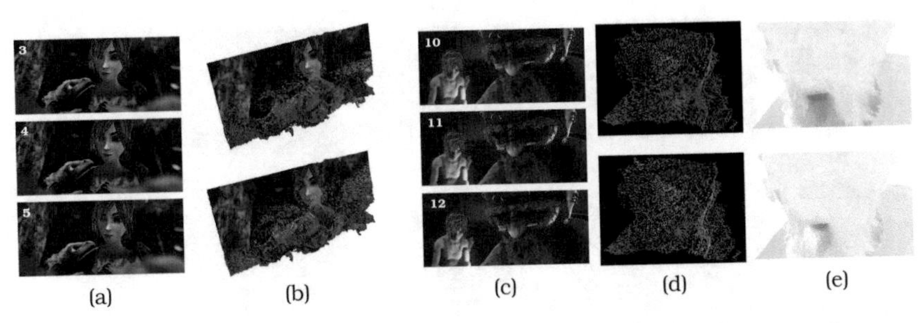

Figure 10.6: Experimental results on the SINTEL dataset [61]: (a) selected frames from the *bandage2 (final)* sequence; (b) whole scene reconstructions with an overlayed scene flow between the corresponding frames on the left; (c) selected frames from the *shaman2 (final)* sequence; (d) enlarged ground truth optical flow between the corresponding frames on the left; (e) scene flow between the corresponding frames in (c).

runtime of NRSfM-Flow for the face sequence amounts to 805 seconds which is split amongst preprocessing, correspondence computation and surface recovery as 2, 771 and 32 seconds, respectively. Note that we also tried the NRSfM-Flow pipeline without preprocessing on the face sequence. In that case, face reconstructions were unnaturally lengthened.
The barn owl sequence [82]. The sequence was acquired outdoors. It depicts a barn owl performing movement peculiar to a predator bird — jerky head turns followed by periods of a focused gaze. Due to the movements, observed surfaces deform non-rigidly. The sequence contains 602 frames in resolution 960×540. There is almost no translation, but there are a lot of redundant frames. In the preprocessing step, 400 out of 602 frames are removed using redundancy removal. An exemplary scene flow field for the barn owl sequence can be seen in Fig. 10.4. The number of points per surface amounts to $2 \cdot 10^5$ and reconstructions look realistic. Scene geometry is correctly explained by rotational and deformational effects. See Fig. 10.5-b for exemplary textured and shaded Poisson surfaces. The runtimes for this sequence amount to 70, 1504, 6300 seconds for the pre-processing, correspondence computation and surface recovery, respectively.
Sintel Flow Dataset [61]. The Sintel Flow Dataset emerged as a response to the growing demand for evaluation of optical flow methods in challenging scenarios. It includes multiple monocular image sequences covering a broad range of realistic scenes varying by type of motion and deformations, environmental conditions and disturbing effects (motion blur, defocus). We tested the proposed framework on several Sintel sequences. Fig. 10.6 depicts selected results on *bandage2 (final)* and

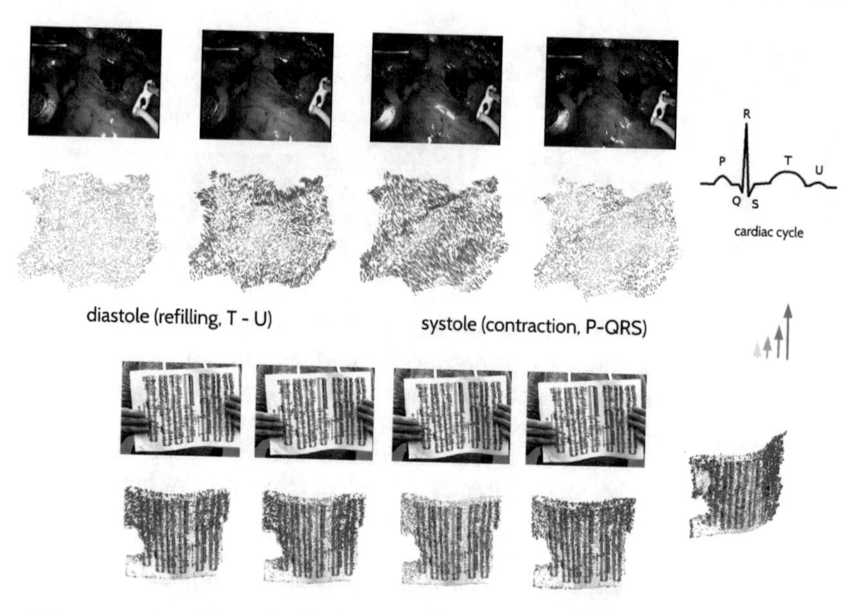

Figure 10.7: Results on the heart [259] (top) and new music notes sequence (bottom). It is possible to recognise different phases of the cardiac cycle in the recovered scene flow. Similarly, the scene flow overlayed with the geometry enhances the perception of the dynamic reconstruction of the music sheet.

shaman2 (final) sequences with 50 non-redundant frames in 1024×436 resolution each. With the first one, we tested the performance of NRSfM-flow in a complex scenario with multiple non-rigid objects. The result discloses few limitations of the proposed approach — without segmentation or shape prior, the variational NRSfM cannot recover relative depths of individual parts correctly, mainly due to the assumed orthographic camera (see Fig. 10.6-a,-b). The relative depths of Sintel and Scales dragon are recovered correctly, but the background is inserted between them. In the case of additional regions (*e.g.,* when the Sintel's hand enters the scene after frame 20), more depth ambiguities occur. Another limitation concerns objects' boundaries — due to the variational nature, NRSfM produces smooth transitions from the foreground objects to the background. Those limitations define the open issues in the area of NRSfM. The *shaman2* sequence shows a slowly moving human face in the foreground (Fig. 10.6-c) and provides optimal conditions for reconstruction with current NRSfM methods. As a result, we were able to obtain accurate scene flow (Fig. 10.6-d) given a foreground-background mask, matching visually well with the ground truth optical flow (Fig. 10.6-e). Both sequences took around 2000 seconds for correspondence computation and 450 seconds for surface recovery.

Discussion. Due to MFOF and linear subspace model of the NRSfM, our approach can handle self- and external occlusions (*e.g.*, occurring in the *bandage2* sequence). Using the proposed framework, it is possible to recover scene flow from monocular image sequences in scenarios not tackleable by existing MSF methods. Concerning NRSfM, we observe a favourable side effect. Serendipitously, MSF allows visualisation of results of a 4D reconstruction better compared to sequentially showing recovered surfaces. MSF also enables one to differentiate between rotational and deformational components in a convenient manner, analyse properties of NRSfM algorithms effectively, tune parameters and uncover directions for further algorithmic improvements.

Our framework inherits limitations peculiar to the current NRSfM methods. Most of them are able to reconstruct scenes which can be easily segmented in background and foreground. Scenes with multiple segments would preferably need additional preprocessing. Due to the orthographic camera model, the proposed framework does not recover absolute depths. Nevertheless, the most appropriate NRSfM method can be chosen depending on requirements (*e.g.*, real-time operation, handling complex composed scenes or support of perspective views). For instance, the method of Russell *et al.* [239] allows joint segmentation and reconstruction complex real-world scenes. The NRSfM-Flow framework will directly benefit from advances in NRSfM methods. As shown experimentally, NRSfM-Flow is not real-time capable in its current form. However, by adopting sequential processing or iterative schemes, it will be possible to achieve real-time MSF recovery performance.

10.1.5 Conclusion

In this section, a new framework for scene flow recovery from monocular image sequences with two preprocessing steps is proposed — the NRSfM-Flow. We introduce a novel analytical framework which allows for relating NRSfM and MSF problems in the continuous domain. We believe that it provides additional insights into both problems and thus will facilitate the development of next-generation algorithms. We would like to draw attention to model-based methods for MSF recovery and to emphasize the importance of differential interpretation of NRSfM.

NRSfM-Flow does not prescribe any particular NRSfM algorithm and inherits advantages and disadvantages of the NRSfM methods. The proposed framework can qualitatively outperform existing MSF methods in the ability to capture 3D motion fields of non-rigidly deforming scenes since less restrictive assumptions about the scene and camera motion are made. For making this conclusion, we consider the results of MSF recovery shown in literature so far and experimental results from this section. One of the central concerns of future work lies in performing comprehensive comparative studies of existing MSF algorithms. As a next step, we plan on using the proposed theoretical apparatus to improve variational NRSfM and to formalise new challenges in the area of NRSfM. NRSfM-Flow will also be used for visualisation purposes supporting the development of augmented reality and medical applications.

10.2 RGB-D Multiframe Scene Flow with Piecewise Rigid Motion

In section, we introduce a new scene flow approach which assumes the scene to be composed of multiple rigid parts which can move independently of coherently relative to each other. The proposed approach considers a short frame of multiple frames. The new approach — Multiframe Scene Flow (MSF) with piecewise rigid motion — jointly optimises the consistency of the patch appearances and their local rigid motions from RGB-D image sequences. In contrast to the competing methods, it takes advantage of an oversegmentation of the reference frame and robust optimisation techniques. MSF is formulated as a global non-linear least squares problem which is iteratively solved by a damped Gauss-Newton approach. As a result, we obtain a qualitatively new level of accuracy in RGB-D based scene flow estimation, and our MSF can potentially run in real-time. MSF can handle challenging cases with rigid, piecewise rigid, articulated and moderate non-rigid motion, and does not rely on prior knowledge about the types of motions and deformations. Extensive experiments on synthetic and real data show that our method outperforms state of the art.

10.2.1 Motivation, Preliminaries and Contributions

Multiframe scene flow (MSF) refers to a flow between a *reference* frame and every other frame of an image sequence (non-reference frames are referred as *current* frames). In the case of depth-augmented or RGB-D images, the input of a scene flow algorithm is a sequence of 2D images with corresponding depth maps. RGB-D based methods use the known depth measurements and inherently provide more accurate estimates compared to other classes.

The accuracy of current RGB-D methods is still not sufficient for many applications, even with predominantly small rigid motions. It deteriorates under *large* scene changes and multi-body transformations (when multiple scene regions transform

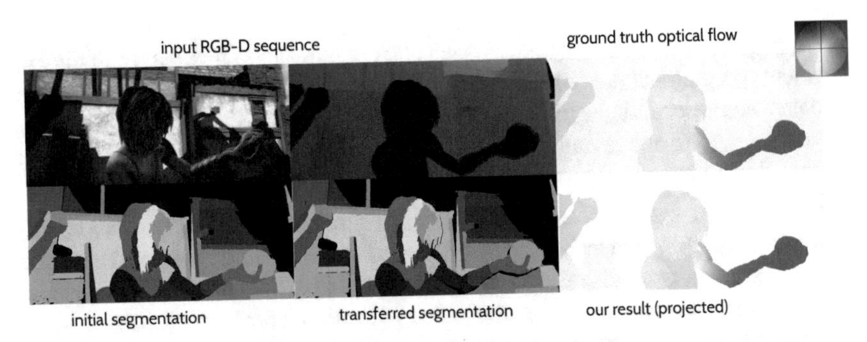

Figure 10.8: The proposed MSF approach accurately estimates scene flow between two or multiple frames. Taking a set of RGB-D frames as an input, it computes an oversegmentation of the reference frame and solves for coherent rigid segment transformations into the set of current frames. The recovered scene flow is highly accurate, with sharp motion boundaries. In the right column, the ground truth optical flow for the given input frames from the MPI SINTEL [61] dataset is visualised using the Middlebury colour scheme [29] (top), together with the projection of the scene flow onto the image plane estimated by our method.

and deform independently). Moreover, the real-time requirement narrows down the choice of suitable optimisation techniques and algorithmic solutions, which often sacrifices the accuracy in favour of the processing speed. Thus, an open question is increasing the accuracy (for small and large motions) without jeopardising the speed. This motivated us to review all aspects of the modern RGB-D based scene flow estimation and find improvements for the bottleneck aspects.

We found *two major realms for improvements*. The first one is the piecewise rigid motion modelling combined with an oversegmentation of the scene. While the idea of piecewise rigidity was explored in the context of scene flow estimation in several ways before [154,225,296], we propose to combine a rigid parameterisation with an over-segmentation of the reference frame, and jointly optimise for the movement of the segment pose pairs in a global manner (for multiple frames). In our model, we assume that object boundaries coincide with a subset of the segment boundaries. In other words, whenever there is an object boundary, a segment boundary must follow it; e converso, segment boundaries inside the objects can run arbitrarily. This is a reasonable assumption in practice, and reliable methods for such oversegmentation exist [7,91]. Moreover, our method performs robustly even if the assumption about conciding boundaries is not entirely fulfilled, *i.e.,* it is robust against inaccuracies in oversegmentation. Another advantage is that segmentation updates result only in segment merging but not splitting. Requiring continuous per-frame segmentation updates would result in a significant increase of the solution space dimensionality, the number of unknowns and the runtime. The proposed assumption allows to avoid those side effects.

The second realm considers the design of an energy functional as well as a choice of a robust optimisation technique. Most current state-of-the art methods make use of total variation regularisers and employ variants of gradient descent or variational optimisation for flow. In contrast, our energy functional is given in terms of sums of squared residuals and optimised using linearisation techniques. This strategy has proven to be efficient for multiple computer graphics and 3D reconstruction problems [149, 334]. In this section, we show that *it is also highly effective for scene flow estimation*. A high-level overview of MSF is shown in Fig. 10.8.

10.2.2 Previous and Related Works in the Area of RGB-D Scene Flow Estimation

Early methods for scene flow estimation required multiple consistent observations of optical flows [147,292,305,326], and additionally solved for the unknown depths or disparities. Piecewise rigid scene flow exploits local rigidity of a scene as an additional constraint for scene flow estimation from stereo images [296]. Vogel *et al.* proposed a sliding window multiframe scene flow approach [295] which imposes consistency of planar patches across stereopairs in all frames.

Since the advent of affordable depth sensors such as Kinect, RGB-D based scene flow estimation became an active research area. In Semi-Rigid Scene Flow (SRSF), Quiroga *et al.* proposed to overparameterize point displacements by individual rigid body motions (twists) [225]. Assuming the rigid motion in the scene is predominant, SRSF estimates scene flow as a sum of a rigid component and a non-rigid residual. The SRSF energy consists of brightness constancy, depth variation and a sum of weighted total variations terms; it is decoupled for an alternating non-linear ROF-model optimisation. SRSF was shown to handle small and simple motions well. A similar concept of rigid body motion parameterisation was proposed in the SphereFlow [146] where the correspondences are searched within a spherical range. Jaimez *et al.* proposed a real-time variational Primal-Dual (PD-)Flow [153]. Its energy includes photometric and geometric consistency terms as well as a spatial total variation regulariser. A GPU implementation of a parallel primal-dual solver enabled real-time processing rates for RGB-D scene flow estimation. Several approaches jointly estimate segmentation and scene flow [154,264]. Motion

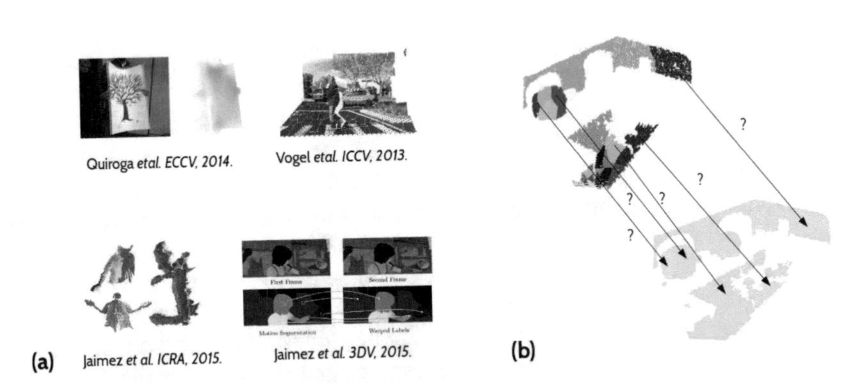

Figure 10.9: An overview of the main related works (Quiroga *et al.* [225], Vogel *et al.* [296], Jaimez *et al.* [153, 154]) in comparison to the proposed idea: given a known segmentation of the reference frame, our method automatically propagates the initial segmentation to one or several frames. Every segment can move rigidly compared to any other segment.

Cooperation (MC-)Flow [154] relies on a linear subspace model, *i.e.,* the velocity of every point is represented as a sum of estimated per-segment rigid motions. In MC-Flow, the scene segmentation is initialised with K-means on the depth channel and the energy is estimated in an alternating manner for motion fields and smooth rigid segmentation. Sun *et al.* rely on depth layering intrinsic to the depth measurements for estimation of scene segmentation and flow [264]. Both latter methods are computationally expensive.

While most approaches rely on the assumption of small motion, only a few approaches explicitly address the scenario with large displacements [324]. In contrast to the previous methods, we assign individual rigid body transformations to the segments retrieved on the depth channel, and use a unified non-linear least-squares optimisation framework (it can potentially run in real-time on a GPU). Instead of a 2D depth reprojection error, we optimise for coordinate differences in 3D space (see Sec. 8.3.2). Moreover, none of the depth-augmented methods performs global optimisation for a subsequence of frames. The global optimisation allows us to estimate larger motions and contributes to the overall accuracy of the method. Another distinguishing property of our MSF approach is kernel lifting. While previous methods use robust kernels (*e.g.*Tukey biweight), we employ kernel lifting, *i.e.,* kernel augmentation with fidelity weights. Our method generates qualitatively more accurate results compared to SRSR, PD-Flow as well as several optical flow methods [267, 323] (after scene flow projection).

RGB-D based scene flow is also used in non-rigid reconstruction [149, 206]. In DynamicFusion [206], depth measurements are warped to a static canonical view by a sparse displacement field and eventually integrated into a volumetric representation. The methods rely on the object texture and feature extraction, can handle moderate motion and support loop-closure but accumulate drift under large deformations. RGB-D scene flow can also be computed by non-rigid point set registration approaches such as *gmmreg* [157]. Point set registration methods recover 3D displacement fields between two or several point clouds. These methods constitute a separate class since differences in appearance and transforming displacement fields between the states can be large, and the consideration is not restricted to surfaces. The idea of scene segmentation has recently found its way in the optical flow estimation. Sevilla-Lara *et al.* showed that accurate object segmentation in 2D can significantly improve the accuracy of the optical flow estimation [253]. Similar to our case, their method requires an accurate object segmentation which follows the object boundaries. Since the depth channel is available in our case, we use a more accurate Felzenszwalb's segmentation algorithm [91] on the depth data.

10.2.3 Multiframe Scene Flow (MSF) with Piecewise Rigid Motion

Our method takes a sequence of N RGB-D image pairs as an input. We denote by $\mathscr{C}_i \in \mathscr{C}$ and $\mathscr{D}_i \in \mathscr{D}$ corresponding colour and depth images, respectively. We assume that all $\mathscr{C}_i$ and $\mathscr{D}_i$ are synchronised temporally as well as spatially (*i.e.*, registered). We work with the corresponding intensity images which are denoted by $\mathscr{I}_i$. The objective of MSF is reconstruction of a 3D displacement field $\rho(\mathbf{x}^t)$ of points $\mathbf{x}$ visible in the *reference* frame $\mathrm{N}_{\mathrm{ref}}$ into all remaining *current* $\mathrm{N}-1$ frames of the observed scene. Thus, MSF relates 3D positions of every 3D point $\mathbf{P} = (P_x, P_y, P_z)$ in all observed frames as

$$\mathbf{P}^t = \mathbf{P}_{\mathrm{ref}} + \rho(\mathbf{x}^t). \tag{10.11}$$

We can handle the case when changes in a scene observed from $\mathrm{N}_{\mathrm{ref}}$ to every other frame are significant and, consequently, $\rho(\mathbf{x}^t)$ could contain large displacements (compared to frame-to-frame cases with small scene changes). A scene can consist of multiple independently moving rigid as well as non-rigid parts. The scene flow estimation in this case is an inherently ill-posed problem as multiple $\rho(\mathbf{x}^t)$ can lead to the same observed scene states.

In our model, every visible 3D Point $\mathbf{P}$ is projected onto the image plane with the projection operator $\pi : \mathbb{R}^3 \to \mathbb{R}^2$:

$$\mathbf{p}(x,y) = \pi(\mathbf{P}) = \left(f_x \frac{P_x}{P_z} + c_x, f_y \frac{P_y}{P_z} + c_y \right)^{\mathsf{T}}, \tag{10.12}$$

with $\{f_x, f_y\}$ the focal lengths of the camera and the principal point $(c_x, c_y)^{\mathsf{T}}$. The inverse projection operator $\pi^{-1} : \mathbb{R}^2 \times \mathbb{R} \to \mathbb{R}^3$ maps a 2D image point to a 3D scene point along the preimage given a depth value z as follows:

$$\pi^{-1}(\mathbf{p}(x,y), z) = \left(z\frac{x - c_x}{f_x}, z\frac{y - c_y}{f_y}, z \right)^{\mathsf{T}}. \tag{10.13}$$

Furthermore, we assume scene transformations to be locally rigid, *i.e.*, the whole scene can be split into K segments $\mathscr{S}_k$, $k \in \{1, \ldots, \mathrm{K}\}$, and every point moves in the segment $\mathscr{S}_k$. By varying the segment granularity, different piecewise rigid and locally non-rigid motions can be accounted for. Movement of every segment $\mathscr{S}_k$ is given by its frame l to frame m rotation $\mathbf{R}_k^{l,m}$ and translation $\mathbf{t}_k^{l,m}$. We denote by $\mathbf{T}_k^{l,m} = (\mathbf{R}_k^{l,m}, \mathbf{t}_k^{l,m})$ the pose of the k-th segment. $\mathbf{T}_k^{l,m}$ has in total 6 DOF, *i.e.*, 3 DOF for $\mathbf{R}_k^{l,m}$ and 3 DOF for $\mathbf{t}_k^{l,m}$. $\mathbf{R}_k^{l,m}$ is parameterised through the angle-axis representation. In the angle-axis representation, a rotation is encoded by a vector $\boldsymbol{\alpha} = (\alpha_x, \alpha_y, \alpha_z)$. The direction of $\boldsymbol{\alpha}$ indicates the axis of rotation $\boldsymbol{\alpha}_n \in \mathfrak{so}(3)$ (it is obtained by normalisation of $\boldsymbol{\alpha}$) — an element of Lie algebra — and the length of $\boldsymbol{\alpha}$ indicates the angle of rotation θ around $\boldsymbol{\alpha}_n$ according to the right-hand rule. To rotate a segment, we convert $\boldsymbol{\alpha}$ to the corresponding rotation matrix $\mathbf{R}$ using a corollary of the Rodrigues' rotation formula leading to the exponential map:

$$\mathbf{R} = \exp(\theta \mathbf{K}) = \mathbf{I} + \sin\theta\, \mathbf{K} + (1 - \cos\theta)\mathbf{K}^2, \tag{10.14}$$

where $\mathbf{K} = \begin{pmatrix} 0 & -\alpha_z & \alpha_y \\ \alpha_z & 0 & \alpha_x \\ -\alpha_y & -\alpha_x & 0 \end{pmatrix} \in \mathfrak{so}(3)$ is given by a skew-symmetric cross-product matrix of $\boldsymbol{\alpha}_n$.

To recover scene flow $\rho(\mathbf{x}^t)$, we jointly solve for all segment poses $\mathbf{T}^{l,m}$ from every frame to every other frame in the input RGB-D sequence. Once the poses are recovered, MSF is estimated by considering per-point correspondences of the segments throughout the sequence. Our MSF approach is based on energy functional minimisation. The energy $\mathfrak{E}$ in the two-frame case consists of four terms:

$$\begin{aligned} \mathfrak{E}(\mathbf{T}, \mathbf{w}) = \alpha\, \mathfrak{E}_{\text{data}}(\mathbf{T}) + \beta\, \mathfrak{E}_{\text{pICP}}(\mathbf{T}) + \\ + \gamma \mathfrak{E}_{\text{l. reg}}(\mathbf{T}, \mathbf{w}) + \eta\, \mathfrak{E}_{\text{r. opt.}}(\mathbf{w}). \end{aligned} \tag{10.15}$$

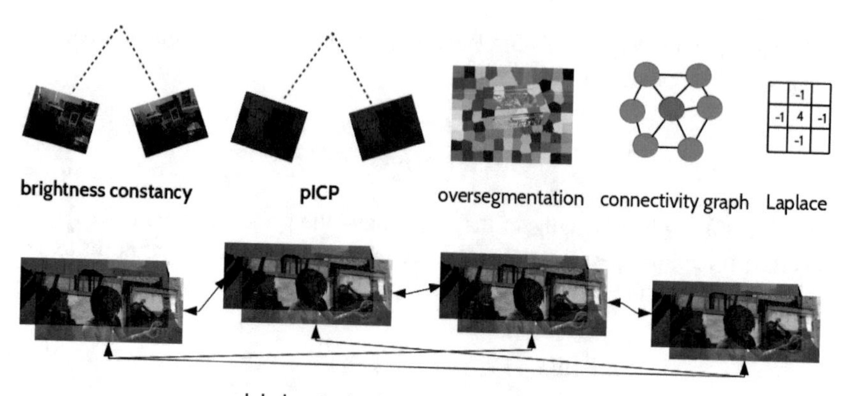

Figure 10.10: An overview of the main components of the proposed MSF energy: brithness constancy term, pICP term, lifting term with oversegmentation, a segment connectivity graph and Laplace regulariser. The multiframe pose concatenation term (shown in the lower part) insures segmant consistency throughout a few frames.

In Eq. (10.15), $\mathbf{T}_k \in \mathbf{T}$ are rigid transformations (rotation and translation) for every segment $k \in \{1,\ldots,K\}$ from the reference frame to the single current frame, $w_{j,h} \in \mathbf{w}$ is a set of lifting weights for the segment pairs optimised coherently. In the following, we describe each of the energy terms from Eq. (10.15) in detail. Prior to the energy-based optimisation, we perform Gaussian smoothing of $\mathscr{I}_i$ to reduce the influence of noise and outliers when computing gradients (see Sec. 10.2.4 for optimisation details).

Data Term

The data term accounts for the brightness constancy of 3D points observed in multiple views $\mathscr{I}_i$. That is, the same 3D point $\mathbf{P}$ must cause the same brightness value in both views (similarly, the reprojection error of the brightness values associated with the same 3D point shall be minimised):

$$\mathfrak{E}_{\text{data}}(\mathbf{T}) = \sum_k \sum_{\mathbf{p}\in\Omega_k} \|(\mathscr{I}_1(\mathbf{p}) - \mathscr{I}_2(\pi(g(\mathbf{T}_k, \pi^{-1}(\mathbf{p}, z_{\mathbf{p}})))))^2\|_\varepsilon, \tag{10.16}$$

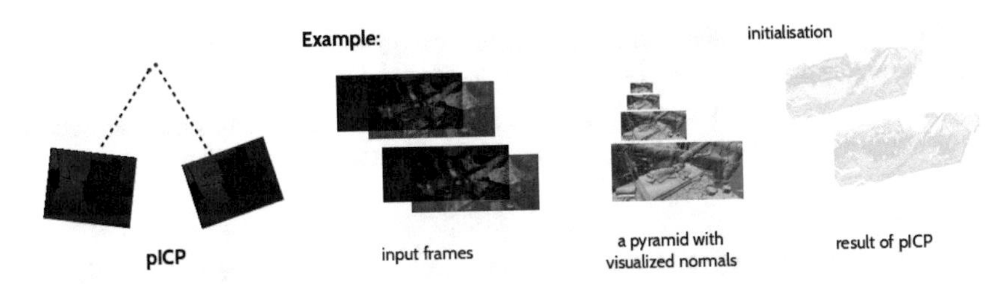

Figure 10.11: Projective ICP term backprojects depth values into 3D with known intrinsics and involves normals for the stabilisation of the registration of planes.

where Ω_k is a set of points in the segment $\mathscr{S}_k$, $g(\cdot,\cdot)$ is the rigid point transformation operator, $z_{\mathbf{p}} = \mathscr{D}_i(\mathbf{p})$ are known and $\|\cdot\|_\varepsilon$ stands for the Huber loss defined as:

$$\|a^2\|_\varepsilon = \begin{cases} \frac{1}{2}a^2, & \text{for } |a| \leq \varepsilon \\ \varepsilon(|a| - \frac{1}{2}\varepsilon), & \text{otherwise,} \end{cases} \tag{10.17}$$

with a non-negative scalar threshold ε. The term in Eq. (10.16) performs dense image alignment of the current and reference intensity frames and contributes to the recovery of the relative segment transformations $\mathbf{T}_k$. The Huber loss works as an ℓ_2 loss for the smallest values of the brightness residuals. If higher than ε, the differences are not squared which is equivalent to an ℓ_1 loss. As a result, the influence of significant non-Gaussian distributed outliers is reduced. The form of the Huber loss in Eq. (10.17) allows, de facto, to use an ℓ_1 norm in the non-linear least squares framework.

Projective ICP Term

The second data term in our target energy is the point-to-plane registration or projective Iterative Closest Point (ICP) term. In contrast to the intensity-based data term — which minimises reprojected intensity values in the image space — the projective ICP minimises Euclidean distances of corresponding points projected onto the normals $\mathbf{n_p}$ of the reference image $\mathrm{N}_{\mathrm{ref}}$ directly in 3D:

$$\mathfrak{E}_{\mathrm{pICP}}(\mathbf{T}) = \sum_k \sum_{\mathbf{p}\in\Omega_k} \|((g(\mathbf{T}_k,\mathbf{p}) - \mathbf{p}^{\mathrm{corr}}) \cdot \mathbf{n_p})^2\|_\varepsilon. \tag{10.18}$$

Every iteration of the projective ICP consists of two alternating steps: first, while segment poses are fixed, the correspondences $\mathbf{p}^{\mathrm{corr}}$ for every point $\mathbf{p}$ are updated and, second, given point correspondences, the new segment poses $\mathbf{T}_k$ are computed.

The optimum is achieved when the difference between two points is small and orthogonal to the normal $\mathbf{n_p}$ of the reference. The normals $\mathbf{n_p}$ are computed on the reference frame. For every point $\mathbf{p}$, the normal $\mathbf{n_p}$ is obtained as a cross product of the central differences d_x and d_y in the x and y directions, respectively:

$$\mathbf{n_p} = d_x \times d_y. \tag{10.19}$$

The projective ICP term is brightness invariant and operates purely on the 3D points computed from the depth map. Thus, spatial diversity of a scene facilitates registrations of a higher accuracy.

Lifted Segment Pose Regulariser

The data terms $\mathfrak{E}_{\text{data}}(\mathbf{T})$ and $\mathfrak{E}_{\text{pICP}}(\mathbf{T})$ alone are not sufficient to accurately align piecewise rigid scenes, especially when an accurate object-background segmentation of the reference frame could be unavailable. Without an explicit regularisation, relatively small segments would be influenced by clustered outliers, noise, missing point correspondences and occlusions. Thus, we propose the following term:

$$\mathfrak{E}_{\text{l. reg}}(\mathbf{T}, \mathbf{w}) = \sum_{w_{j,h}:\Psi[j,h]=1} \left\| (w_{j,h}^2 (\mathbf{T}_j - \mathbf{T}_h))^2 \right\|_2, \tag{10.20}$$

where j and h define a pair of segments with an imposed coherent movement and $\|\cdot\|_2$ is an ℓ_2 norm. The segment pose regulariser favors coherent transformation of the neighboring segments.

The pairs of the segments which are demanded to move coherently are determined based on the segment vicinities. We maintain a segment adjacency matrix, *i.e.*, a sparse $\mathrm{K} \times \mathrm{K}$ matrix which contains ones if segments j and h move coherently, and zeroes otherwise. The number of segments can be large, and we store only non-zero elements of Ψ. We parameterize $\Psi = \psi_{j,h}$ with the number of adjacent segments n_ψ for each segment. For every $\mathscr{S}_k$, a corresponding row of Ψ contains n_ψ segments with the closest centroids. For every pose pair, Eq. (10.20) defines a residual. Altogether, there are $\sum \psi_{j,h}$ residuals in the segment pose regulariser.

For the sake of the robustness to disturbing effects, we opt for lifting of the pose regulariser. Kernel lifting was shown to outperform robust costs (such as iterative reweighting and Triggs correction) in avoiding local minima [322]. The idea of lifting consists in augmenting the energy with confidence weights. Thus, we introduce $w_{j,h}$ which in our formulation account for the strength of the segment connections and allow to continuously adjust segment coherencies. Recall that the scene can exhibit multiple independent motions and deformations, and $w_{j,h}$ allow to weaken or break the connections if two segments move independently.

Kernel lifting works in combination with the robust weight optimiser term which we discuss next.

Robust Weight Optimiser

The lifted segment pose regulariser term alone is not sufficient to influence segment pair connectivities. The robust weight optimiser thus prevents highly coherent segment pairs from weakening their weights:

$$\mathfrak{E}_{\text{r. opt.}}(\mathbf{w}) = \sum_{\forall w_{j,h}:\Psi[j,h]=1} \|(1-w_{j,h}^2)^2\|_2. \tag{10.21}$$

The operator introducing $\mathbf{w}$ into the energy is often called a lifting function. We choose the lifting function

$$\mathscr{F}(\cdot,\mathbf{w}) = \sum(w_i^2(\cdot)+(1-w_i^2)) \tag{10.22}$$

as it was shown to simplify the overall energy landscape well and also proven to be the most robust among other lifting kernel choices in our experiments. If optimally balanced, the segment pose regulariser and the robust weight optimiser can maintain strong connections by keeping $w_{j,h}$ high, even if the partial energy of the segment pose regulariser remains high.

10.2.3.1 Multiframe Formulation

We generalise the energy proposed in Eq. (10.15) for the case of multiple frames. Consider a temporal window of N frames; we perform pairwise alignment of all frame pairs with individual data, projective ICP and segment pose regulariser terms but common weights $w_{j,h}$. This allows to further constrain the problem and enhance the accuracy (reduce multiframe registration drift) by adding more observations and performing the joint minimisation. Let $\mathfrak{i}_{\text{ref}}$ and $\mathfrak{i}_{\text{target}}$ denote indexes of the first and the last frame of the temporal window and let Z be the set of pairwise frame combinations. In total, there are ${}^{\text{N}}C_2$ frame pairs, *i.e.*, a number of 2-combinations in a set of N frames. Moreover, let $\zeta \in \text{Z}$ be a one-dimensional variable indexing frame-to-frame transformations from frame l to frame m, *i.e.*, $\mathbf{T}^{\zeta}$ is a shortcut for $\mathbf{T}^{l,m} \in \text{Z}$. Thus, the energy functional for the multiframe case reads

$$\begin{aligned}\mathfrak{E}(\mathbf{T}^1,\mathbf{T}^2,\ldots,\mathbf{T}^{|\text{Z}|},\mathbf{w}) &= \sum_{\zeta\in\text{Z}} \alpha_{\zeta}\,\mathfrak{E}_{\text{data}}(\mathbf{T}^{\zeta}) + + \sum_{\zeta\in\text{Z}} \beta_{\zeta}\,\mathfrak{E}_{\text{pICP}}(\mathbf{T}^{\zeta}) + \\ &+ \gamma_{\zeta} \sum_{\zeta\in\text{Z}} \mathfrak{E}_{\text{l.reg.}}(\mathbf{T}^{\zeta},\mathbf{w}) + + \eta\,\mathfrak{E}_{\text{r.opt.}}(\mathbf{w}) + \sum_{\zeta=3}^{|\text{Z}|} \lambda_{\zeta}\,\mathfrak{E}_{\text{c.}}(\mathbf{T}^{\zeta}).\end{aligned} \tag{10.23}$$

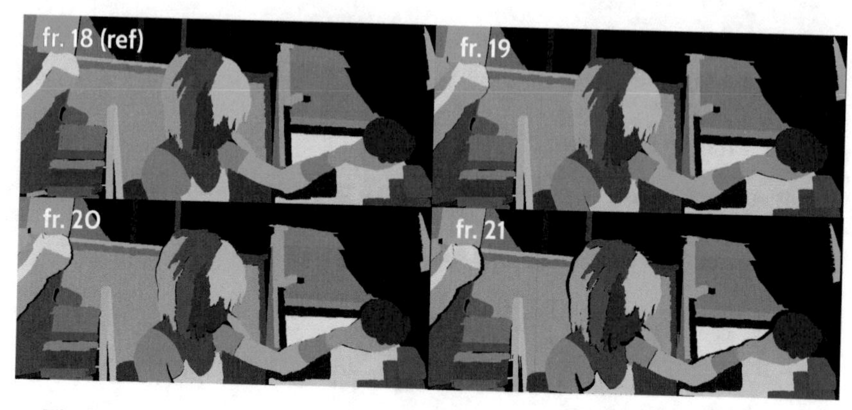

Figure 10.12: Segmentation transfer from the reference frame to three other frames in *alley1*. In this experiment, scene flow was computed in the global manner (every frame was optimised jointly with every other). In all frames, the same colours denote same segments, and the shapes are obtained by projection of the initial segment to the respective frames.

The global energy in Eq. (10.23) contains multiple data and projective ICP terms, lifted segment pose regulariser and robust weight optimiser terms (*cf.* the two-frame energy in Eq. (10.15)) as well as an additional pose concatenation term.

Multiframe Pose Concatenation Term

Since we also optimise for poses between several frames with large displacements, the accumulated changes can be intractable for direct optimisation. Additionally to the terms in the energy functional in the two-frame case, we add a pose concatenation term involving pose transformations to non-adjacent frames:

$$\mathfrak{E}_{\text{c.}}(\mathbf{T}^{l,m}) = \sum_{k} \left\| (\mathbf{T}_k^{l,m} - \mathbf{T}_k^{m-1,m} \cdot \ldots \cdot \mathbf{T}_k^{l,l+1})^2 \right\|_2, \tag{10.24}$$

with $\cdot$ denoting the pose concatenation operator. Thus, the pose concatenation term enforces the transformations between non-adjacent frames to be close to the concatenation of transformations which sequentially lead from frame l to frame m.

10.2.4 Energy Optimisation

Our target energy consists of several terms, and every term contains a sum of squared residuals. Due to the perspective 3D to 2D and inverse 2D to 3D projections, the objective is non-linear and minimisation of Eq. (10.23) is a non-linear least squares problem. We minimise it with a Gauss-Newton method. In total, there are M residuals:

$$\mathrm{M} = {}^{\mathrm{N}}C_2\,(n_{\mathscr{C}} + n_{\mathscr{D}}) + ({}^{\mathrm{N}}C_2 + 1)\,n_{\mathrm{pp}} + n_{\mathrm{c}}, \tag{10.25}$$

where $n_{\mathscr{C}}$ is the number of residuals in the two-frame data term ($\leq$ number of pixels in an image), $n_{\mathscr{D}}$ is the number of residuals in the two-frame projective ICP term (maximum the number of non-zero depth measurements); both $n_{\mathscr{C}}$ and $n_{\mathscr{D}}$ are appearing ${}^{\mathrm{N}}C_2$ times. $n_{\mathrm{pp}} = {}^{\mathrm{N}}C_2 \sum \psi_{k,l}$ is the number of pose pairs (number of non-zero elements in the segment adjacency matrix) for every frame-to-frame combination. Additionally, there are n_{pp} residuals in the robust regulariser (there are n_{pp} weights in total). Finally, $n_{\mathrm{c}} = \mathrm{K}\,(\mathrm{N}-2)$ is the number of residuals in the pose concatenation term, *i.e.*, a number of non-adjacent frame combinations for given N per segment (if $\mathrm{N} = 2$, then $n_{\mathrm{c}} = 0$).

In the following, x is a shorthand symbol for the set of unknowns in the target energy $\mathfrak{E}$. We denote the scaled residuals by the compact notation $f_r(\mathrm{x})$ and stack them into a single multivariate vector-valued function $\mathbf{F}(\mathrm{x}) : \mathbb{R}^{\mathrm{K}^{\mathrm{N}}C_2 + n_{\mathrm{pp}}} \rightarrow \mathbb{R}^{\mathrm{M}}$:

$$\mathbf{F}(\mathrm{x}) = [f_1(\mathrm{x}), f_2(\mathrm{x}), \ldots, f_r(\mathrm{x})]^{\mathsf{T}}. \tag{10.26}$$

The total number of parameters is composed of K poses for each two-frame combination out of N and n_{pp} weights. The objective function $\mathfrak{E}(x)$ can now be written in the new symbols as

$$\mathfrak{E}(x) = \|\mathbf{F}(\mathrm{x})\|_2^2. \tag{10.27}$$

We aim at an optimal parameter set x' minimising $\mathfrak{E}(x)$:

$$\mathrm{x}' = \arg\min_{\mathrm{x}} \|\mathbf{F}(\mathrm{x})\|_2^2. \tag{10.28}$$

As the problem in Eq. (10.28) is non-linear, we iteratively linearize the objective around the current solution x_t and find an update through minimisation of the linear objective. Next, starting from the first-order Taylor expansion of (10.28), the NLLS optimisation follows the steps described in Sec. 2.1.4.

While optimising $\mathfrak{E}$, we allow for non-monotonic steps. In the classic implementation, LM reverses to the previous state if a current update leads to an increased energy. Allowing suboptimal steps could result in the surpassing of local minima in the long term. After computing the update $\Delta\mathrm{x}$ in each iteration, we iteratively

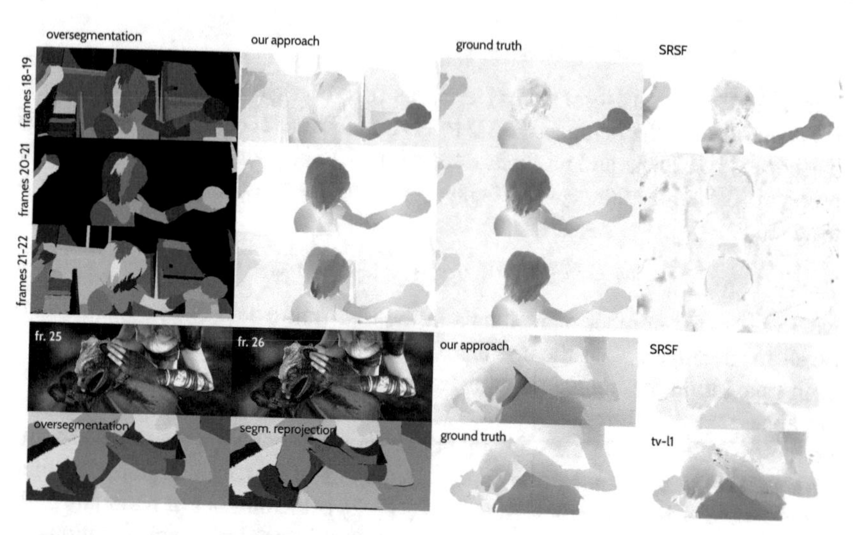

Figure 10.13: Experimental results on the SINTEL *alley1* (top three rows) and *bandage1* (bottom rows) sequences. *alley1*, from the left to the right: oversegmentation for each reference frame, results of our approach, ground truth optical flow and results of SRSF [225]; *bandage1*, on the left: the input frames, the initial oversegmentation and its re-projection into the current frame; on the right: results of our approach, SMSR [225], ground truth and the tv-l1 [323]. Due to optimally chosen segment sizes, our approach recovers an accurate scene flow compared to SRSF and tv-l1.

update the current solution as $x_{t+1} = x_t + \Delta x$. If the difference between several consecutive energy values continuously falls under the ε value, the optimisation is considered as converged and the lowest energy value corresponding to the optimal x is returned. Once the energy is minimised, we compute correspondences between points of the reference and the current frames by 3D point reprojection. The scene flow ρ is recovered as displacements between corresponding points. As a side effect, our approach can perform segmentation from motion. By grouping the recovered $\mathbf{T}_k$, we are able to determine independent rigid motions and deformations throughout the scene.

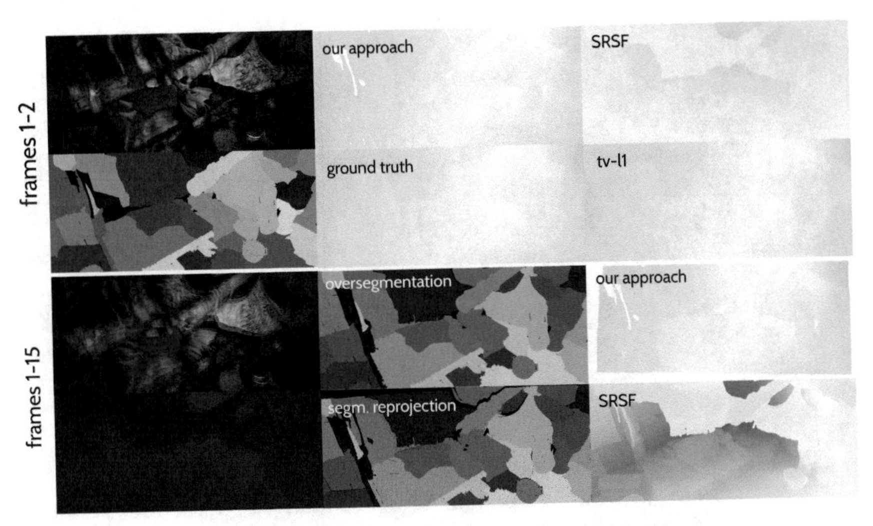

Figure 10.14: Experimental results on a static scene observed by a moving camera from the SINTEL dataset (*sleeping2*). First column: input frames, an oversegmentation, large displacement frames and corresponding depth maps. Both SRSF [225] and MSF can incorporate prior knowledge of the rigidity while solving for the 3D flow. Both for the case of small (one frame) and large (14 frames) displacements, our approach outputs a very accurate result. We notice from the input images that in the case of large displacements, the camera preserves the trajectory; thus, direction of the flow does not change significantly in this scene. This effect is reflected in the projection of the scene flow estimated by our MSF approach.

10.2.4.1 Energy Initialisation and Settings

To alleviate the influence of noise during computation of image gradients in the brightness term, we perform Gaussian smoothing of $\mathscr{I}_i$. We do not filter depth since the number of residuals in $\mathfrak{E}$ exceeds the number of parameters (segment poses and pose combination weights) by several orders of magnitude.

We initialise segments with graph-based segmentation on the depth values [91]. At the beginning, when no motion segmentation is available, the segment pairs are initialised based on the vicinity of the centroids (and so is the adjacency matrix Ψ). First, the center coordinates of the segments $\mathscr{S}_k$ are computed. Then, for

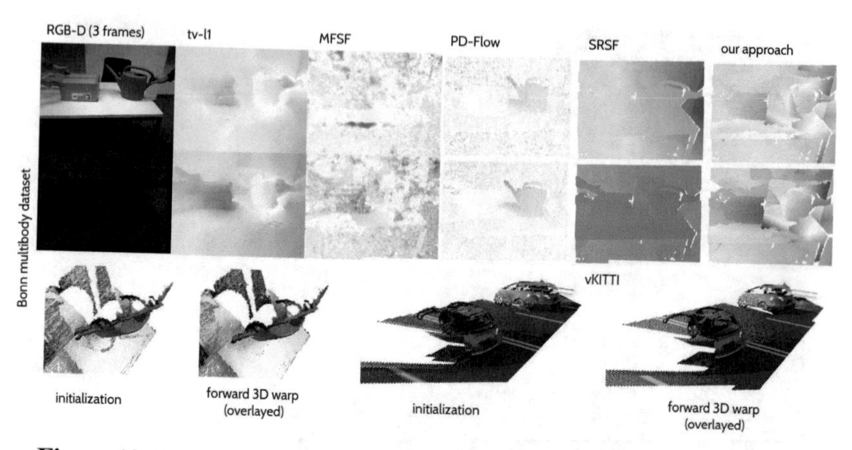

Figure 10.15: Results of several RGB-D scene flow and optical flow approaches on the Bonn multibody dataset [261]. Top row: results for the frames 1-2; middle row: results for the frames 1-3; bottom row: initialised (not aligned) and warped (aligned) RGB-D frames visualised as point clouds (left, processed by MSF); an example of an urbane driving scene processing form the vKITTI dataset [101] by MSF (right).

every segment (row in the adjacency matrix), ones are set for K nearest segments (corresponding rows). We initialise the poses by aligning all neighboring frame pairs using the two-frame energy formulation. Transformations between non-adjacent frames are initialised with respective pose concatenations. We normalise every term in $\mathfrak{E}$ w.r.t. the total number of residuals. Thus, we set $\frac{\alpha\zeta}{\mathrm{M}_{\alpha\zeta}}, \frac{\beta\zeta}{\mathrm{M}_{\beta\zeta}}, \ldots$ as the term weights, where $\mathrm{M}_{\alpha\zeta}, \mathrm{M}_{\beta\zeta}, \ldots$ are the respective numbers of residuals.

10.2.5 Experimental Evaluation

In this section, we describe the experimental evaluation of the proposed technique. All experiments are performed under Ubuntu 16.04 on a system with 32 GB RAM and Intel Core i7-6700K CPU running at 4GHz. We implement MSF as a standalone framework in C++. As a non-linear least squares solver, we use *ceres* [242]. For the sake of accuracy and speed, we opt for automatic differentiation of cost functions. Since every flow vector is parameterised by 6 DOF, every corresponding residual depending on rigid transformation is a 6-vector. After every successful solver step, we update point correspondences for the projective ICP term. The algorithm converges in average after $15-20$ iterations.

10.2.5.1 Experiments on Synthetic Data

We use two synthetic datasets for the quantitative evaluation — MPI SINTEL [61] and v-(irtual)KITTI [101]. SINTEL dataset represents several sequences of synthetically rendered images of an animated movie with an additional depth channel (the resolution is 1024×436 pixel). The imaging process (motion blur, defocus) and the atmospheric effects are accurately simulated so that the images look naturalistic. vKITTI represents a synthetic dataset of a frequently encountered urbane driving scenes. We compare MSF with SRSF [225] as well as tv-l1 optical flow method of Zach *et al.* [323]. Comparison with PD-Flow is performed on the real data (Sec. 10.2.5.2), as it is hard-coded for Kinect recordings. Unfortunately, the source code for MC-Flow [154] is not publicly available.

We set the minimum number of vertices per segment to $2 \cdot 10^3$ (if less, a segment is discarded) and the threshold to 0.5 in the Felzenszwalb's algorithm [91]. The oversegmentation is used for parameterisation of the scene flow by a set of rigid 6-DOF transformations. When the number and size of the segments are optimal, parametrisation explains observed rigid motion and approximates non-rigid deformations well so that our results consistently outperform the competing methods. In this experiment, we have empirically determined the neccesary condition on the initial segmentation: *the segments boundaries must follow the object boundaries*. The more completely this condition is fulfilled, the more accurate can be the result. On the other side, if segmentation is not accurate, the accuracy decays insignificantly, up to a breakpoint. Fig. 10.13 shows representative results (*i.e.,* not biased) on the *alley1* and *bandage1* sequences. Projection of the scene flow and optical flow are visualised with the Middlebury colour code [29].

Next, we compare MSF on the scene undergoing a purely rigid transformation — *sleeping2*. Similar to our approach, SRSF allows parameterising the whole displacement field by a single rigid transformation, given prior knowledge of a scene. Fig. 10.14 shows results of the comparison for the case of small (one frame) and large (14 frames) differences. In the case of one frame difference, the ground truth is available. Our approach recovers a similar flow to the ground truth, whereas SRSF misses the direction by several degrees. In the case of large displacements, no ground truth is available, but the flow preserves the direction (this can be noticed from the images). MSF recovers the scene flow accurately also in this case, while the discrepancy of SRSF increases.

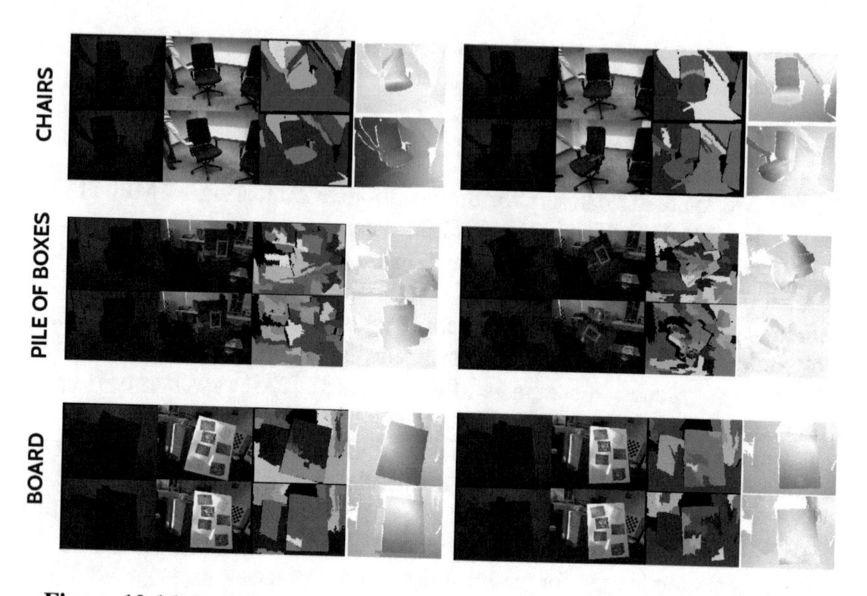

Figure 10.16: Results on several real datasets, *i.e.,* Chairs [261], Pile of Boxes and a Board. The order in every line: depth map, RGB image, segmentation of the reference frame, projection of the scene flow between the shown frame and the next frame of the sequence.

Table 10.2: Quantitative comparison between scene flow projections and the ground truth optical flow on the MPI SINTEL [29] training sequences using Average EPE errors (in pixel).

	alley1	*bandage1*	*sleeping2, rigid*
SRSF [225]	2.46122/2.40833	2.47801/2.46389	1.13584
MSF	0.740127	1.69865	0.307526

Table 10.3: Average runtime comparison of the proposed MSF and SRSF [225] for different configurations, in seconds. Legend: "(r)" stands for rigid scene, and "l.d." stands for large displacements (14 frames).

	2 fr.	3 fr.	4 fr.	2 fr. (r)	l. d. (r)
SRSF [225]	274	n.a.	n.a.	**87.5**	84.5
MSF	**49**	221	541	**90**	254

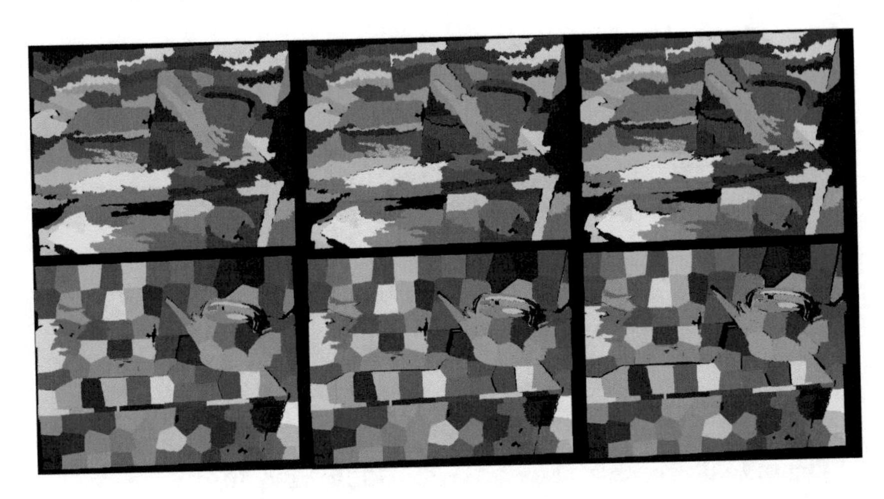

Figure 10.17: Segmentation transfer on the Bonn watering can sequence [261] with Felzenszwalb segmentation [91] (top row) and SLIC segmentation [7] (bottom row).

We evaluate accuracy of MSF quantitatively by comparing projections of the scene flow with the ground truth optical flow on the SINTEL dataset. Table 10.2 reports the average End Point Error (EPE)[2] for SRSF [225] and our approach for several image sequences. The obtained metrics agree with the qualitative results. Table 10.3 lists average runtimes for SRSF [225] and MSF on the SINTEL dataset, for different launch configurations. Additionally, we report EPE of our MSF method for *shaman2* (0.354061), *shaman3* (1.00192), *mountain1* (1.6898) and *bandage2* (1.0715) sequences.

With the example of the *alley1* sequence, we study the influence of the Huber loss on the scene flow result. Using the ℓ_1 norm instead of the Huber loss (a Huber loss with a zero threshold) leads to the decrease in the runtime and accuracy by 5% and 4%, respectively. Choosing the ℓ_2 norm leads to an $\approx 40\%$ increase of EPE. Some excerpts from the *alley1*, *bandage1*, *sleeping1*, *sleeping2*, *shaman2*, *shaman3*, *mountain1* sequences are given in Fig. 10.19–10.21, see Fig. 10.18 for the order of frames in these figures.

Evaluation on the vKITTI dataset demonstrated the applicability of MSF for driving scenarios. Fig. 10.15-(bottom row) shows two consecutive frames from a driving scene as point clouds overlayed and warped to the current frame after scene

[2]EPE is defined as $\|(u-u_{\text{GT}}),(v-v_{\text{GT}})\|$, where $(u,v)^{\mathsf{T}}$ is a projected flow vector and $(u_{\text{GT}},v_{\text{GT}})^{\mathsf{T}}$ is a ground truth vector.

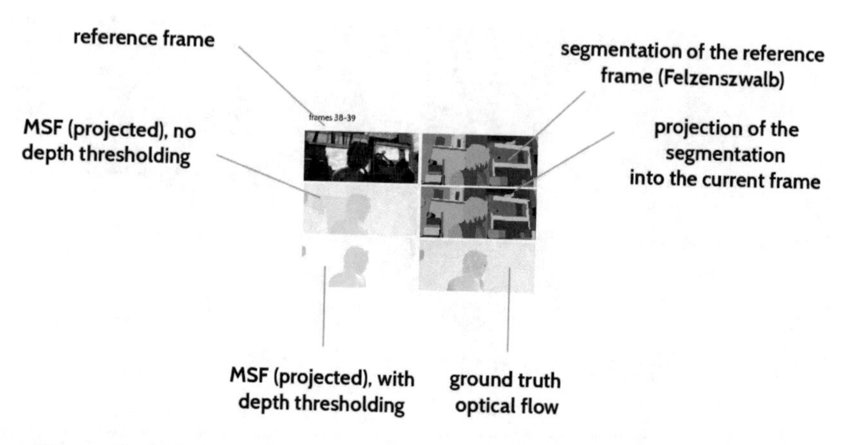

Figure 10.18: The order of frames in Figs. 10.19, 10.20, 10.21.

flow estimation. Although the motion of the cars differs in direction and magnitude, both scene subflows are recovered accurately.

10.2.5.2 Experiments on Real Data

We also evaluate the proposed approach on several real datasets, *i.e.,* rigid multibody dataset [261] recorded by a Kinect. We compare MSF with SRSF [225], PD-Flow [153] as well as tv-l1 optical flow [323] and Multi-Frame Subspace Flow (MFSF) [267]. Selected results are shown in Fig. 10.15. Since no ground truth is available for this real dataset, several qualitative observations can be made. First, the watering pot moves towards the observer and the motion between three frames is rather small. MFSF, PD-Flow and SRSR generate more noisy output than tv-l1 and our approach. MSF preserves object boundaries, but some spurious boundaries are introduced which are more apparent than in the case of the SINTEL sequences. Obtaining accurate segmentation on noisy real world data requires additional pre-processing and can be readily done. Fig. 10.15-(bottom left) shows initialised and warped overlayed RGB-D frames visualised as point clouds.

10.2.6 Discussion

The experiments show advantages of the new scene flow formulation. We believe that the segmentation along with the requirement of sharp object boundaries in combination with suitable segment sizes reason the notable accuracy of MSF. In some cases, however, SLIC [7] or a simple regular tiled segmentation can be more

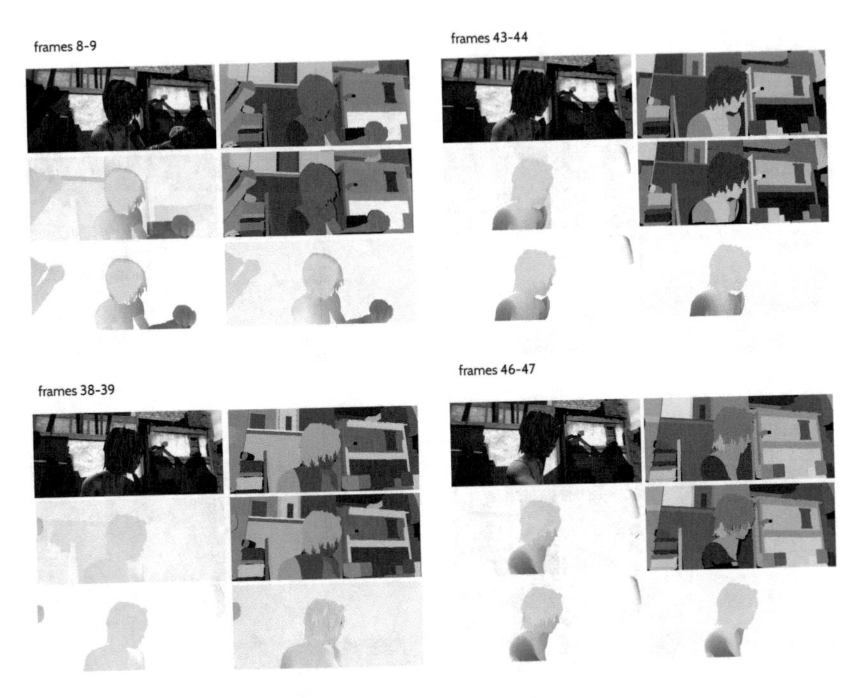

Figure 10.19: Additional visualisations of the *alley1* sequence. For every block: reference RGB frame – segmentation of the reference frame (top row), our result, projection of the scene flow into the image plane (no depth thresholding) – projection of the segmentation into the current frame (second row), our result, projection of the segmentation into the current frame (depth thresholded to two meters) – ground truth optical flow (bottom row). The scheme is once again visualised in Fig. 10.18.

suitable than Felzenszwalb though this is case-dependent. Another reason could be the projective point-to-plane ICP term which endows MSF with a property of an articulated point set registration algorithm. Finally, an improved lifted energy landscape is perhaps a part of the answer.

Yet, MSF has some limitations. If a scene does not provide sufficient cues for the data terms (brightness and varying depth values), segments which include such regions can be influenced by the moving parts in the vicinity and erroneously involved in the motion (cf. Fig. 10.13, our result on frames 21-22 — the segments

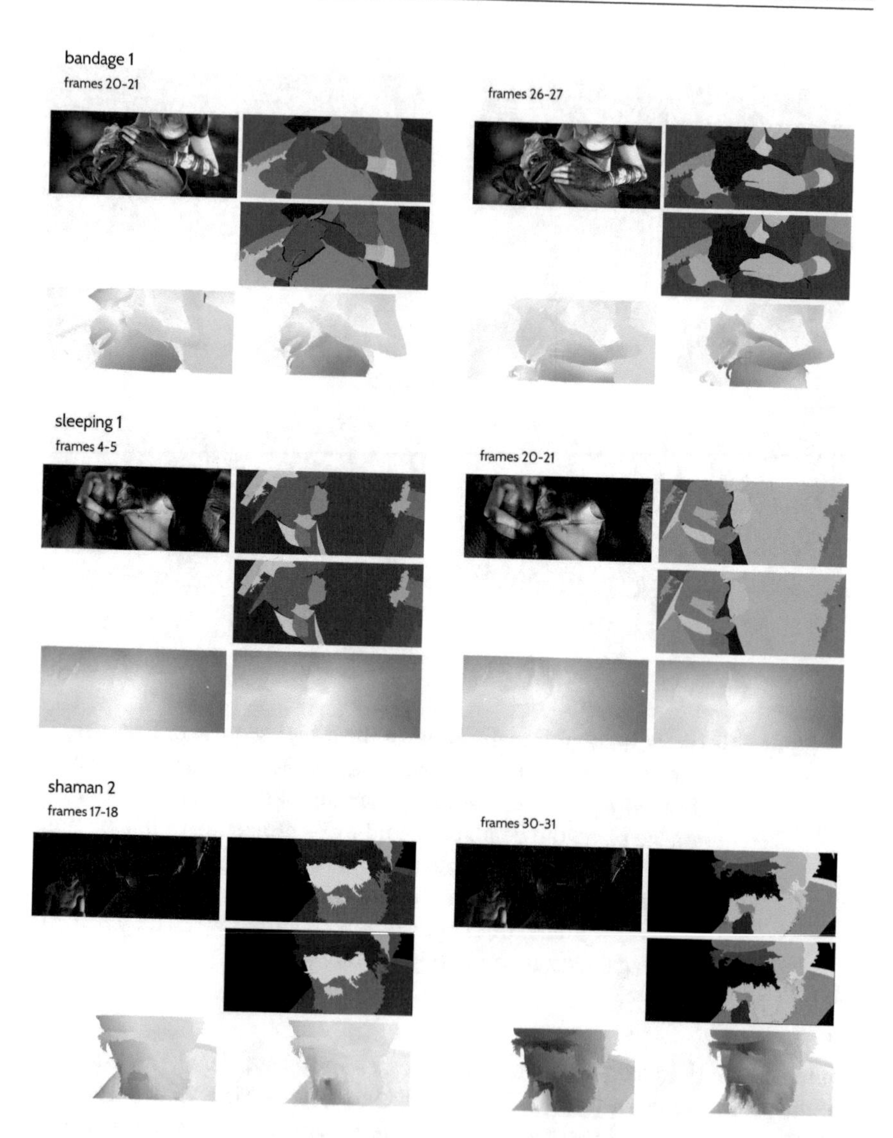

Figure 10.20: Additional visualisations of the *bandage1*, *sleeping1* and *shaman2* sequences. See Fig. 10.18 for the legend.

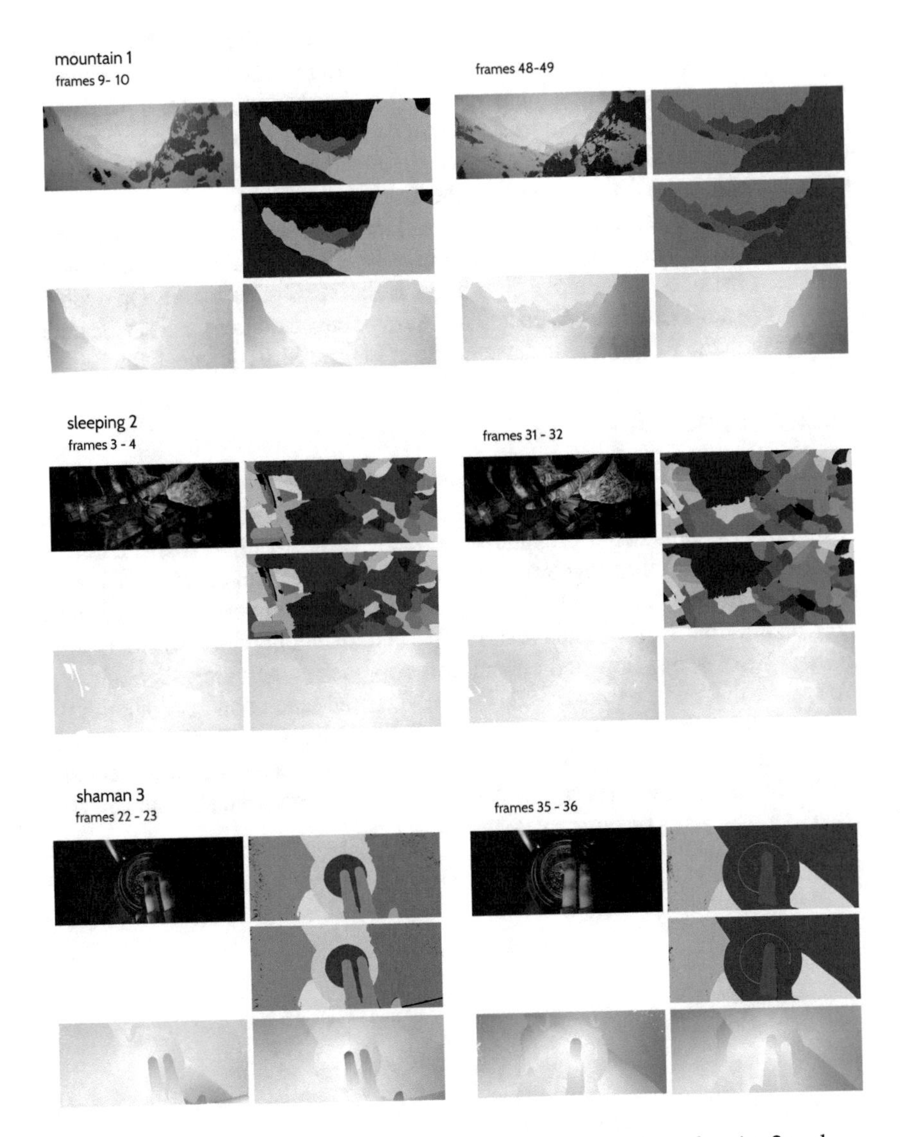

Figure 10.21: Additional visualisations of the *mountain1*, *sleeping2* and *shaman3* sequences. See Fig. 10.18 for the legend.

near the girl's face get spuriously involved in the motion). The method also requires a balance between the lifted Laplacian pose regulariser and robust weight regulariser terms. If all weights are coming close to 1.0 in the absolute value, η needs to be *decreased*. Parameters should be set so that the weights w_{ij} are distributed around 0.0 and 1.0 or -1.0 with a low variance. In all experiments, γ and η were fixed, while α and β were set depending on the data properties.

Sometimes, Felzenszwalb segmentation does not generate sufficiently accurate segmentation of a reference frame. In this case, other initialisation methods can be used. One of the alternative algorithms is the SLIC superpixels approach [7]. Fig. 10.17 shows examples of the reference frame segmentations and segmentation transfer with the Felzenszwalb segmentation [91] and SLIC segmentation.

10.2.7 Conclusion

We propose a novel multiframe scene flow approach. Our method relies on an over-segmentation of the reference frame with sharp object boundaries. The underlying energy functional includes brightness constancy, projective ICP, robust Laplacian as well as chained pose regulariser terms, and each of them consists of sums of squared residuals. As a result, we obtain a robust method which performs accurately on multiple synthetic and real datasets and favourably compares to multiple RGB-D based scene flow and optical flow approaches including state of the art. We believe that more improvements in RGB-D based scene flow estimation can be achieved within the proposed framework. In future work, we will focus on real-time performance aspects, since MSF is well parallelizable and can be implemented on a GPU. Moreover, we will evaluate the proposed model regarding the suitability for visual odometry and occlusion detection problems.

11 Summary, Conclusions and Outlook

IN this thesis, we proposed several new methods for monocular non-rigid 3D reconstruction and point set registration. This final chapter concludes the work and outlines future directions in these areas.

Chapters 1, 2 and 3 introduced both addressed research fields and covered the core preliminaries and related works. In Chapter 4, we proposed two scalable methods for dense NRSfM — *Accelerated Metric Projections* (AMP) and *Scalable Monocular Surface Reconstruction* (SMSR). Through the efficient minimisation of a quadratic function on a set of orthonormal matrices, we obtained a fast method which can factorise batches of dense measurements in seconds on a single CPU. AMP can be used in combination with the fast optical flow for interactive monocular non-rigid 3D reconstruction. We have reported the method's accuracy and runtime in different scenarios. The primary attribute of SMSR is the steady accuracy across different scenarios and datasets exhibiting various motions and deformations.

In Chapter 5, we introduced a method with enhanced robustness in situations with large and long occlusions in input images, or, equivalently, significant inaccuracies in dense point tracks. The energy functional of *Shape Prior Variational Approach* (SPVA) includes a geometrical prior term weighted by an occlusion tensor for the stabilisation of reconstructions in the regions where the measurements are highly uncertain. SPVA can operate in a pipeline for estimation of the shape prior from several unoccluded frames of the sequence. In turn, the occlusion tensor helps to determine whether the frames are occluded or not. The *Dynamic Shape Prior Reconstruction* (DSPR) approach generalised the notion of the static shape prior to a dynamic shape prior. Instead of selecting a single initially reconstructed state as a prior, DSPR takes advantage of multiple representative states of a sequence and adaptively uses them, *i.e.,* dynamically selects the most suitable prior state related to the current state by a rigid transformation.

Lifted Coherent Depth Fields (L-CDF) is a new dense NRSfM technique introduced in Chapter 6 with a new spatial regulariser based on the motion coherence theory. Thanks to the coherency term, moderate occlusions can now also be handled without a shape prior term and an occlusion tensor. L-CDF includes a parameter

V. Golyanik, *Robust Methods for Dense Monocular Non-Rigid 3D Reconstruction and Alignment of Point Clouds*,

expressing the prior assumption on the degree of inaccuracies in the correspondences. Besides, the high dimensional space model allows L-CDF to leverage the factorisation-based formulation of NRSfM for shape compression.

In Chapter 7, we proposed a deep neural network based framework for monocular surface recovery from a single image. The *Hybrid Deformation Model Network* (HDM-Net) was trained on a new synthetic dataset encompassing multiple states of a thin surface with a known shape at rest, and the deformation model was automatically learned from the data. HDM-Net regresses a point cloud of the observed surface from an input image, and processes up to 200 images per second. Improving HDM-Net is a promising research direction for augmented reality.

Chapter 8 addressed *embedding of prior correspondences* into a probabilistic rigid and non-rigid point set registration and demonstrated applications of the proposed techniques for generation of animatable human avatars. If a sparse set of correspondences is known in advance, *Extended Coherent Point Drift* (ECPD) can be used to guide non-rigid registration and enhance registration accuracy in the region of interest. In the rigid case, a joint transformation estimation and point set alignment resulted in a broader convergence basin and a higher registration accuracy, especially with a low number of prior matches.

A new *physics-based gravitational class of methods* for point set registration was introduced in Chapter 9. *Gravitational approaches* were shown in this thesis to achieve the superior accuracy in scenarios with highly noisy and missing data as well as clustered outliers. In the rigid case, acceleration techniques enabled globally multiply-linked registration of large point sets. In the non-rigid case, we successfully solved the problem of appearance transfer from a highly noisy 3D scan, whereas other point set registration methods were not able to differentiate between the data and noise.

Finally, in Chapter 10 we showed how NRSfM and point set registration methods can be used to recover *scene flow from monocular and RGB-D image sequences*. We demonstrated that monocular scene flow estimation could benefit from models developed for NRSfM so far, and derived the mathematical relation between these two problems. We interpreted scene flow estimation from RGB-D images as a piecewise rigid point set registration with known point intensities and combined photometric and geometric terms optimised by a non-linear least squares algorithm.

To conclude, the developed NRSfM and monocular surface recovery methods advance the field regarding the accuracy, runtime and robustness to occlusions, thus broadening the spectrum of the available modelling techniques, and, in overall, contributing to a more profound problem understanding. The methods for point set registration became more robust through the integration of additional prior knowledge and the introduction of the new gravitational model.

11.1 Future Directions

The new methods open different directions for further investigation. Future work in dense NRSFM could continue analysing temporally-disjoint rigidity effects. One of the still-unsolved problems is the shape completion under non-rigid deformations when the camera observes only parts of a scene at a time.

Recently, supervised learning techniques started to find their way into monocular non-rigid reconstruction, with several preliminary methods — including HDM-Net — already shown in the literature. The promising research direction is semi-supervised learning for monocular shape recovery and combining neural architectures with model-based elements and prior knowledge.

It is foreseeable that point set registration with correspondence priors — both probabilistic and gravitational — will be further investigated in the next few years, with conceivable extensions for multiple point sets, point clusters and image registration, among others. Those will bring more insights about the convergence and new possible application scenarios of the methods. Next, principles of particle interaction under gravitational forces could be adapted for other related problems such as finding correspondences on meshes. Moreover, registration of multiple point sets and dynamic mass assignment in gravitational approaches are of high interest and undoubtedly worth to investigate. There is also the first evidence that semi-supervised learning techniques can be applied in piece-wise rigid and non-rigid point set registration. We also believe that the gravitational potential energy functional can be advantageous as a gravitational loss in machine learning approaches and in mapping correspondence problems to quantum hardware.

Bibliography

[1] ARPACK - Arnoldi Package. http://www.caam.rice.edu/software/ARPACK/. Accessed: 23.03.2015. 198

[2] *OpenCV Web-Page.* http://opencv.org/, 2016. [online; accessed on 11.05.2016]. 284

[3] AANAES, H., AND KAHL, F. Estimation of deformable structure and motion. In *ECCV Workshop on Vision and Modelling of Dynamic Scenes* (2002). 47

[4] AARSETH, S. J. Dynamical evolution of clusters of galaxies, I. *Monthly Notices of the Royal Astronomical Society (MNRAS) 126* (1963), 223. 229

[5] AARSETH, S. J. *Gravitational N-body Simulations Tools and Algorithms.* Cambridge University Press, 2003. 38

[6] AARSETH, S. J., TOUT, C. A., AND MARDLING, R. A., Eds. *The Cambridge N-body lectures.* Springer, 2008. 38, 39, 234, and 235

[7] ACHANTA, R., SHAJI, A., SMITH, K., LUCCHI, A., FUA, P., AND SUSSTRUNK, S. Slic superpixels compared to state-of-the-art superpixel methods. *Transactions on Pattern Analysis and Machine Intelligence (T-PAMI) 34*, 11 (2012), 2274–2282. 291, 307, 308, and 312

[8] AGRAWAL, M., AND DAVIS, L. S. Camera calibration using spheres: a semi-definite programming approach. In *International Conference on Computer Vision (ICCV)* (2003), pp. 782–789 vol.2. 60 and 69

[9] AGUDO, A., AGAPITO, L., CALVO, B., AND MONTIEL, J. M. M. Good vibrations: A modal analysis approach for sequential non-rigid structure from motion. In *Computer Vision and Pattern Recognition (CVPR)* (2014), pp. 1558–1565. 47, 105, and 278

[10] AGUDO, A., CALVO, B., AND MONTIEL, J. Finite element based sequential bayesian non-rigid structure from motion. In *Computer Vision and Pattern Recognition (CVPR)* (2012), pp. 1418–1425. 47

V. Golyanik, *Robust Methods for Dense Monocular Non-Rigid 3D Reconstruction and Alignment of Point Clouds*,

[11] AGUDO, A., MONTIEL, J. M. M., AGAPITO, L., AND CALVO, B. Online dense non-rigid 3d shape and camera motion recovery. In *British Machine Vision Conference (BMVC)* (2014). 47, 49, 60, 61, and 137

[12] AGUDO, A., MONTIEL, J. M. M., CALVO, B., AND MORENO-NOGUER., F. Mode-shape interpretation: Re-thinking modal space for recovering deformable shapes. In *Winter Conference on Applications of Computer Vision (WACV)* (2016). 47, 91, 105, and 150

[13] AGUDO, A., AND MORENO-NOGUER, F. Learning shape, motion and elastic models in force space. In *International Conference on Computer Vision (ICCV)* (2015). 47, 70, 91, 114, 148, and 154

[14] AGUDO, A., AND MORENO-NOGUER, F. Simultaneous pose and non-rigid shape with particle dynamics. In *Conference on Computer Vision and Pattern Recognition (CVPR)* (2015). 47 and 91

[15] AGUDO, A., AND MORENO-NOGUER, F. Global model with local interpretation for dynamic shape reconstruction. In *Winter Conference on Applications of Computer Vision (WACV)* (2017), pp. 264–272. 150

[16] AGUDO, A., AND MORENO-NOGUER, F. Force-based representation for non-rigid shape and elastic model estimation. *Transactions on Pattern Analysis and Machine Intelligence (TPAMI)* (2018). 148 and 154

[17] AGUDO, A., MORENO-NOGUER, F., CALVO, B., AND MONTIEL, J. M. M. Sequential non-rigid structure from motion using physical priors. *Transactions on Pattern Analysis and Machine Intelligence (TPAMI)* (2016). 47 and 146

[18] AHMAD, A., AND COHEN, L. A numerical integration scheme for the n-body gravitational problem. *Journal of Computational Physics 12* (1973), 389–402. 39 and 234

[19] AIDIBE, A., AND TAHAN, A. Adapting the coherent point drift algorithm to the fixtureless dimensional inspection of compliant parts. *The International Journal of Advanced Manufacturing Technology 79*, 5 (2015), 831–841. 137

[20] AKHTER, I., SHEIKH, Y., AND KHAN, S. In defense of orthonormality constraints for nonrigid structure from motion. In *Computer Vision and Pattern Recognition (CVPR)* (2009), pp. 1534–1541. 46 and 47

[21] AKHTER, I., SHEIKH, Y., KHAN, S., AND KANADE, T. Nonrigid structure from motion in trajectory space. *Neural Information Processing Systems (NIPS)* (2008). *47, 79, 81, 82, 83, 84, 85, 87, 145, 146, and 160*

[22] AKHTER, I., SHEIKH, Y., KHAN, S., AND KANADE, T. Trajectory space: A dual representation for nonrigid structure from motion. *Transactions on Pattern Analysis and Machine Intelligence (TPAMI) 33*, 7 (2011), 1442–1456. *47, 82, 88, 115, 120, 121, 148, 154, 160, and 278*

[23] ALLEN, B., CURLESS, B., AND POPOVIĆ, Z. Articulated dody deformation from range scan data. *ACM SIGGRAPH 2002 21*, 3 (2002), 612–619. *193 and 204*

[24] AMBERG, B., ROMDHANI, S., AND VETTER, T. Optimal step nonrigid icp algorithms for surface registration. In *Computer Vision and Pattern Recognition (CVPR)* (June 2007), pp. 1–8. *55, 266, 268, 269, and 271*

[25] ANDERSON, E., BAI, Z., BISCHOF, C., BLACKFORD, L. S., DEMMEL, J., DONGARRA, J. J., DU CROZ, J., HAMMARLING, S., GREENBAUM, A., MCKENNEY, A., AND SORENSEN, D. *LAPACK Users' Guide (Third Ed.)*. Society for Industrial and Applied Mathematics, 1999. *68 and 69*

[26] ANGUELOV, D., SRINIVASAN, P., CHEUNG PANG, H., KOLLER, D., THRUN, S., AND DAVIS, J. The correlated correspondence algorithm for unsupervised registration of nonrigid surfaces. In *Advances in Neural Information Processing Systems (NIPS)* (2005), pp. 33–40. *52*

[27] ANSARI, M., GOLYANIK, V., AND STRICKER, D. Scalable dense monocular surface reconstruction. In *International Conference on 3D Vision (3DV)* (2017). *114, 117, 119, 120, 121, and 122*

[28] ASTHANA, A., ZAFEIRIOU, S., CHENG, S., AND PANTIC, M. Incremental face alignment in the wild. In *CVPR* (2014), pp. 1859–1866. *206 and 214*

[29] BAKER, S., SCHARSTEIN, D., LEWIS, J. P., ROTH, S., BLACK, M. J., AND SZELISKI, R. A database and evaluation methodology for optical flow. *International Journal of Computer Vision (IJCV) 92*, 1 (2011). *73, 130, 203, 280, 285, 286, 291, 305, and 306*

[30] BALLESTER, C., GARRIDO, L., LAZCANO, V., AND CASELLES, V. A tv-l1 optical flow method with occlusion detection. In *Pattern Recognition* (Berlin, Heidelberg, 2012), pp. 31–40. *34*

[31] BARNES, J., AND HUT, P. A hierarchical O(N log N) force-calculation algorithm. *Nature 324* (1986), 446–449. *39, 234, 245, and 248*

[32] BARNUM, H., SAKS, M., AND SZEGEDY, M. Quantum query complexity and semi-definite programming. In *Conference on Computational Complexity (CCC)* (2003), pp. 179–193. *59*

[33] BARTOLI, A., GAY-BELLILE, V., CASTELLANI, U., PEYRAS, J., OLSEN, S., AND SAYD, P. Coarse-to-fine low-rank structure-from-motion. In *Computer Vision and Pattern Recognition (CVPR)* (2008). *47 and 151*

[34] BARTOLI, A., GÉRARD, Y., CHADEBECQ, F., AND COLLINS, T. On template-based reconstruction from a single view: Analytical solutions and proofs of well-posedness for developable, isometric and conformal surfaces. In *Computer Vision and Pattern Recognition (CVPR)* (2012), pp. 2026–2033. *106*

[35] BECK, A., AND TEBOULLE, M. A fast iterative shrinkage-thresholding algorithm for linear inverse problems. *SIAM Journal on Imaging Sciences 2*, 1 (2009), 183–202. *75 and 77*

[36] BEELER, T., HAHN, F., BRADLEY, D., BICKEL, B., BEARDSLEY, P., GOTSMAN, C., SUMNER, R. W., AND GROSS, M. High-quality passive facial performance capture using anchor frames. *ACM Transactions on Graphics (TOG) 30*, 4 (2011), 75. *80, 84, 85, 87, 137, 142, 143, and 145*

[37] BELONGIE, S., MALIK, J., AND PUZICHA, J. Shape matching and object recognition using shape contexts. *Transactions on Pattern Analysis and Machine Intelligence (TPAMI) 24*, 4 (2002), 509–522. *194*

[38] BENITEZ-QUIROZ, C. F., GÖKGÖZ, K., WILBUR, R. B., AND MARTINEZ, A. M. Discriminant features and temporal structure of nonmanuals in american sign language. *PLoS ONE 9* (2014), 1–17. *107, 108, 111, and 112*

[39] BERGSTRÖM, P., AND EDLUND, O. Robust registration of point sets using iteratively reweighted least squares. *Computational Optimization and Applications 58*, 3 (2014), 543–561. *53*

[40] BESL, P. J., AND MCKAY, N. D. A method for registration of 3-d shapes. *Transactions on Pattern Analysis and Machine Intelligence (TPAMI) 14*, 2 (1992), 239–256. *13, 36, 52, 54, 183, 212, 227, 228, 252, 253, and 255*

[41] BIRKBECK, N., COBZAŞ, D., AND JÄGERSAND, M. Basis constrained 3d scene flow on a dynamic proxy. In *International Conference on Computer Vision (ICCV)* (2011), pp. 1967–1974. *277, 278, and 285*

[42] BIRKBECK, N., COBZAŞ, D., AND JÄGERSAND, M. Depth and scene flow from a single moving camera. In *3DPVT* (2010). *277, 278, 285, and 286*

[43] BLANZ, V., AND VETTER, T. A morphable model for the synthesis of 3d faces. In *ACM Trans. Graphics (TOG)* (1999), pp. 187–194. *114 and 166*

[44] BLENDER FOUNDATION. blender, v. 2.79a. open source 3d creation. `https://www.blender.org/`, 2018. *173*

[45] BOGO, F., ROMERO, J., LOPER, M., AND BLACK, M. J. Faust: Dataset and evaluation for 3d mesh registration. In *Computer Vision and Pattern Recognition (CVPR)* (2014), pp. 3794–3801. *55, 218, 219, and 220*

[46] BOOKSTEIN, F. L. Principal warps: Thin-plate splines and the decomposition of deformations. *Transactions on Pattern Analysis and Machine Intelligence (TPAMI) 11*, 6 (June 1989), 567–585. *55 and 194*

[47] BORCHERS, B. CSDP 2.3 user's guide. *Optimization Methods and Software 11*, 1 (1999), 597–611. *69*

[48] BORCHERS, B. CSDP, a C library for semidefinite programming. *Optimization Methods and Software 11*, 1 (1999), 613–623. *59 and 69*

[49] BOYD, S., PARIKH, N., CHU, E., PELEATO, B., AND ECKSTEIN, J. Distributed optimization and statistical learning via the alternating direction method of multipliers. *Foundations and Trends® in Machine Learning 3*, 1 (2011), 1–122. *75*

[50] BRACHMANN, E., KRULL, A., MICHEL, F., GUMHOLD, S., SHOTTON, J., AND ROTHER, C. Learning 6D object pose estimation using 3D object coordinates. In *European Conference on Computer Vision (ECCV)* (2014), Springer, pp. 536–551. *226*

[51] BRAND, M. Morphable 3d models from video. In *Computer Vision and Pattern Recognition (CVPR)* (2001), vol. 2, pp. II–456–II–463. *46*

[52] BRAND, M. A direct method for 3d factorization of nonrigid motion observed in 2d. In *Computer Vision and Pattern Recognition (CVPR)* (2005), vol. 2, pp. 122–128. *46 and 154*

[53] BREGLER, C., HERTZMANN, A., AND BIERMANN, H. Recovering non-rigid 3d shape from image streams. In *Computer Vision and Pattern Recognition (CVPR)* (2000), pp. 690–696. *13, 22, 25, 45, 46, 61, 90, 137, 148, and 278*

[54] BROWN, B., AND RUSINKIEWICZ, S. Global non-rigid alignment of 3-D scans. *ACM Transactions on Graphics (TOG) 26*, 3 (2007). *52*

[55] BRUNTON, A., SALAZAR, A., BOLKART, T., AND WUHRER, S. Review of statistical shape spaces for 3d data with comparative analysis for human faces. *Computer Vision and Image Understanding (CVIU) 128* (2014), 1–17. *261 and 271*

[56] BUE, A. D. A factorization approach to structure from motion with shape priors. In *Computer Vision and Pattern Recognition (CVPR)* (2008). *47 and 91*

[57] BUE, A. D., AANAES, H., JENSEN, S. N., AND SHEIKH, Y. Non-rigid structure from motion challenge 2017. `http://nrsfm2017.compute.dtu.dk/benchmark/`, 2017. *86*

[58] BUE, A. D., SMERALDI, F., AND AGAPITO, L. Non-rigid structure from motion using ranklet-based tracking and non-linear optimization. *Image and Vision Computing 25*, 3 (2007), 297 – 310. Articulated and Non-rigid motion. *47*

[59] BURKARDT, J. PLY repository of the Florida State University. `http://people.sc.fsu.edu/~jburkardt/data/ply/ply.html`. [accessed on 30.10.2015]. *226*

[60] BURTSCHER, M., AND PINGALI, K. An efficient CUDA implementation of the tree-based barnes hut n-body algorithm. *GPU Computing Gems Emerald Edition* (2011), 75–92. *235*

[61] BUTLER, D. J., WULFF, J., STANLEY, G. B., AND BLACK, M. J. A naturalistic open source movie for optical flow evaluation. In *European Conference on Computer Vision (ECCV)* (2012), pp. 611–625. *69, 70, 147, 161, 163, 202, 203, 256, 257, 258, 275, 287, 291, and 305*

[62] CALVETTI, D., REICHEL, L., AND SORENSEN, D. An implicitly restarted lanczos method for large symmetric eigenvalue problems. *ETNA*, 2 (1994), 1–21. *198*

[63] CAMPBELL, D., AND PETERSSON, L. An adaptive data representation for robust point-set registration and merging. In *International Conference on Computer Vision (ICCV)* (2015). *54*

[64] Chambolle, A., and Pock, T. A first-order primal-dual algorithm for convex problems with applications to imaging. *Journal of Mathematical Imaging and Vision 40*, 1 (2011), 120–145. *34, 35, and 49*

[65] Chen, Q., and Koltun, V. Robust nonrigid registration by convex optimization. In *International Conference on Computer Vision (ICCV)* (2015). *211*

[66] Chen, Y., and Medioni, G. Object modelling by registration of multiple range images. *Image and Vision Computing 10*, 3 (1992), 145 – 155. *37 and 52*

[67] Chhatkuli, A., Pizarro, D., Collins, T., and Bartoli, A. Inextensible non-rigid structure-from-motion by second-order cone programming. *Transactions on Pattern Analysis and Machine Intelligence (TPAMI) PP*, 99 (2018). *47, 60, and 167*

[68] Choy, C. B., Xu, D., Gwak, J., Chen, K., and Savarese, S. 3D-R2N2: A unified approach for single and multi-view 3d object reconstruction. In *European Conference on Computer Vision (ECCV)* (2016). *167*

[69] Chui, H., and Rangarajan, A. A feature registration framework using mixture models. In *Workshop on Mathematical Methods in Biomedical Image Analysis (MMBIA)* (2000), pp. 190–197. *53 and 55*

[70] Chui, H., and Rangarajan, A. A new point matching algorithm for non-rigid registration. *Computer Vision and Image Understanding 89*, 2-3 (2003), 114–141. *54, 55, 137, 199, 201, 266, and 269*

[71] Combettes, P. L., and Pesquet, J.-C. Proximal splitting methods in signal processing. In *Fixed-Point Algorithms for Inverse Problems in Science and Engineering*, vol. 49 of *Springer Optimization and Its Applications*. Springer New York, 2011, pp. 185–212. *49*

[72] Cook, R. L., and Torrance, K. E. A reflectance model for computer graphics. *ACM Trans. Graph. (TOG) 1*, 1 (1982), 7–24. *173*

[73] Costeira, J. P., and Kanade, T. A multibody factorization method for independently moving objects. *International Journal of Computer Vision (IJCV) 29*, 3 (1998), 159–179. *48*

[74] Cruz, F. A., Layton, S. K., and Barba, L. A. How to obtain efficient gpu kernels: an illustration using FMM & FGT. *Comput. Phys. Commun.*, 10 (2011), 2084–2098. *198*

[75] DAI, Y., DENG, H., AND HE, M. Dense non-rigid structure-from-motion made easy – a spatial-temporal smoothness based solution. In *International Conference on Image Processing (ICIP)* (2017), pp. 4532–4536. *114, 120, and 121*

[76] DAI, Y., LI, H., AND HE, M. A simple prior-free method for non-rigid structure-from-motion factorization. *International Journal of Computer Vision 107*, 2 (2014), 101–122. *47, 60, 71, 75, 78, 79, 82, 84, 85, 88, 91, and 92*

[77] DEMMEL, J., AND KAHAN, W. Accurate singular values of bidiagonal matrices. *SIAM J. Sci. Stat. Comput. 11*, 5 (1990), 873–912. *22*

[78] DENG, Y., RANGARAJAN, A., EISENSCHENK, S., VEMURI, AND BABA, C. A riemannian framework for matching point clouds represented by the schrödinger distance transform. In *Computer Vision and Pattern Recognition (CVPR)* (2014). *54 and 55*

[79] DEY, T. K., FU, B., WANG, H., AND WANG, L. Automatic posing of a meshed human model using point clouds. *Computers & Graphics 46* (2015), 14 – 24. *212*

[80] DIACU, F. The solution of then-body problem. *The Mathematical Intelligencer 18*, 3 (1996), 66–70. *37*

[81] DIACU, F. The classical n-body problem in the context of curved space. *Canadian Journal of Mathematics 69* (2017), 790–806. *38*

[82] DINNING, P. *Barn Owl at Screech Owl Sanctuary*. `https://www.youtube.com/watch?v=xmou8t-DHh0`, 2014. [online; accessed 12.05.2016; usage rights obtained]. *69, 70, 80, 83, 84, 85, 87, 147, 286, and 287*

[83] DODIG, M., STOŠIĆ, M., AND XAVIER, J. On minimizing a quadratic function on stiefel manifold. *Linear Algebra and Its Applications 475* (2015), 251–264. *62*

[84] ECKART, B., KIM, K., TROCCOLI, A., KELLY, A., AND KAUTZ, J. Mlmd: Maximum likelihood mixture decoupling for fast and accurate point cloud registration. In *International Conference on 3D Vision (3DV)* (2015). *53*

[85] EDELMAN, A., ARIAS, T. A., AND SMITH, S. T. The geometry of algorithms with orthogonality constraints. *SIAM Journal on Matrix Analysis and Applications 20*, 2 (1999), 303–353. *62*

[86] EGGERT, D., LORUSSO, A., AND FISHER, R. Estimating 3-d rigid body transformations: a comparison of four major algorithms. *Machine Vision and Applications 9*, 5-6 (1997), 272–290. 183

[87] EINSTEIN, A. Die grundlage der allgemeinen relativitätstheorie. *Annalen der Physik 354*, 7 (1916), 769 – 822. 38

[88] FAN, H., SU, H., AND GUIBAS, L. J. A point set generation network for 3d object reconstruction from a single image. In *Computer Vision and Pattern Recognition (CVPR)* (2017). 167

[89] FAYAD, J., AGAPITO, L., AND DEL BUE, A. Piecewise quadratic reconstruction of non-rigid surfaces from monocular sequences. In *European Conference on Computer Vision (ECCV)* (2010). 47 and 150

[90] FAYAD, J., RUSSELL, C., AND AGAPITO, L. Automated articulated structure and 3d shape recovery from point correspondences. In *International Conference on Computer Vision (ICCV)* (2011), pp. 431–438. 48

[91] FELZENSZWALB, P. F., AND HUTTENLOCHER, D. P. Efficient graph-based image segmentation. *International Journal of Computer Vision (IJCV) 59*, 2 (2004), 167–181. 291, 294, 303, 305, 307, and 312

[92] FISCHER, B., AND MODERSITZKI, J. Curvature based image registration. *Journal of Mathematical Imaging and Vision 18*, 1 (2003), 81–85. 260

[93] FITZGIBBON, A. W. Robust registration of 2D and 3D point sets. In *British Machine Vision Conference (BMVC)* (2001), pp. 662–670. 53, 55, 183, 199, and 201

[94] FORD, W. *Numerical Linear Algebra with Applications: Using MATLAB*. Elsevier Science, 2014. 19, 20, and 21

[95] FOUCART, S., AND RAUHUT, H. *A Mathematical Introduction to Compressive Sensing*. Birkhäuser Basel, 2013. 150

[96] FRAGKIADAKI, K., SALAS, M., ARBELAEZ, P., AND MALIK, J. Grouping-based low-rank trajectory completion and 3d reconstruction. In *Neural Information Processing Systems (NIPS)* (2014). 145 and 147

[97] FREEDEN, W. *Lecture Notes on Inverse Problems in Geosciences*. 2015. 16

[98] FRIGO, M., AND JOHNSON, S. G. The design and implementation of FFTW3. *Proceedings of the IEEE. Special issue on "Program Generation, Optimization, and Platform Adaptation" 93*, 2 (2005), 216–231. 142

[99] FURUKAWA, Y., AND PONCE, J. Accurate, dense, and robust multiview stereopsis. *IEEE Transactions on Pattern Analysis and Machine Intelligence (TPAMI) 32*, 8 (2010), 1362–1376. *189, 190, 206, 218, 238, and 239*

[100] GAËL, G., BENOÎT, J., ET AL. Eigen v3. http://eigen.tuxfamily.org, 2010. *68 and 142*

[101] GAIDON, A., WANG, Q., CABON, Y., AND VIG, E. Virtual worlds as proxy for multi-object tracking analysis. In *Computer Vision and Pattern Recognition (CVPR)* (2016). *275, 304, and 305*

[102] GAL, R., AND COHEN-OR, D. Salient geometric features for partial shape matching and similarity. *ACM Transactions on Graphics 25*, 1 (2006), 130–150. *260*

[103] GARG, R., PIZARRO, L., RUECKERT, D., AND AGAPITO, L. Dense multi-frame optic flow for non-rigid objects using subspace constraints. In *Asian Conference on Computer Vision (ACCV)*. 2011, pp. 460–473. *91, 92, 101, 103, 105, 108, 109, 110, 111, 112, and 113*

[104] GARG, R., ROUSSOS, A., AND AGAPITO, L. Dense variational reconstruction of non-rigid surfaces from monocular video. In *Computer Vision and Pattern Recognition (CVPR)* (2013), pp. 1272–1279. *47, 49, 60, 64, 70, 71, 72, 75, 79, 80, 83, 84, 85, 87, 91, 92, 101, 103, 108, 109, 110, 111, 112, 113, 114, 117, 119, 120, 121, 123, 127, 131, 137, 138, 142, 143, 144, 145, 146, 147, 155, 158, 159, 160, 161, 163, 173, 176, 177, 278, 279, 283, and 284*

[105] GARG, R., ROUSSOS, A., AND AGAPITO, L. A variational approach to video registration with subspace constraints. *International Journal of Computer Vision (IJCV) 104*, 3 (2013), 286–314. *33, 34, 35, 48, 61, 64, 70, 72, 83, 115, 126, 128, 131, 138, 142, 144, 145, 279, 283, and 284*

[106] GARRIDO, P., ZOLLHÖFER, M., CASAS, D., VALGAERTS, L., VARANASI, K., PEREZ, P., AND THEOBALT, C. Reconstruction of personalized 3d face rigs from monocular video. In *ACM Trans. Graph. (TOG)* (2016), vol. 35, pp. 28:1–28:15. *166*

[107] GE, S., AND FAN, G. Non-rigid articulated point set registration with local structure preservation. In *Computer Vision and Pattern Recognition Workshops (CVPRW)* (2015), pp. 126–133. *55 and 212*

[108] GE, S., FAN, G., AND DING, M. Non-rigid point set registration with global-local topology preservation. In *Computer Vision and Pattern Recognition (CVPR) Workshops* (June 2014). *55 and 212*

[109] GEIGER, A., LENZ, P., STILLER, C., AND URTASUN, R. Vision meets robotics: The kitti dataset. *International Journal of Robotics Research (IJRR)* (2013). 260

[110] GEROGIANNIS, D., NIKOU, C., AND LIKAS, A. The mixtures of student's t-distributions as a robust framework for rigid registration. *Image and Vision Computing 27*, 9 (2009), 1285 – 1294. 183

[111] GIANNAROU, S., VISENTINI-SCARZANELLA, M., AND YANG, G. Z. Probabilistic tracking of affine-invariant anisotropic regions. *Transactions on Pattern Analysis and Machine Intelligence (TPAMI) 35*, 1 (2013), 130–143. 173 and 179

[112] GIROSI, F., JONES, M., AND POGGIO, T. Regularization theory and neural networks architectures. *Neural Computation 7* (1995), 219–269. 138 and 209

[113] GOLD, S., RANGARAJAN, A., PING LU, C., AND MJOLSNESS, E. New algorithms for 2D and 3D point matching: Pose estimation and correspondence. *Pattern Recognition 31* (1997), 957–964. 53

[114] GOLUB, G. H., AND VAN LOAN, C. F. *Matrix computations. Third Edition.* The John Hopkins University Press, 2012. 78

[115] GOLYANIK, V., ALI, A. S., AND STRICKER, D. Gravitational approach for point set registration. In *Computer Vision and Pattern Recognition (CVPR)* (2016). 252 and 253

[116] GOLYANIK, V., FETZER, T., AND STRICKER, D. Accurate 3d reconstruction of dynamic scenes from monocular image sequences with severe occlusions. In *Winter Conference on Applications of Computer Vision (WACV)* (2017). 60, 114, 115, 120, 121, 126, 129, 131, 136, 137, 144, 145, and 147

[117] GOLYANIK, V., FETZER, T., AND STRICKER, D. Introduction to coherent depth fields for dense monocular surface recovery. In *British Machine Vision Conference (BMVC)* (2017). 149

[118] GOLYANIK, V., KIM, K., MAIER, R., NIESSNER, M., STRICKER, D., AND KAUTZ, J. Multiframe scene flow with piecewise rigid motion. In *International Conference on 3D Vision (3DV)* (2017). 203 and 204

[119] GOLYANIK, V., MATHUR, A. S., AND STRICKER, D. Nrsfm-flow: Recovering non-rigid scene flow from monocular image sequences. In *British Machine Vision Conference (BMVC)* (2016). 70, 85, 86, and 147

[120] GOLYANIK, V., REIS, G., TAETZ, B., AND STRIEKER, D. A framework for an accurate point cloud based registration of full 3d human body scans. In *International Conference on Machine Vision Applications (MVA)* (2017). *270*

[121] GOLYANIK, V., AND STRICKER, D. Dense batch non-rigid structure from motion in a second. In *Winter Conference on Applications of Computer Vision (WACV)* (2017). *79, 101, 103, 108, 112, 142, 143, 144, 146, 175, 176, 177, and 179*

[122] GOLYANIK, V., TAETZ, B., REIS, G., AND STRICKER, D. Extended coherent point drift algorithm with correspondence priors and optimal subsampling. In *Winter Conference on Applications of Computer Vision (WACV)* (2016). *105, 160, and 193*

[123] GOLYANIK, V., TAETZ, B., AND STRICKER, D. Joint pre-alignment and robust rigid point set registration. In *International Conference on Image Processing (ICIP)* (2016), pp. 4503–4507. *XX, 252, 254, and 255*

[124] GOLYANIK, V., AND THEOBALT, C. Optimising for scale in globally multiply-linked gravitational point set registration leads to singularities. In *International Conference on 3D Vision (3DV)* (2019). *231*

[125] GOTARDO, P. F. U., AND MARTINEZ, A. M. Computing smooth time trajectories for camera and deformable shape in structure from motion with occlusion. *Transactions on Pattern Analysis and Machine Intelligence (TPAMI) 33*, 10 (2011), 2051–2065. *47, 75, 79, 82, 83, 84, 85, 88, and 145*

[126] GOTARDO, P. F. U., AND MARTINEZ, A. M. Kernel non-rigid structure from motion. In *International Conference on Computer Vision (ICCV)* (2011), pp. 802–809. *47, 82, 108, 115, 138, 151, and 278*

[127] GOTARDO, P. F. U., AND MARTINEZ, A. M. Non-rigid structure from motion with complementary rank-3 spaces. In *Computer Vision and Pattern Recognition (CVPR)* (2011), pp. 3065–3072. *79, 82, 83, 84, 85, 88, and 278*

[128] GOWER, J. C. Generalized procrustes analysis. *Psychometrika 40*, 1 (1975), 33–51. *80*

[129] GRANGER, S., AND PENNEC, X. Multi-scale em-icp: A fast and robust approach for surface registration. In *European Conference on Computer Vision (ECCV)* (2002), pp. 418–432. *52 and 53*

[130] GREENGARD, L., AND ROKHLIN, V. A fast algorithm for particle simulations. *Journal of Computational Physics 73*, 2 (1987), 325–348. 39 and 234

[131] GREENGARD, L., AND STRAIN, J. The fast gauss transform. *SIAM Journal on Scientific and Statistical Computing 12*, 1 (1991), 79–94. 234

[132] GREENSPAN, M., AND GODIN, G. A nearest neighbor method for efficient icp. In *International Conference on 3-D Digital Imaging and Modeling (3DIM)* (2001), pp. 161–168. 54

[133] GUAN, P., WEISS, A., BĂLAN, A. O., AND BLACK, M. J. Estimating human shape and pose from a single image. In *International Conference on Computer Vision (ICCV)* (2009), pp. 1381–1388. 166

[134] GUMEROV, N., ZANDIFAR, A., DURAISWAMI, R., AND DAVIS, L. S. Structure of applicable surfaces from single views. In *European Conference on Computer Vision (ECCV)* (2004), pp. 482–496. 167

[135] HADAMARD, J. Sur les problèmes aux dérivés partielles et leur signification physique. *Princeton University Bulletin 13* (1902), 49–52. 15

[136] HAGER, W., AND ZHANG, H. A survey of nonlinear conjugate gradient method. *Pacific journal of Optimization* 2, 1 (01 2006), 35–58. 19

[137] HAMSICI, O. C., GOTARDO, P. F. U., AND MARTINEZ, A. M. Learning spatially-smooth mappings in non-rigid structure from motion. In *European Conference on Computer Vision (ECCV)* (2012). 108

[138] HARTLEY, R., AND VIDAL, R. Perspective nonrigid shape and motion recovery. In *European Conference on Computer Vision (ECCV)* (2008), pp. 276–289. 47

[139] HE, K., ZHANG, X., REN, S., AND SUN, J. Deep residual learning for image recognition. In *Computer Vision and Pattern Recognition (CVPR)* (2016). 168

[140] HELMBERG, C., RENDL, F., VANDERBEI, R. J., AND WOLKOWICZ, H. An interior-point method for semidefinite programming. *SIAM Journal on Optimization 6*, 2 (1996), 342–361. 59 and 69

[141] HINTON, G. E., WILLIAMS, C. K. I., AND REVOW, M. D. Adaptive elastic models for hand-printed character recognition. In *NNeural Information Processing Systems (NIPS)*. 1992, pp. 512–520. 55

[142] HIROSE, O. Dependent landmark drift: robust point set registration based on the gaussian mixture model with a statistical shape model. *CoRR abs/1711.06588* (2017). 194

[143] HIRSHBERG, D. A., LOPER, M., RACHLIN, E., AND BLACK, M. J. Coregistration: Simultaneous alignment and modeling of articulated 3d shape. In *European Conference for Computer Vision (ECCV)*, vol. 7577 of *Lecture Notes in Computer Science*. 2012, pp. 242–255. 55

[144] HORN, B. K. P. Closed-form solution of absolute orientation using unit quaternions. *Journal of the Optical Society of America A 4*, 4 (1987), 629–642. 265 and 267

[145] HORN, B. K. P., AND SCHUNCK, B. G. Determining optical flow. *ARTIFICAL INTELLIGENCE 17* (1981), 185–203. 32, 277, and 281

[146] HORNÁCEK, M., FITZGIBBON, A., AND ROTHER, C. Sphereflow: 6 dof scene flow from rgb-d pairs. In *Computer Vision and Pattern Recognition (CVPR)* (2014), pp. 3526–3533. 292

[147] HUGUET, F., AND DEVERNAY, F. A variational method for scene flow estimation from stereo sequences. In *International Conference on Computer Vision (ICCV)* (2007). 275 and 292

[148] HUMPHREY, G. The psychology of the gestalt. *Journal of Educational Psychology 15*, 7 (1924), 401–412. 137

[149] INNMANN, M., ZOLLHÖFER, M., NIESSNER, M., THEOBALT, C., AND STAMMINGER, M. Volumedeform: Real-time volumetric non-rigid reconstruction. In *European Conference on Computer Vision (ECCV)* (2016). 292 and 294

[150] ITTI, L., AND KOCH, C. A saliency-based search mechanism for overt and covert shifts of visual attention. *Vision Research 40*, 10–12 (2000), 1489 – 1506. 137

[151] J. LAWIN, F., DANELLJAN, M., S. KHAN, F., FORSSÉN, P.-E., AND FELSBERG, M. Density adaptive point set registration. In *Computer Vision and Pattern Recognition (CVPR)* (2018). 251

[152] JADERBERG, M., SIMONYAN, K., ZISSERMAN, A., AND KAVUKCUOGLU, K. Spatial transformer networks. In *Advances in Neural Information Processing Systems (NIPS)* (2015), pp. 2017–2025. 172

[153] JAIMEZ, M., SOUIAI, M., GONZÁLEZ-JIMÉNEZ, J., AND CREMERS, D. A primal-dual framework for real-time dense rgb-d scene flow. In *International Conference on Robotics and Automation (ICRA)* (2015). *292, 293, and 308*

[154] JAIMEZ, M., SOUIAI, M., STÜCKLER, J., GONZÁLEZ-JIMÉNEZ, J., AND CREMERS, D. Motion cooperation: Smooth piece-wise rigid scene flow from rgb-d images. In *International Conference on 3D Vision (3DV)* (2015), pp. 64–72. *291, 292, 293, and 305*

[155] JENSEN, S. H. N., BUE, A. D., DOEST, M. E. B., AND AANÆS, H. A benchmark and evaluation of non-rigid structure from motion. In *arXiv.org* (2018). *88*

[156] JIAN, B., AND VEMURI, B. C. A robust algorithm for point set registration using mixture of gaussians. In *International Conference on Computer Vision (ICCV)* (2005), pp. 1246–1251. *53, 54, 55, 266, 268, 269, and 271*

[157] JIAN, B., AND VEMURI, B. C. Robust point set registration using gaussian mixture models. *Transactions on Pattern Analysis and Machine Intelligence (TPAMI) 33*, 8 (2011), 1633–1645. *53, 54, 137, 252, 253, 255, and 294*

[158] KABSCH, W. A solution for the best rotation to relate two sets of vectors. *Acta Crystallographica Section A 32*, 5 (1976), 922 – 923. *35, 187, 226, 232, 263, 265, and 267*

[159] KERL, C., STURM, J., AND CREMERS, D. Dense visual slam for rgb-d cameras. In *International Conference on Intelligent Robot Systems (IROS)* (2013). *52*

[160] KET NG, S., KRISHNAN, T., AND MCLACHLAN, G. The em algorithm. *Handbook of Computational Statistics: Concepts and Methods* (01 2004). *41*

[161] KEUCHEL, J. Multiclass image labeling with semidefinite programming. In *European Conference on Computer Vision (ECCV)* (2006), pp. 454–467. *60*

[162] KEUCHEL, J., SCHNÖRR, C., SCHELLEWALD, C., AND CREMERS, D. Binary partitioning, perceptual grouping, and restoration with semidefinite programming. *Transactions on Pattern Analysis and Machine Intelligence (TPAMI) 25*, 11 (2003), 1364–1379. *60*

[163] KIM, K., LEE, D., AND ESSA, I. Gaussian process regression flow for analysis of motion trajectories. In *International Conference on Computer Vision (ICCV)* (2011). *275*

[164] KOFFKA, K. *Principles of Gestalt psychology*. Harcourt, 1935. 137

[165] KOLESOV, I., LEE, J., SHARP, G., VELA, P., AND TANNENBAUM, A. A stochastic approach to diffeomorphic point set registration with landmark constraints. *Transactions on Pattern Analysis and Machine Intelligence (TPAMI) 38*, 2 (2016), 238–251. 194

[166] KONG, C., AND LUCEY, S. Prior-less compressible structure from motion. In *Computer Vision and Pattern Recognition (CVPR)* (2016). 114 and 150

[167] KOVNATSKY, A., BRONSTEIN, M. M., BRONSTEIN, A. M., GLASHOFF, K., AND KIMMEL, R. Coupled quasi-harmonic bases. *CoRR abs/1210.0026* (2012). 56

[168] KRIZHEVSKY, A., SUTSKEVER, I., AND HINTON, G. E. Imagenet classification with deep convolutional neural networks. In *Advances in Neural Information Processing Systems (NIPS)*. 2012, pp. 1097–1105. 168

[169] KRUSKAL, J. B. On the Shortest Spanning Subtree of a Graph and the Traveling Salesman Problem. In *Proceedings of the American Mathematical Society, 7* (1956). 215

[170] KUMAR, S., CHERIAN, A., DAI, Y., AND LI, H. Scalable dense non-rigid structure-from-motion: A grassmannian perspective. In *Computer Vision and Pattern Recognition (CVPR)* (2018). 114, 120, 121, and 122

[171] KUMAR, S., DAI, Y., AND LI, H. Spatio-temporal union of subspaces for multi-body non-rigid structure-from-motion. *Pattern Recognition* (2017). 48 and 150

[172] LEE, M., CHO, J., CHOI, C.-H., AND OH, S. Procrustean normal distribution for non-rigid structure from motion. In *Computer Vision and Pattern Recognition (CVPR)* (June 2013), pp. 1280–1287. 47

[173] LEE, M., CHO, J., AND OH, S. Consensus of non-rigid reconstructions. In *Computer Vision and Pattern Recognition (CVPR)* (2016). 47, 75, 79, 82, 84, 85, 88, and 150

[174] LEHOUCQ, R., AND SORENSEN, D. C. Deflation techniques for an implicitly re-started arnoldi iteration. *SIAM J. Matrix Anal. Appl 17* (1996), 789–821. 20

[175] LEVENBERG, K. A method for the solution of certain non-linear problems in least squares. *Quarterly Journal of Applied Mathmatics II*, 2 (1944), 164–168. 18 and 251

[176] LI, X., LI, H., JOO, H., LIU, Y., AND SHEIKH, Y. Structure from recurrent motion: From rigidity to recurrency. In *Computer Vision and Pattern Recognition (CVPR)* (2018). *114 and 115*

[177] LI, Z., LIU, J., AND TANG, X. Constrained clustering via spectral regularization. In *Computer Vision and Pattern Recognition (CVPR)* (2009), pp. 421–428. *60 and 69*

[178] LIN, W., CHEONG, L., TAN, P., DONG, G., AND LIU, S. Simultaneous camera pose and correspondence estimation with motion coherence. *International Journal of Computer Vision 96*, 2 (2012), 145–161. *137*

[179] LIN, Z., CHEN, M., AND MA, Y. The augmented lagrange multiplier method for exact recovery of corrupted low-rank matrices. *Technical Report UILU-ENG-09-2215, University of Illinois at Urbana-Champaign* (2009). *75 and 79*

[180] LIU-YIN, Q., YU, R., AGAPITO, L., FITZGIBBON, A., AND RUSSELL, C. Better together: Joint reasoning for non-rigid 3d reconstruction with specularities and shading. In *British Machine Vision Conference (BMVC)* (2016). *92*

[181] LLADÓ, X., BUE, A. D., AND AGAPITO, L. Non-rigid metric reconstruction from perspective cameras. *Image and Vision Computing 28*, 9 (2010), 1339 – 1353. *47*

[182] LOPER, M., MAHMOOD, N., ROMERO, J., PONS-MOLL, G., AND BLACK, M. J. Smpl: A skinned multi-person linear model. *ACM TOG 34* (2015), 248:1–248:16. *212*

[183] LUCAS, B. D., AND KANADE, T. An iterative image registration technique with an application to stereo vision. In *International Joint Conference on Artificial Intelligence (IJCAI)* (1981), pp. 674–679. *202, 203, and 284*

[184] M. SETHI, A. RANGARAJAN, K. G. The schrödinger distance transform (sdt) for point-sets and curves. In *Computer Vision and Pattern Recognition (CVPR)* (2012). *54*

[185] MA, J., ZHAO, J., JIANG, J., AND ZHOU, H. Non-rigid point set registration with robust transformation estimation under manifold regularization. In *Conference on Artificial Intelligence (AAAI)* (2017), pp. 4218–4224. *54*

[186] MA, J., ZHAO, J., AND YUILLE, A. L. Non-rigid point set registration by preserving global and local structures. *Transactions on Image Processing 25*, 1 (2016), 53–64. *194 and 260*

[187] MALTI, A., HARTLEY, R., BARTOLI, A., AND KIM, J. H. Monocular template-based 3d reconstruction of extensible surfaces with local linear elasticity. In *Computer Vision and Pattern Recognition (CVPR)* (2013), pp. 1522–1529. *45*

[188] MARKLEY, L., CHENG, Y., CRASSIDIS, J., AND OSHMAN, Y. Averaging quaternions. 1193–1196. *30 and 31*

[189] MARQUARDT, D. W. An algorithm for least-squares estimation of nonlinear parameters. *Journal of the Society for Industrial and Applied Mathematics 11*, 2 (1963), 431–441. *18 and 251*

[190] MATHWORKS. File exchange: Iterative closest point. `http://www.mathworks.com/matlabcentral/fileexchange/27804-iterative-closest-point`. [accessed on 30.10.2015]. *235*

[191] MAYER, N., ILG, E., HAUSSER, P., FISCHER, P., CREMERS, D., DOSOVITSKIY, A., AND BROX, T. A large dataset to train convolutional networks for disparity, optical flow, and scene flow estimation. In *Computer Vision and Pattern Recognition (CVPR)* (2016). *69 and 70*

[192] MCLACHLAN, G. J., AND KRISHNAN, T. *The EM algorithm and extensions*, 2. ed ed. 2008. *41*

[193] MENZE, M., AND GEIGER, A. Object scene flow for autonomous vehicles. In *Computer Vision and Pattern Recognition (CVPR)* (2015), pp. 3061–3070. *275*

[194] MIKLOS, B., GIESEN, J., AND PAULY, M. Discrete scale axis representations for 3d geometry. *ACM TOG 29*, 4 (2010), 101:1–101:10. *215*

[195] MITICHE, A., MATHLOUTHI, Y., AND BEN AYED, I. Monocular concurrent recovery of structure and motion scene flow. *Frontiers in ICT 2* (2015), 16. *277 and 285*

[196] MOGHARI, M., AND ABOLMAESUMI, P. Point-based rigid-body registration using an unscented kalman filter. *IEEE Transactions on Medical Imaging 26*, 12 (2007), 1708 – 1728. *54*

[197] MONAGHAN, J. J. Smoothed particle hydrodynamics. *Annual Review of Astronomy and Astrophysics 30*, 1 (1992), 543 – 574. *266*

[198] MORADI, M., AND ABOLMAESUMI, P. Medical image registration based on distinctive image features from scale-invariant (sift) key-points. In *Inter'l Congress and Exhib. on Computer Assisted Radiology and Surgery* (2005), vol. 1281. *56*

[199] MORITA, T., AND KANADE, T. A sequential factorization method for recovering shape and motion from image streams. *Transactions on Pattern Analysis and Machine Intelligence (TPAMI) 19*, 8 (1997), 858–867. *24, 27, and 77*

[200] MOURNING, C., NYKL, S., XU, H., CHELBERG, D. M., AND LIU, J. GPU acceleration of robust point matching. In *ISVC* (2010), pp. 417–426. *55*

[201] MYRONENKO, A. Coherent point drift (cpd) project page. `https://sites.google.com/site/myronenko/research/cpd`. [accessed on 30.10.2015]. *235*

[202] MYRONENKO, A., AND SONG, X. On the closed-form solution of the rotation matrix arising in computer vision problems. *Computing Research Repository (CoRR)* (2009). *31, 94, 186, 187, and 226*

[203] MYRONENKO, A., AND SONG, X. Point-set registration: Coherent point drift. *Transactions on Pattern Analysis and Machine Intelligence (TPAMI)* (2010). *13, 41, 53, 55, 137, 181, 183, 188, 195, 196, 197, 199, 201, 204, 210, 226, 227, 247, 252, 253, 255, 258, 260, 266, 269, and 271*

[204] MYRONENKO, A., SONG, X., AND CARREIRA-PERPIÑÁN, M. A. Non-rigid point set registration: Coherent point drift. In *Neural Information Processing Systems (NIPS)* (2006), pp. 1009–1016. *55, 183, 196, and 209*

[205] NEUMANN, T., VARANASI, K., WENGER, S., WACKER, M., MAGNOR, M., AND THEOBALT, C. Sparse localized deformation components. *ACM Trans. Graph. 32*, 6 (2013), 179:1–179:10. *150*

[206] NEWCOMBE, R. A., FOX, D., AND SEITZ, S. M. Dynamicfusion: Reconstruction and tracking of non-rigid scenes in real-time. In *Computer Vision and Pattern Recognition (CVPR)* (2015). *275 and 294*

[207] NEWCOMBE, R. A., IZADI, S., HILLIGES, O., MOLYNEAUX, D., KIM, D., DAVISON, A. J., KOHLI, P., SHOTTON, J., HODGES, S., AND FITZGIBBON, A. Kinectfusion: Real-time dense surface mapping and tracking. In

International Symposium on Mixed and Augmented Reality (ISMAR) (2011), pp. 127–136. *52 and 221*

[208] NIESSNER, M., ZOLLHÖFER, M., IZADI, S., AND STAMMINGER, M. Real-time 3d reconstruction at scale using voxel hashing. *ACM Transactions on Graphics (TOG) 32*, 6 (2013), 169:1–169:11. *52*

[209] NUECHTER, A., LINGEMANN, K., AND HERTZBERG, J. Cached k-d tree search for icp algorithms. In *Sixth International Conference on 3-D Digital Imaging and Modeling (3DIM 2007)* (2007), pp. 419–426. *53*

[210] NVIDIA CORPORATION. NVIDIA CUDA C programming guide, 2018. Version 9.0. *100 and 175*

[211] NYLAND, L., HARRIS, M., AND PRINS, J. Fast n-body simulation with CUDA. In *GPU Gems 3* (2007), pp. 677–795. *235*

[212] OPENMP ARCHITECTURE REVIEW BOARD. OpenMP application program interface version 3.0, May 2008. *198*

[213] P. YANG, X. Q. Direct computing of surface curvatures for point-set surfaces. In *Eurographics Symposium on Point-Based Graphics (SPBG)* (2007). *263*

[214] PALADINI, M., BARTOLI, A., AND AGAPITO, L. Sequential non-rigid structure-from-motion with the 3d-implicit low-rank shape model. In *European Conference on Computer Vision (ECCV)* (2010), pp. 15–28. *47 and 278*

[215] PALADINI, M., DEL BUE, A., STOSIC, M., DODIG, M., XAVIER, J., AND AGAPITO, L. Factorization for non-rigid and articulated structure using metric projections. In *Computer Vision and Pattern Recognition (CVPR)* (2009). *79, 82, 83, 84, 85, and 88*

[216] PALADINI, M., DEL BUE, A., XAVIER, J., AGAPITO, L., STOSIĆ, M., AND DODIG, M. Optimal metric projections for deformable and articulated structure-from-motion. *International Journal of Computer Vision (IJCV) 96*, 2 (2012), 252–276. *45, 46, 47, 48, 57, 58, 59, 60, 61, 62, 63, 70, 91, 116, 120, 121, 138, 144, 145, 146, 160, and 278*

[217] PALÁGYI, K., AND KUBA, A. A 3d 6-subiteration thinning algorithm for extracting medial lines. *Pattern Recognition Letters 19*, 7 (1998), 613 – 627. *215*

[218] PAPAZOV, C., AND BURSCHKA, D. Deformable 3D shape registration based on local similarity transforms. *Computer Graphics Forum 30*, 5 (2011), 1493–1502. *226*

[219] PARRILO, P. A., AND LALL, S. Semidefinite programming relaxations and algebraic optimization in control. *European Journal of Control 9* (2003), 2–3. *59*

[220] PASZKE, A., GROSS, S., CHINTALA, S., CHANAN, G., YANG, E., DEVITO, Z., LIN, Z., DESMAISON, A., ANTIGA, L., AND LERER, A. Automatic differentiation in pytorch. In *Advances in Neural Information Processing Systems Workshops (NIPS-W)* (2017). *173*

[221] PASZKE, A., GROSS, S., MASSA, F., AND CHINTALA, S. pytorch. `https://github.com/pytorch`, 2018. *173*

[222] PELLEGRINI, S., SCHINDLER, K., AND NARDI, D. A generalisation of the icp algorithm for articulated bodies. In *BMVC* (2008). *212*

[223] PERRIOLLAT, M., HARTLEY, R., AND BARTOLI, A. Monocular template-based reconstruction of inextensible surfaces. *International Journal of Computer Vision (IJCV) 95*, 2 (2011), 124–137. *106, 114, and 167*

[224] PUMAROLA, A., AGUDO, A., PORZI, L., SANFELIU, A., LEPETIT, V., AND MORENO-NOGUER, F. Geometry-aware network for non-rigid shape prediction from a single view. In *Computer Vision and Pattern Recognition (CVPR)* (2018). *114*

[225] QUIROGA, J., BROX, T., DEVERNAY, F., AND CROWLEY, J. L. Dense semi-rigid scene flow estimation from RGBD images. In *European Conference on Computer Vision (ECCV)* (2014), pp. 567–582. *291, 292, 293, 302, 303, 305, 306, 307, and 308*

[226] RABAUD, V., AND BELONGIE, S. Re-thinking non-rigid structure from motion. In *Computer Vision and Pattern Recognition (CVPR)* (2008). *47, 150, 151, and 278*

[227] RABAUD, V., AND BELONGIE, S. Linear embeddings in non-rigid structure from motion. In *Computer Vision and Pattern Recognition (CVPR)* (2009), pp. 2427–2434. *47 and 151*

[228] REHAN, A., ZAHEER, A., AKHTER, I., SAEED, A., HARIS USMANI, M., MAHMOOD, B., AND KHAN, S. Nrsfm using local rigidity. In *Applications*

of Computer Vision (WACV), 2014 IEEE Winter Conference on (March 2014), pp. 69–74. *47, 91, and 114*

[229] Revaud, J., Weinzaepfel, P., Harchaoui, Z., and Schmid, C. Epic-Flow: Edge-Preserving Interpolation of Correspondences for Optical Flow. In *Computer Vision and Pattern Recognition* (2015). *45*

[230] Revow, M., Williams, C. K. I., and Hinton, G. E. Using generative models for handwritten digit recognition. *Transactions on Pattern Analysis and Machine Intelligence (TPAMI) 18* (1996), 592–606. *55*

[231] Ricco, S., and Tomasi, C. Dense lagrangian motion estimation with occlusions. *Computer Vision and Pattern Recognition (CVPR)* (2012), 1800–1807. *91 and 98*

[232] Ricco, S., and Tomasi, C. Video motion for every visible point. In *International Conference on Computer Vision (ICCV)* (2013). *91*

[233] Rocchini, C., Cignoni, P., Montani, C., Pingi, P., and Scopigno, R. A low cost 3 d scanner based on structured light. In *Eurographics* (2001). *259*

[234] Rother, C., Kolmogorov, V., and Blake, A. Grabcut -interactive foreground extraction using iterated graph cuts. *ACM Transactions on Graphics (SIGGRAPH)* (2004). *101 and 283*

[235] Roussos, A., Garg, R., and Agapito., L. *Multi-Frame Subspace Flow (MFSF)*. `http://www0.cs.ucl.ac.uk/staff/lagapito/subspace_flow/`, 2015. [online; accessed on 12.02.2016]. *105 and 284*

[236] Rusinkiewicz, S., and Levoy, M. Efficient variants of the ICP algorithm. In *International Conference on 3-D Imaging and Modeling (3DIM)* (2001), pp. 145–152. *54, 56, and 183*

[237] Russell, C., Fayad, J., and Agapito, L. Energy based multiple model fitting for non-rigid structure from motion. In *Computer Vision and Pattern Recognition (CVPR)* (2011), pp. 3009–3016. *127, 131, 138, 146, 159, and 161*

[238] Russell, C., Fayad, J., and Agapito, L. Dense non-rigid structure from motion. In *International Conference on 3D Imaging, Modeling, Processing, Visualization and Transmission (3DIMPVT)* (2012), pp. 509–516. *47, 49, 60, 61, 91, 137, and 278*

[239] RUSSELL, C., YU, R., AND AGAPITO, L. Video pop-up: Monocular 3d reconstruction of dynamic scenes. In *European Conference on Computer Vision (ECCV)* (2014), pp. 583–598. *47, 49, and 289*

[240] RUSU, R. B., BLODOW, N., AND BEETZ, M. Fast point feature histograms (fpfh) for 3d registration. In *International Conference on Robotics and Automation (ICRA)* (2009), pp. 3212–3217. *52 and 194*

[241] RUSU, R. B., BLODOW, N., MARTON, Z. C., AND BEETZ, M. Aligning Point Cloud Views using Persistent Feature Histograms. In *IROS* (2008). *183, 189, 191, and 204*

[242] S. AGARWAL, K. MIERLE AND OTHERS. Ceres solver. `http://ceres-solver.org`. *252 and 304*

[243] SALAS-MORENO, R. F., NEWCOMBE, R. A., STRASDAT, H., KELLY, P. H. J., AND DAVISON, A. J. Slam++: Simultaneous localisation and mapping at the level of objects. In *Computer Vision and Pattern Recognition (CVPR)* (2013), pp. 1352–1359. *52*

[244] SALOMON, D. *Data Compression: The Complete Reference*. Springer-Verlag New York, Inc., 2006. *150*

[245] SALZMANN, M., AND FUA, P. Reconstructing sharply folding surfaces: A convex formulation. *Conference on Computer Vision and Pattern Recognition (CVPR)* (2009), 1054–1061. *106*

[246] SALZMANN, M., URTASUN, R., AND FUA, P. Local deformation models for monocular 3d shape recovery. In *Computer Vision and Pattern Recognition (CVPR)* (2008). *45*

[247] SANCHEZ GIRALDO, L. G., HASANBELLIU, E., RAO, M., AND PRINCIPE, J. C. Group-wise point-set registration based on Renyi's second order entropy. In *Computer Vision and Pattern Recognition (CVPR)* (2017). *55*

[248] SANDHU, R., DAMBREVILLE, S., AND TANNENBAUM, A. Particle filtering for registration of 2D and 3D point sets with stochastic dynamics. In *Computer Vision and Pattern Recognition (CVPR)* (2008). *54*

[249] SANROMA, G., ALQUÉZAR, R., SERRATOSA, F., AND HERRERA, B. Smooth point-set registration using neighboring constraints. *Pattern Recognition Letters 33*, 15 (2012), 2029 – 2037. *137*

[250] SAVAL-CALVO, M., LÓPEZ, J. A., GUILLÓ, A. F., VILLENA-MARTINEZ, V., AND FISHER, R. B. 3d non-rigid registration using color: Color coherent point drift. *Computer Vision and Image Understanding 169* (2018), 119–135. *52 and 193*

[251] SCOTT, G. L., AND LONGUET-HIGGINS, H. C. An algorithm for associating the features of two images. *Proceedings of the Royal Society of London B: Biological Sciences 244*, 1309 (1991), 21–26. *53*

[252] SELA, M., RICHARDSON, E., AND KIMMEL, R. Unrestricted facial geometry reconstruction using image-to-image translation. In *International Conference on Computer Vision (ICCV)* (2017). *166*

[253] SEVILLA-LARA, L., SUN, D., JAMPANI, V., AND BLACK, M. J. Optical flow with semantic segmentation and localized layers. In *Computer Vision and Pattern Recognition (CVPR)* (2016). *294*

[254] SIPIRAN, I., AND BUSTOS, B. Harris 3d: A robust extension of the harris operator for interest point detection on 3d meshes. *The Visual Computer 27*, 11 (2011), 963–976. *183 and 189*

[255] SLOAN DIGITAL SKY SURVEY. `http://sdssorgdev.pha.jhu.edu/dr1/en/proj/kids/constellation/images/orionstars_tiny.jpg`. [accessed on 25.10.2015]. *240*

[256] SONG, X., MYRONENKO, A., AND SAHN, D. J. Speckle tracking in 3d echocardiography with motion coherence. In *Computer Vision and Pattern Recognition (CVPR)* (2007). *137*

[257] SORKINE, O., AND ALEXA, M. As-rigid-as-possible surface modeling. In *Eurographics Symposium on Geometry Processing (SGP)* (2007), pp. 109–116. *115*

[258] STAY & PLAY ROTORUA LTD. *A hot balloon*. `http://stayandplaynz.com/rotorua/the-real-new-zealand-experience/`. [Online; accessed June 29, 2018]. *179*

[259] STOYANOV, D. Stereoscopic scene flow for robotic assisted minimally invasive surgery. In *Medical Image Computing and Computer-Assisted Intervention (MICCAI)* (2012), pp. 479–486. *34, 70, 73, 80, 83, 87, 107, 108, 109, 112, 127, 132, 146, 161, and 288*

[260] STRANG, G. *Linear algebra and its applications*. Thomson, Brooks/Cole, 2006. *19 and 21*

[261] STUECKLER, J., AND BEHNKE, S. Efficient dense rigid-body motion segmentation and estimation in rgb-d video. *International Journal of Computer Vision (IJCV)* (2015). *275, 304, 306, 307, and 308*

[262] STURM, J. F. Using sedumi 1.02, a matlab toolbox for optimization over symmetric cones. *Optimization Methods and Software 11*, 1-4 (1999), 625–653. *59*

[263] SUN, D., ROTH, S., AND BLACK, M. J. Secrets of optical flow estimation and their principles. In *Computer Vision and Pattern Recognition (CVPR)* (2010), pp. 2432–2439. *126, 127, 132, 203, and 204*

[264] SUN, D., SUDDERTH, E. B., AND PFISTER, H. Layered rgbd scene flow estimation. In *Computer Vision and Pattern Recognition (CVPR)* (2015). *292 and 293*

[265] SUN, J., OVSJANIKOV, M., AND GUIBAS, L. A concise and provably informative multi-scale signature based on heat diffusion. In *Symposium on Geometry Processing (SGP)* (2009), pp. 1383–1392. *260*

[266] SUWAJANAKORN, S., KEMELMACHER-SHLIZERMAN, I., AND SEITZ, S. M. Total moving face reconstruction. In *European Conference on Computer Vision (ECCV)* (2014). *166*

[267] TAETZ, B., BLESER, G., GOLYANIK, V., AND STRICKER, D. Occlusion-aware video registration for highly non-rigid objects. In *Winter Conference on Applications of Computer Vision (WACV)* (2016). *33, 34, 48, 72, 83, 91, 92, 98, 101, 103, 105, 108, 109, 112, 113, 132, 133, 136, 137, 138, 144, 145, 147, 161, 175, 279, 283, 284, 294, and 308*

[268] TAETZ, B., BLESER, G., AND MIEZAL, M. Towards self-calibrating inertial body motion capture. In *FUSION* (2016), pp. 1751–1759. *223*

[269] TAM, G. K. L., CHENG, Z.-Q., LAI, Y.-K., LANGBEIN, F. C., LIU, Y., MARSHALL, D., MARTIN, R. R., SUN, X., AND ROSIN, P. L. Registration of 3d point clouds and meshes: A survey from rigid to nonrigid. *IEEE Trans. Vis. Comput. Graph. 19*, 7 (2013), 1199–1217. *52 and 260*

[270] TAO, L., AND MATUSZEWSKI, B. J. Non-rigid structure from motion with diffusion maps prior. In *Computer Vision and Pattern Recognition (CVPR)* (2013), pp. 1530–1537. *47, 91, and 151*

[271] TAO, L., MATUSZEWSKI, B. J., AND MEIN, S. J. Non-rigid structure from motion with incremental shape prior. In *International Conference on Image Processing (ICIP)* (2012), pp. 1753–1756. *47*

[272] TAO, L., MEIN, S. J., QUAN, W., AND MATUSZEWSKI, B. J. Recursive non-rigid structure from motion with online learned shape prior. *Computer Vision and Image Understanding 117*, 10 (2013), 1287 – 1298. *47*

[273] TAYLOR, J., JEPSON, A. D., AND KUTULAKOS, K. N. Non-rigid structure from locally-rigid motion. In *Computer Vision and Pattern Recognition (CVPR)* (2010), pp. 2761–2768. *47 and 150*

[274] TEVS, A., BOKELOH, M., WAND, M., SCHILLING, A., AND SEIDEL, H. P. Isometric registration of ambiguous and partial data. In *Computer Vision and Pattern Recognition (CVPR)* (2009), pp. 1185–1192. *260*

[275] TEWARI, A., ZOLLHOFER, M., KIM, H., GARRIDO, P., BERNARD, F., PEREZ, P., AND THEOBALT, C. Mofa: Model-based deep convolutional face autoencoder for unsupervised monocular reconstruction. In *International Conference on Computer Vision (ICCV)* (2017). *167*

[276] TEXTURES.COM. *WrincklesHanging0037*. `https://www.textures.com/browse/hanging/112398`. [Online; accessed June 29, 2018]. *179*

[277] THE MATHWORKS INC. *MATLAB version 9.0 (R2016a)*. `http://mathworks.com/products/matlab/`, 2016. *79, 188, 191, and 284*

[278] THE STANFORD 3D SCANNING REPOSITORY. `http://graphics.stanford.edu/data/3Dscanrep/`. [accessed on 30.10.2015]. *235, 236, 237, 252, and 261*

[279] THE UK DARK SKY DISCOVERY PARTNERSHIP. `http://www.darkskydiscovery.org.uk/the_night_sky/orion_the_huntersmall.jpg`. [accessed on 25.10.2015]. *240*

[280] TIEN NGO, D., PARK, S., JORSTAD, A., CRIVELLARO, A., YOO, C. D., AND FUA, P. Dense image registration and deformable surface reconstruction in presence of occlusions and minimal texture. In *International Conference on Computer Vision (ICCV)* (2015). *106*

[281] TOMASI, C., AND KANADE, T. Shape and motion from image streams under orthography: a factorization method. *International Journal of Computer Vision (IJCV) 9* (1992), 137–154. *13, 22, 23, 26, 46, 64, 70, 71, 90, 96, 100, 115, 118, 142, 143, 145, 152, 159, 176, and 278*

[282] TORRESANI, L., HERTZMANN, A., AND BREGLER, C. Nonrigid structure-from-motion: Estimating shape and motion with hierarchical priors. *Transactions on Pattern Analysis and Machine Intelligence (TPAMI) 30*, 5 (2008), 878–892. *46, 47, 81, 82, 87, 91, 138, and 278*

[283] TORRESANI, L., YANG, D., ALEXANDER, E., AND BREGLER, C. Tracking and modeling non-rigid objects with rank constraints. In *Computer Vision and Pattern Recognition (CVPR)* (2001), vol. 1, pp. I–493–I–500. *27 and 46*

[284] TRENTI, M., AND HUT, P. Gravitational N-body simulations. *Scholarpedia 3*, 5 (2008). *37, 229, and 234*

[285] TRESADERN, P., AND REID, I. Articulated structure from motion by factorization. In *Conference on Computer Vision and Pattern Recognition (CVPR)* (2005), vol. 2, pp. 1110–1115. *48*

[286] TSIN, Y., AND KANADE, T. A correlation-based approach to robust point set registration. In *European Conference on Computer Vision (ECCV)*. Springer, 2004, pp. 558–569. *53, 55, and 247*

[287] TSOLI, A., LOPER, M., AND BLACK, M. J. Model-based anthropometry: Predicting measurements from 3d human scans in multiple poses. In *WACV* (2014), pp. 83–90. *213*

[288] VALGAERTS, L., BRUHN, A., ZIMMER, H., WEICKERT, J., STOLL, C., AND THEOBALT, C. Joint estimation of motion, structure and geometry from stereo sequences. In *European Conference on Computer Vision (ECCV)* (2010), pp. 568–581. *285 and 286*

[289] VALGAERTS, L., WU, C., BRUHN, A., SEIDEL, H.-P., AND THEOBALT, C. Lightweight binocular facial performance capture under uncontrolled lighting. *ACM Trans. Graph. (TOG) 31*, 6 (2012), 187:1–187:11. *126*

[290] VALMADRE, J., AND LUCEY, S. General trajectory prior for non-rigid reconstruction. In *Computer Vision and Pattern Recognition (CVPR)* (2012), pp. 1394–1401. *48 and 278*

[291] VAROL, A., SHAJI, A., SALZMANN, M., AND FUA, P. Monocular 3d reconstruction of locally textured surfaces. *Transactions on Pattern Analysis and Machine Intelligence (TPAMI) 34*, 6 (2012), 1118–1130. *167*

[292] VEDULA, S., BAKER, S., RANDER, P., COLLINS, R., AND KANADE, T. Three-dimensional scene flow. *Transactions on Pattern Analysis and Machine Intelligence (T-PAMI) 27*, 3 (2005), 475–480. 292

[293] VICENTE, S., AND AGAPITO, L. Soft inextensibility constraints for template-free non-rigid reconstruction. In *European Conference on Computer Vision (ECCV)* (2012), pp. 426–440. 47 and 61

[294] VICSEK, T., CZIRÓK, A., BEN-JACOB, E., COHEN, I., AND SHOCHET, O. Novel type of phase transition in a system of self-driven particles. *Physical Review Letters (PRL) 75* (1995), 1226–1229. 260 and 266

[295] VOGEL, C., ROTH, S., AND SCHINDLER, K. View-consistent 3d scene flow estimation over multiple frames. In *European Conference on Computer Vision (ECCV)* (2014), pp. 263–278. 292

[296] VOGEL, C., SCHINDLER, K., AND ROTH, S. Piecewise rigid scene flow. In *International Conference on Computer Vision (ICCV)* (2013). 291, 292, and 293

[297] ŠKERL, D., LIKAR, B., FITZPATRICK, J. M., AND PERNUŠ, F. Comparative evaluation of similarity measures for the rigid registration of multi-modal head images. *Physics in Medicine and Biology 52*, 18 (2007). 261 and 270

[298] WANDT, B., ACKERMANN, H., AND ROSENHAHN, B. 3d reconstruction of human motion from monocular image sequences. *Transactions on Pattern Analysis and Machine Intelligence (TPAMI) 38*, 8 (2016), 1505–1516. 166

[299] WANG, G., AND CHEN, Y. Fuzzy correspondences guided gaussian mixture model for point set registration. *Knowledge-Based Systems 136* (2017), 200 – 209. 194

[300] WANG, G., WANG, Z., CHEN, Y., LIU, X., REN, Y., AND PENG, L. Learning coherent vector fields for robust point matching under manifold regularization. *Neurocomputing 216* (2016), 393 – 401. 260

[301] WANG, G., ZHOU, Q., AND CHEN, Y. Robust non-rigid point set registration using spatially constrained gaussian fields. *IEEE Transactions on Image Processing 26*, 4 (April 2017), 1759–1769. 193

[302] WANG, P., WANG, P., QU, Z., GAO, Y., AND SHEN, Z. A refined coherent point drift (cpd) algorithm for point set registration. *Science China Information Sciences 54*, 12 (2011), 2639–2646. 53, 55, and 137

[303] WASENMÜLLER, O., MEYER, M., AND STRICKER, D. CoRBS: Comprehensive rgb-d benchmark for slam using kinect v2. In *Winter Conference on Applications of Computer Vision (WACV)* (2016). 239, 243, and 244

[304] WASENMÜLLER, O., PETERS, J. C., GOLYANIK, V., AND STRICKER, D. Precise and automatic anthropometric measurement extraction using template registration. In *3DB-ST* (2015). 213

[305] WEDEL, A., BROX, T., VAUDREY, T., RABE, C., FRANKE, U., AND CREMERS, D. Stereoscopic scene flow computation for 3d motion understanding. *International Journal of Computer Vision (IJCV) 95*, 1 (2011), 29–51. 275 and 292

[306] WEINBERGER, K. Q., AND SAUL, L. K. Unsupervised learning of image manifolds by semidefinite programming. *International Journal of Computer Vision (IJCV) 70*, 1 (2006), 77–90. 69

[307] WEISS, Y. Smoothness in layers: Motion segmentation using nonparametric mixture estimation. In *Computer Vision and Pattern Recognition (CVPR)* (1997). 137

[308] WELLS III, W. M. Statistical approaches to feature-based object recognition. *International Journal of Computer Vision 21*, 1-2 (1997), 63–98. 53

[309] WHITE, R., CRANE, K., AND FORSYTH, D. A. Capturing and animating occluded cloth. *ACM Trans. Graph. (TOG) 26* (2007). 71, 83, 84, and 87

[310] WILL, C. M. *Gravity: Newtonian, Post-Newtonian, and General Relativistic.* Springer International Publishing, 2016, pp. 9 – 72. 38

[311] WILM, J. Iterative closest point. `http://www.mathworks.com/matlabcentral/fileexchange/27-804-iterative-closest-point`. version 1.14. 252

[312] XIAO, D., YANG, Q., YANG, B., AND WEI, W. Monocular scene flow estimation via variational method. *Multimedia Tools and Applications (An International Journal)* (2015), 1–23. 277 and 285

[313] XIAO, J., CHAI, J., AND KANADE, T. A closed-form solution to non-rigid shape and motion recovery. *International Journal of Computer Vision (IJCV) 67*, 2 (2006), 233–246. 47

[314] XIAO, J., AND KANADE, T. Uncalibrated perspective reconstruction of deformable structures. In *Tenth IEEE International Conference on Computer Vision (ICCV '05)* (October 2005), vol. 2, pp. 1075 – 1082. 47

[315] YAN, J., AND POLLEFEYS, M. A factorization-based approach for articulated nonrigid shape, motion and kinematic chain recovery from video. *IEEE Transactions on Pattern Analysis and Machine Intelligence 30*, 5 (2008), 865–877. *48*

[316] YANG, R., AND ALLEN, P. K. Registering, integrating, and building cad models from range data. In *International Conference on Robotics and Automation (ICRA)* (1998), vol. 4, pp. 3115–3120. *52*

[317] YIN, L., WEI, X., SUN, Y., WANG, J., AND ROSATO, M. J. A 3d facial expression database for facial behavior research. In *7th International Conference on Automatic Face and Gesture Recognition (FGR06)* (2006), pp. 211–216. *261, 270, and 271*

[318] YOKOTA, R., AND BARBA, L. A. Hierarchical n-body simulations with autotuning for heterogeneous systems. *Computing in Science and Engineering 14*, 3 (2012), 30–39. *235*

[319] YU, R., RUSSELL, C., CAMPBELL, N. D. F., AND AGAPITO, L. Direct, dense, and deformable: Template-based non-rigid 3d reconstruction from rgb video. In *International Conference on Computer Vision (ICCV)* (2015). *92, 106, 108, 115, 136, 173, 174, 176, 177, and 179*

[320] YUILLE, A., AND GRZYWACZ, N. The motion coherence theory. In *International Conference on Computer Vision (ICCV)* (1988). *55, 136, 137, and 149*

[321] YUILLE, A. L., AND GRZYWACZ, N. M. A mathematical analysis of the motion coherence theory. *International Journal of Computer Vision 3*, 2 (1989), 155–175. *55, 136, and 137*

[322] ZACH, C. Robust bundle adjustment revisited. In *European Conference on Computer Vision (ECCV)* (2014), pp. 772–787. *298*

[323] ZACH, C., POCK, T., AND BISCHOF, H. A duality based approach for realtime tv-l1 optical flow. In *In Annual Symposium of the German Association for Pattern Recognition* (2007), pp. 214–223. *73, 281, 286, 294, 302, 305, and 308*

[324] ZANFIR, A., AND SMINCHISESCU, C. Large displacement 3d scene flow with occlusion reasoning. In *International Conference on Computer Vision (ICCV)* (2015). *293*

[325] ZHANG, R., ZHOU, W., LI, Y., YU, S., AND XIE, Y. Nonrigid registration of lung ct images based on tissue features. *56*

[326] ZHANG, Y., AND KAMBHAMETTU, C. On 3d scene flow and structure estimation. In *Computer Vision and Pattern Recognition (CVPR)* (2001), pp. 778–785. *292*

[327] ZHANG, Z. A flexible new technique for camera calibration. *Transactions on Pattern Analysis and Machine Intelligence (TPAMI) 22*, 11 (2000), 1330–1334. *226*

[328] ZHONG, Y. Intrinsic shape signatures: A shape descriptor for 3d object recognition. In *ICCV Workshops* (2009), pp. 689–696. *183, 189, 191, and 204*

[329] ZHOU, Q.-Y., AND KOLTUN, V. Dense scene reconstruction with points of interest. *ACM Transactions on Graphics (TOG) 32*, 4 (2013), 112:1–112:8. *239, 242, and 243*

[330] ZHOU, Z., ZHENG, J., DAI, Y., ZHOU, Z., AND CHEN, S. Robust non-rigid point set registration using student's-t mixture model. *PLoS ONE 9*, 3 (03 2014), e91381. *55*

[331] ZHU, S., ZHANG, L., AND SMITH, B. M. Model evolution: An incremental approach to non-rigid structure from motion. In *Computer Vision and Pattern Recognition (CVPR)* (2010), pp. 1165–1172. *47 and 150*

[332] ZHU, Y., HUANG, D., DE LA TORRE, F., AND LUCEY, S. Complex non-rigid motion 3d reconstruction by union of subspaces. In *Computer Vision and Pattern Recognition (CVPR)* (2014). *114 and 150*

[333] ZHU, Y., AND LUCEY, S. Convolutional sparse coding for trajectory reconstruction. In *Transactions on Pattern Analysis and Machine Intelligence (TPAMI)* (2014). *278*

[334] ZOLLHÖFER, M., NIESSNER, M., IZADI, S., REHMANN, C., ZACH, C., FISHER, M., WU, C., FITZGIBBON, A., LOOP, C., THEOBALT, C., AND STAMMINGER, M. Real-time non-rigid reconstruction using an rgb-d camera. *ACM Transactions on Graphics (TOG) 33*, 4 (2014), 156:1–156:12. *275 and 292*

Publication List

Core Publications (2016-2019)

i. **Vladislav Golyanik**, Torben Fetzer and Didier Stricker. Accurate 3D Reconstruction of Dynamic Scenes from Monocular Image Sequences with Severe Occlusions. In *Winter Conference on Applications of Computer Vision (WACV)*, 2017, Santa Rosa, USA.

ii. **Vladislav Golyanik**, Torben Fetzer and Didier Stricker. Introduction to Coherent Depth Fields for Dense Monocular Surface Recovery. In *British Machine Vision Conference (BMVC)*, 2017, London, UK.

iii. **Vladislav Golyanik** and Didier Stricker. High Dimensional Space Model for Dense Monocular Surface Recovery. In *International Conference on 3D Vision (3DV)*, 2017, Qingdao, China.

iv. **Vladislav Golyanik** and Didier Stricker. Dense Batch Non-Rigid Structure from Motion in a Second. In *Winter Conference on Applications of Computer Vision (WACV)*, 2017, Santa Rosa, USA.

v. Mohammad Dawud Ansari, **Vladislav Golyanik** and Didier Stricker. Scalable Dense Monocular Surface Reconstruction. In *International Conference on 3D Vision (3DV)*, 2017, Qingdao, China.

vi. **Vladislav Golyanik**, Soshi Shimada, Kiran Varanasi and Didier Stricker. HDM-Net: Monocular Non-Rigid 3D Reconstruction with Learned Deformation Model. In *EuroVR*, 2018, London, UK.

vii. **Vladislav Golyanik**, Aman Shankar Mathur and Didier Stricker. NRSfM-Flow: Recovering Non-Rigid Scene Flow from Monocular Image Sequences. In *British Machine Vision Conference (BMVC)*, 2016, York, UK.

viii. **Vladislav Golyanik***, Bertram Taetz*, Gerd Reis and Didier Stricker. Extended Coherent Point Drift Algorithm with Correspondence Priors and Optimal Subsampling. In *Winter Conference on Applications of Computer Vision (WACV)*, 2016, Lake Placid, USA.

$*$ marks approximately equal contributions

V. Golyanik, *Robust Methods for Dense Monocular Non-Rigid 3D Reconstruction and Alignment of Point Clouds*,

ix. **Vladislav Golyanik**, Bertram Taetz and Didier Stricker. Joint Pre-Alignment and Robust Rigid Point Set Registration. In *International Conference on Image Processing (ICIP)*, 2016, Phoenix, USA.

x. **Vladislav Golyanik**, Gerd Reis, Bertram Taetz and Didier Stricker. A Framework for an Accurate Point Cloud Based Registration of Full 3D Human Body Scans. In *International Conference on Machine Vision Applications (MVA)*, 2017, Nagoya, Japan.

xi. **Vladislav Golyanik**, Sk Aziz Ali and Didier Stricker. Gravitational Approach for Point Set Registration. In *Computer Vision and Pattern Recognition (CVPR)*, 2016, Las Vegas, USA.

xii. **Vladislav Golyanik**, Christian Theobalt and Didier Stricker. Accelerated Gravitational Point Set Alignment with Altered Physical Laws. In *International Conference on Computer Vision (ICCV)*, 2019, Seoul, South Korea.

xiii. Sk Aziz Ali, **Vladislav Golyanik** and Didier Stricker. NRGA: Gravitational Approach for Non-Rigid Point Set Registration. In *International Conference on 3D Vision (3DV)*, 2018, Verona, Italy.

xiv. **Vladislav Golyanik**, Kihwan Kim, Robert Maier, Matthias Nießner, Didier Stricker and Jan Kautz. Multiframe Scene Flow with Piecewise Rigid Motion. In *International Conference on 3D Vision (3DV)*, 2017, Qingdao, China.

xv. **Vladislav Golyanik**, André Jonas, Didier Stricker and Christian Theobalt. Intrinsic Dynamic Shape Prior for Dense Monocular Non-Rigid 3D Reconstruction and Compression of Sequences with Temporally-Disjoint Rigidity. *arXiv.org*, 2019.

Additional Publications as of 2020

xvi. Bertram Taetz, Gabriele Bleser, **Vladislav Golyanik** and Didier Stricker. Occlusion-Aware Video Registration for Highly Non-Rigid Objects. In *Winter Conference on Applications of Computer Vision (WACV)*, 2016, Lake Placid, USA.

xvii. Oliver Wasenmüller, Jan C. Peters, **Vladislav Golyanik** and Didier Stricker. Precise and Automatic Anthropometric Measurement Extraction using Template Registration. In *3D Body Scanning Technologies (3DBST)*, 2015, Lugano, Switzerland.

xviii. **Vladislav Golyanik**, Mitra Nasri and Didier Stricker. Towards Scheduling Hard Real-Time Image Processing Tasks on a Single GPU. In *International Conference on Image Processing (ICIP)*, 2017, Beijing, China.

xix. Tomonari Yoshida, **Vladislav Golyanik**, Oliver Wasenmüller and Didier Stricker. Improving Time-of-Flight Sensor for Specular Surfaces with Shape From Polarization. In *International Conference on Image Processing (ICIP)*, 2018, Athens, Greece.

xx. Jyothish K. James, Georg Puhlfürst, **Vladislav Golyanik** and Didier Stricker. Classification of LIDAR Sensor Contaminations with Deep Neural Networks. In *Computer Science in Cars Symposium (CSCS)*, 2018, Munich, Germany.

xxi. **Vladislav Golyanik**, André Jonas and Didier Stricker. Consolidating Segmentwise Non-Rigid Structure from Motion. In *International Conference on Machine Vision Applications (MVA)*, 2019, Tokio, Japan.

xxii. Soshi Shimada, **Vladislav Golyanik**, Christian Theobalt and Didier Stricker. IsMo-GAN: Adversarial Learning for Monocular Non-Rigid 3D Reconstruction. In *Computer Vision and Pattern Recognition (CVPR) Workshops*, 2019, Long Beach, USA.

xxiii. Jilliam Maria Diaz Barros, **Vladislav Golyanik**, Kiran Varanasi and Didier Stricker. FACE IT!: A Pipeline For Real-Time Performance-Driven Facial Animation. In *International Conference on Image Processing (ICIP)*, 2019, Taipei, Taiwan.

xxiv. **Vladislav Golyanik** and Christian Theobalt. Optimising for Scale in Globally Multiply-Linked Gravitational Point Set Registration Leads to Singularities. In *International Conference on 3D Vision (3DV)*, 2019, Québec, Canada.

xxv. Soshi Shimada, **Vladislav Golyanik**, Edgar Tretschk, Didier Stricker and Christian Theobalt. DispVoxNets: Non-Rigid Point Set Alignment with Supervised Learning Proxies. In *International Conference on 3D Vision (3DV)*, 2019, Québec, Canada.

xxvi. Yongzhi Su, **Vladislav Golyanik**, Nareg Minaskan, Sk Aziz Ali and Didier Stricker. A Shape Completion Component for Monocular Non-Rigid SLAM. In *International Symposium on Mixed and Augmented Reality (ISMAR)*, 2019, Beijing, China.

xxvii. Onorina Kovalenko, **Vladislav Golyanik**, Jameel Malik, Ahmed Elhayek and Didier Stricker. Structure from Articulated Motion: Accurate and Stable Monocular 3D Reconstruction without Training Data. Sensors, 19(20), 2019.

xxviii. Edgar Tretschk, Ayush Tewari, Michael Zollhöfer, **Vladislav Golyanik** and Christian Theobalt. DEMEA: Deep Mesh Autoencoders for Non-Rigidly Deforming Objects. *arXiv.org*, 2019.

xxix. Lan Xu, Weipeng Xu, **Vladislav Golyanik**, Marc Habermann, Lu Fang and Christian Theobalt. EventCap: Monocular 3D Capture of High-Speed Human Motions using an Event Camera. In *Computer Vision and Pattern Recognition (CVPR)*, 2020, Seattle, USA.

xxx. Jameel Malik, Ibrahim Abdelaziz, Ahmed Elhayek, Soshi Shimada, Sk Aziz Ali, **Vladislav Golyanik**, Christian Theobalt and Didier Stricker. HandVoxNet: Deep Voxel-Based Network for 3D Hand Shape and Pose Estimation from a Single Depth Map. In *Computer Vision and Pattern Recognition (CVPR)*, 2020, Seattle, USA.

xxxi. **Vladislav Golyanik** and Christian Theobalt. A Quantum Computational Approach to Correspondence Problems on Point Sets. In *Computer Vision and Pattern Recognition (CVPR)*, 2020, Seattle, USA.

Printed in the United States
By Bookmasters